THE DIGITAL ECONOMY

Information technologies (IT) shape economic space, but we have no clear map of the cyber economy since the "digital revolution" began in the early 1970s. The convergence of computers, advanced telecommunications, digital media, and the rise of the Internet have generated tremendous changes in the geographies of business organization, with consequences for places and communities worldwide. Business has become network-centric, global, and flexible and firms and workers are encouraged to be increasingly sophisticated consumers of space.

The Digital Economy offers an up-to-date, critical synthesis that links the various aspects of the digital or cyber economy from the perspective of real firms. A geographic approach emphasizes how IT has made businesses less dependent on locational constraints, and the tangible effects on places and regions are placed at the core of the analysis. Case studies of companies, including Amazon, Dell, Li & Fung, and Volvo, demonstrate that the geography of digitally-driven production is the outcome of both dispersion and agglomeration dynamics. Global corporations are shown to have footprints that ignore—to some degree—distance and time, yet creative and coordinating activities remain anchored in urban innovative ecosystems such as Silicon Valley and Bangalore. These trends have been made possible by the development of a worldwide and integrated telecommunications network, whose unequal presence dictates the capabilities of places and communities to be connected to the global economy. However, the threat of the digital divide must not be overstated. In cities, rural areas, and emerging countries, local development is wrapped up in human capital, rather than technology.

This engaging and accessible text describes and explains the patterns and dynamics of today's digital economic space. The effects on places and regions and the people in them are at the core of the authors' analysis, illustrated by many real examples. This book will be useful to anyone studying Business and Management, Geography, and Information and Communication studies.

Edward J. Malecki is Professor of Geography at The Ohio State University. He is author of over 100 published papers, and of *Technology and Economic Development* (Addison-Wesley Longman 1997) and Associate Editor of *Entrepreneurship and Regional Development*.

Bruno Moriset is Associate Professor and Head, Department of Geography and Planning at the University Jean Moulin of Lyon in France. His research focuses notably on the link between information technologies and regional development.

THE DIGITAL ECONOMY

Business organization, production processes, and regional developments

Edward J. Malecki and Bruno Moriset

LONDON AND NEW YORK

First published 2008
by Routledge
2 Park Square, Milton Park, Abingdon, Oxon OX14 4RN

Simultaneously published in the USA and Canada
by Routledge
270 Madison Ave, New York, NY 10016

Routledge is an imprint of the Taylor & Francis Group

Typeset in Times New Roman by
Florence Production Ltd, Stoodleigh, Devon
Printed and bound in Great Britain by
TJ International Ltd, Padstow, Cornwall

British Library Cataloguing in Publication Data
A catalogue record for this book is available from the British Library

Library of Congress Cataloging in Publication Data
Malecki, Edward J., 1949–
The digital economy: business organization,
production processes and regional developments/
by Edward J. Malecki and Bruno Moriset.
p. cm.
Includes bibliographical references and index.
1. Information technology—Economic aspects. 2. High
technology industries—Management. 3. Business enterprises—
Technological innovations—Case studies. 4. Electronic
commerce. 5. Telecommunication—Technological innovations.
I. Moriset, Bruno, 1960–. II. Title.
HC79.I55M355 2007
303.48′33–dc22 2007022680

ISBN10: 0–415–39695–6 (hbk)
ISBN10: 0–415–39696–4 (pbk)
ISBN10: 0–203–93363–X (ebk)

ISBN13: 978–0–415–39695–0 (hbk)
ISBN13: 978–0–415–39696–7 (pbk)
ISBN13: 978–0–415–93363–3 (ebk)

CONTENTS

FIGURES

TABLES

BOXES

ABBREVIATIONS

3G	third generation
3PL	third party logistics provider
4PL	fourth party logistics provider
ACD	automatic call distribution
ADSL	asymmetric digital subscriber line
ANX	American Network eXchange
ARPA	Advanced Research Projects Agency
AS	autonomous system
ASP	application service provider
ASTEC	Aire de Services et de Transferts de Technologie
ATM	automatic teller machine
ATM	asyncronous transfer mode
AURP	Association of University Research Parks
BBA	British Bankers Association
BLS	Bureau of Labor Statistics
BPO	business process outsourcing
BT	British Telecom
BtoB (B2B)	business-to-business
BtoC (B2C)	business-to-consumer
CAD	computer-aided design
C&D	connect and develop
CAE	computer-aided engineering
CAGR	compound annual growth rate
CAM	computer-aided manufacturing
CBD	central business district
CC	call center / contact center
CEA	Commissariat á l'Energie Atomíque
CENG	Centre d'Etudes Nucléaires de Grenoble
CHIPS	Clearing House Interbank Payment System
CM	contractual manufacturer
CO	central office
CPE	customer premises equipment
CRITO	Counter for Research on Information Technology and Organizations

CRM	customer relationship management
CTI	computer telephony integration
CtoB	consumer-to-business
CtoC (C2C)	consumer-to-consumer
DOMS	Dell Order Management System
DRAM	dynamic random access memory
DSL	digital subscriber line
DSLAM	digital subscriber line access multiplexer
DTP	desktop publishing
DVR	digital video recorder
DWDM	dense wavelength division multiplexing
EAN	European Article Numbering
ECN	electronic communication network
EDI	electronic data interchange
EMEA	Europe, Middle East and Africa
EMIX	Emirates Internet Exchange
EMS	electronic manufacturing services
ENX	European Network eXchange
EO-FDI	export-oriented foreign direct investment
EPC	electronic product code
ERP	enterprise resource planning
ERS	electronic reservation system
ESRF	European Synchrotron Radiation Facility
FDI	foreign direct investment
FIRE	Finance, Insurance and Real Estate
FT	France Télécom
FTE	full time equivalent
FTP	file transfer protocol
FTTB	fiber to the building; fiber to the business
FTTC	fiber to the curb
FTTH	fiber to the home
FTTN	fiber to the node
FTTP	fiber to the premises
GDP	gross domestic product
GIS	geographic information system
GIX	global Internet eXchange point
GM	General Motors
GNP	gross national product
GPS	global positioning system
GPT	general purpose technology
GSA	General Services Administration
GSM	Global System for Mobile Communications
HP	Hewlett Packard
HTML	hypertext mark-up language
HVAC	heating, ventilating and air-conditioning

IASP	International Association of Science Parks
ICT	information and communication technology
ICU	intensive care unit
IMD	International Institute for Management Development
IOS	inter-organizational systems
IP	Internet protocol
IPO	initial public offering
ISDN	integrated service digital network
ISOC	Internet Society
ISP	Internet service provider
IT	information technology
ITAC	International Telework Association and Council
ITES	information technology-enabled services
ITPL	International Technology Park Limited
ITS	intelligent transportation system
ITU	International Telecommunications Union
IVR	interactive voice response
IXP	Internet exchange point
J2EE	Java 2 Enterprise Edition
JPEG	Joint Photographic Experts Group
KIBS	knowledge-intensive business services
KPO	knowledge process outsourcing
LAN	local area network
LETI	Laboratory of Information Technology Electronics
LSP	logistics service providers
M&A	mergers and aquisitions
MAE	metropolitan area exchange
MAN	metropolitan area network
MGI	McKinsey Global Institute
MIT	Massachusetts Institute of Technology
MMDS	microwave multipoint distribution system
MNC	multinational corporation
MPEG	Moving Picture Experts Group
MP3	MPEG-1 Audio Layer 3
MSC	Multimedia Super Corridor
NAFTA	North American Free Trade Agreement
NAICS	North American Industry Classification System
NAP	network access point
NASSCOM	National Association of Software and Service Companies
NASDAQ	National Association of Securities Dealers Automated Quotation
NHTS	National Household Travel Survey
NYSE	New York Stock Exchange
OCR	optical character regognition
ODM	original design manufacturer
OECD	Organisation for Economic Co-operation and Development

OEM	original equipment manufacturer
OLPC	One Laptop per Child
OSI	Open Systems Interconnection
OS	operating system
P2P	peer-to-peer
P&G	Procter & Gamble
PABX	Private Automatic Branch eXchange
PC	personal computer
PCCI	Premium Contact Center International
PDF	Portable Document Format
PDM	product date management
PLM	product lifecycle management
POP	point of presence
POTS	plain old telephone service
PSTN	public switched telecommunication networks
PTT	Post, Telegraph and Telecommunications
QOS	quality of service
R&D	research and development
RAM	random access memory
RFID	Radio Frequency Identification
RIS	regional innovation system
ROI	return on investment
SABRE	Semi-Automated Business Research Environment
SCM	supply chain management
SDH	syncronous digital hierarchy
SIBIS	Statistical Indicators Benchmarking the Information Society
SIC	standard industrial classification
SLC	supplier logistics center
SME	small and medium-sized enterprise
SOHO	small office-home office
SOI	silicon on insulator
SONET	synchronous optical network
SSC	shared service center
SUV	sport utility vehicle
SWIFT	Society for Worldwide Interbank Financial Telecommunication
TCS	Tata Consultancy Services
TFP	total factor productivity
TNC	transnational corporation
TSF	Télécoms sans Frontières
TTB	travel time budget
UCLA	University of California at Los Angeles
UKSPA	UK Science Park Association
UMTS	universal mobile telecommunications system
UNE	unbundled network elements
UPC	universal product code

USB	universal serial bus
VAN	value-added network
VAS	Virtual-Agent Services
VMT	vehicle miles traveled
VoIP	voice over IP
VPN	virtual private network
VSAT	very small aperture terminal
WAN	wide area network
WAP	wireless application protocol
W-CDMA	wideband code division multiple access
WDM	wavelength division multiplexing
WEF	World Economic Forum
WiFi	wireless fidelity
WiMAX	worldwide interoperability for microwave access
WMTC	Washington Metropolitan Telework Centers
WTO	World Trade Organization
XML	Extensible Mark-up Language

ACKNOWLEDGMENTS

Figure 1.1 is reprinted from *Telecommunications Policy*, Volume 27, Dan Steinbock, "Globalization of the wireless value system: from geographic to strategic advantages," Figure 7, page 226, with permission from Elsevier.

Figure 6.4 is reprinted from *A.T. Kearney's 2004 Offshore Location Attractiveness Index: Making Offshore Decisions*, Figure 2, page 4, with permission from A.T. Kearney.

Ed Malecki is grateful to a trio of students who led him into the Internet era: Carlton Boush, Sean Gorman, and Angela McIntee. As our work progressed, comments by Peter Hugill were insightful. Colleagues in the Department of Geography at The Ohio State University provided encouragement as the manuscript took shape. Ed acknowledges his gratitude to the US National Science Foundation, through grants SBE-9911222 and BCS-03117878, the Mershon Center, and the Center for Urban and Regional Analysis at The Ohio State University, for their support of this research. Finally, my love and appreciation go to my wife, Kathleen, who has endured a great deal as the book has taken shape.

Bruno Moriset thanks all the colleagues in the Department of Geography and Planning at the University Jean Moulin of Lyon, or in other institutions, who encouraged him since the start of this project. Also, I do apologize and express thanks to my wife Véronique, and to my children, Apolline and Virgile, whom I too often neglected during the writing of this book, but whose presence proved a critical support.

Both the authors thank warmly Marie-Laure Trémélo, cartography engineer, platform ISIG, CNRS laboratory UMR 5600 "Environnement, Ville, Société" (Lyon), who kindly drew the maps of the present book.

INTRODUCTION
The digital economy and the splintering of economic space

1

In 2001, the first major transatlantic telesurgical operation was carried out: doctors in the United States removed a gall bladder from a 68-year-old woman in France by remotely controlling a surgical robot arm. The two medical teams were linked by a high-speed fiber-optic loop. The time delay between the surgeon's movements and the return video image displayed on screen was less than 200 milliseconds (BBC News 2001; Marescaux *et al.* 2001).

The Dalles, Oregon, a community of 12,000 inhabitants on the Columbia River, has become a key nexus of the World Wide Web. Google has built there a major data center—thousands of interconnected servers—which benefits, like aluminum plants, from the abundant and cheap electricity supply (Spellman 2005).

In Ann Arbor, Michigan, operators at ProQuest, a digital archive company, scan microfilms of nine major US newspapers. Then, the copies are e-mailed to Chennai (Madras), India, where 850 employees of Ninestars Information Technology Ltd cut up and sort the images into articles suitable for Web search and reading. Salesforce.com Inc., a customer-relationship management (CRM) company based in San Francisco, subcontracts some typing to Digital Divide Data, a non-profit organization based in Laos and Cambodia, which has trained poor, and sometimes disabled people (Helm and Kripalani 2006).

These examples are a brief aerial survey of the splintering, but interconnected, economic space which is emerging in the Internet era. There remains virtually no place on earth, neither rural Oregon nor Cambodia, which cannot intrinsically get in touch—maybe in a subordinate way—with the globalized economy. This is the main objective of the present book: to inventory, describe, and explain the original features and dynamics—from both organizational and geographic viewpoints—which derive from the all-out intrusion of digital technology in the contemporary economy.

Splintered, yet connected

In many different ways, businesses are operating in a digital, interconnected space of flows (Castells 1996), that has permitted a fine-grained spatial division of labor, taking advantage of optimal combinations of (high) skills and (low) wages in various places around the world (Grossman and Rossi-Hansberg 2006; Scott 2006). This splintered economic space encompasses not only the production of tangible products but also many office-based tasks as paper is replaced by digital files (Blinder 2006). These files can contain food orders from drive-through lanes of distant restaurants (Richtel 2006) or scans of a patient in the US read by a radiologist in India (Wachter 2006b).

> Until recently, the need to take a patient's history and perform a physical examination, apply complex techniques or procedures, and share information quickly has made medicine a local affair. . . . To examine the heart, the cardiologist could be no farther from the patient than his or her stethoscope allowed, and data gathering required face-to-face discussions with patients and sifting through paper files. But as health care becomes digitized, many activities, ranging from diagnostic imaging to the manipulation of laparoscopic instruments, are rendered borderless. The offshore interpretation of radiologic studies is proof that technology and the political climate will now permit the outsourcing of medical care.
>
> (Wachter 2006a: 661)

As it takes advantage of a growing array of digital technologies, the geography of any business becomes more complex and difficult to locate. Managers and entrepreneurs who seek the ideal location, as well as analysts, scholars, and policymakers who seek to understand and predict them, confront a more complex geography of the global economy. As digital processes have permeated the entire value chain, they have permitted a countless series of tasks amenable to remote treatment (Blinder 2006). This "great unbundling" of tasks has sparked fears that nearly all jobs will disappear from high-wage countries (Baldwin 2006).

Even high-end jobs are not secure. Several hundred US hospitals use overseas teleradiology services, such as Teleradiology Solutions, NightHawk Radiology Services, and Virtual Radiologic (Wachter 2006b). In electronic intensive care units (ICUs), "off-site intensivists monitor patients by closed-circuit television. Streams of physiological data appear in real time on a remote screen, allowing the off-site physician to advise local providers, sometimes even entering orders remotely into the hospital's computer system" (Wachter 2006a: 663). Wachter (2006b: 663) quotes a US radiologist: "Who needs to pay us $350,000 a year if they can get a cheap Indian radiologist for $25,000 a year?"

However, digitalization has not created an economy that has become completely footloose in which any task can be done anywhere. Yet this is the hype that has been promised—or threatened—by many pundits. A state of confusion surrounds the digital economy. Oversold and characterized by hype, the long-awaited paperless office, picture-phones in every home and office, and the global village have not materialized. In part, this is because many different views have coalesced around the information society, the digital economy, the network economy and other labels for the new, technology-based world. Among the recent examples of hype are two books whose titles are frequently mentioned to capture the phenomenon: *The Death of Distance* (Cairncross 2001) and *Global Financial Integration: The End of Geography* (O'Brien 1992).

A more nuanced analysis of the social, economic, and geographical changes underway leads Kotkin (2000) to observe that "the new geography" is highlighted by the growing importance of "sophisticated consumers of place" (Knight 1989: 237). In fact, more than place or places are consumed, as networks and the "space of flows" also are evaluated and consumed in sophisticated ways (Castells 1989). We prefer to generalize the phenomenon to companies as well as people, all of whom have become *sophisticated consumers of space*.

Even though downloaded files (such as music, video, and information from web pages) are perceived as "free" to many users, because it is very easy to retrieve information across long distances almost instantaneously (depending on the last mile link to one's computer or mobile phone), distance is not "dead." Cairncross (2001), whose book is cited routinely in this regard, recognizes that the story is more subtle. "The death of distance loosens the grip of geography. It does not destroy it" (Cairncross 2001: 5). It is not a case of technological determinism, with information technologies (IT) creating a global village. The falling cost of communication has not been felt equally everywhere. Large cities continue to dominate both in network connections and in the agglomeration of face-to-face activities for which "the tyranny of proximity" has replaced the tyranny of distance (Duranton 1999). Poor people and regions continue to lag behind as new technologies flow first into wealthy regions. For the individual, the ability of the telephone to permit voice and image communication over distance has evolved—thanks to both wireless technologies and the Internet—into new "personal mobilities" (Kellerman 2006) and new consumer power (Markillie 2005).

The diffusion and the convergence of digital technologies

It is difficult to avoid hype or exaggeration in the discussion and analysis of the continuous series of events which constitute the so-called "digital revolution." The evolution of telecommunications, discussed in greater detail in Chapter 3, has been a continued improvement of communications technologies since the telegraph and the radio in the nineteenth century (Arnold and Guy 1989; Cukier 2007). The computer was invented in the 1940s, and has benefitted since from a continuous improvement process. But these technologies remained separate until the early 1990s, at least in the public's eyes.

To understand the hype that arose in the 1990s around the concept of IT (information technology)—after all, telegraph and Gutenberg's printing press were also IT—we must consider the close interlinkage of three processes:

- the computerization of the society, thanks to the production of cheap personal computers (PCs) operating user-friendly interfaces (itself a consequence of the materialization of "Moore's law" in microprocessors);[1]
- the convergence of computers and telecommunications (Steinbock 2003; Yoffie 1997);
- the irresistible diffusion of the Internet and the spread of the World Wide Web, its most popular application.

Actually, at least four formerly distinct industries—computers, communications, software, media and entertainment—have converged into a galaxy of overlapping, digitally enabled industries which share common technical standards and channels. But a few business sectors, we suggest in a later section, actually ignore the digital convergence and remain largely un-digitalized.

The analysis of this evolution faces typical "chicken-and-egg" problems. The Internet would not have become so popular if businesses and households would not have gotten access *en masse* to cheap computers and network technologies such as cable modems and digital subscriber lines (DSL). Media and commerce would not have become electronic—in part—without the commoditization of PCs and the Internet. In turn, the richness of the content and applications available now via the Internet is a powerful driver towards more computers and more connections at home, in schools, in enterprises. The system of interlinkage should be extended further. Computers, together with other digital devices such as flat-panel TVs, cameras, and wireless handsets, have become commodities because they are the outcomes of a globalized production system, which includes low cost countries, and could not be coordinated without a wide range of IT applications. The digital convergence should be regarded as the emergence of a huge, worldwide value network (see Chapter 4), which sees a continuous process of creation of new companies, new products, and new niches of value-added creation.

The phenomenon of convergence is clearly visible in the rise of "triple play" bundled subscriptions, which give customers access to the Internet, telephony (Voice over Internet Protocol or VoIP), television, and a roster of home media and entertainment services, from the same copper or fiber-optic line. This tendency materializes in "device convergence": desktop and laptop computers, mobile phones, television sets, and game consoles are all able to access the Internet, display messages and video, and store information (Standage 2006). The latest versions of advanced mobile telephone handsets, which incorporate computing capacity, Internet access, e-mail, photo and video camera, and even video display, perfectly epitomize both the convergence and the unprecedented degree of informational ubiquity it gives to "connected" people (Figure 1.1).

In some way, the latest technology has made real some images and metaphors which have long been associated with science fiction. The design, marketing and servicing of these technologies have generated an enormous industry, with real impacts on producers, consumers, and actual economic effects on places and regions.

Digital technology is spreading throughout the entire economy. Computers and telecommunications are mainly enabling technologies, which serve downstream industries in both manufacturing (e.g. automobiles, aerospace, textiles, and electrical equipment) and services sectors (transport and travel, finance, retail). Even industries in the primary sector, such as farming, fishing, and mining, rely now on computers, digital apparatus and software, and advanced telecommunication services. Fishermen and farmers use radar, sonar, and global positioning systems (GPS). Farming and forestry are important users of remote sensing and geographic information systems (GIS). Oil prospecting could not be performed without powerful computing capacity.

The digital economy: a definition

> The digital economy represents the pervasive use of IT (hardware, software, applications and telecommunications) in all aspects of the economy, including internal operations of organizations (business, government and non-profit);

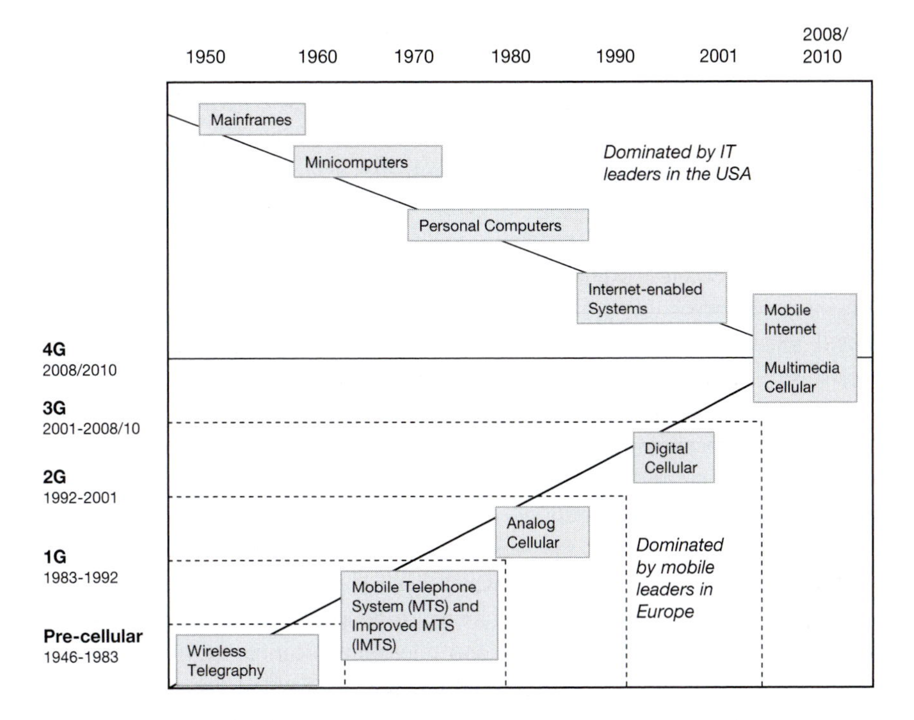

Figure 1.1 ***Convergence in wireless communications***

Source: Steinbock 2003

> transactions between organizations; and transactions between individuals, acting both as consumers and citizens, and organizations. . . . The technologies underlying the digital economy also go far beyond the Internet and personal computers. IT is embedded in a vast array of products, and not just technology products like cell phones, GPS units, PDAs, MP3 players, and digital cameras. IT is in everyday consumer products like washing machines, cars, and credit cards, and industrial products like computer numerically-controlled machine tools, lasers, and robots. Indeed, in 2006, 70 percent of microprocessors did not go into computers but rather went into cars, planes, HDTVs, etc., enabling their digital functionality and connectivity.
>
> (Atkinson and McKay 2007: 7)

As IT has become cheaper, faster, better, and easier to use, organizations continuously find new and expanded uses for IT, creating digital applications connected—increasingly wirelessly—that have become ubiquitous and central to economic and social activity (Atkinson and McKay 2007; Cukier 2007). However, not all industries and regions worldwide have reached the same level of IT adoption. Farmers and fishermen in developing countries are not computerized. However, digital devices (embedding "chips and bits") are becoming almost ubiquitous. In sub-Saharan Africa, where most of the world's poorest countries are located, mobile telephones are becoming a commodity, and are commonly used by merchants to conduct commercial transactions. In some way, we are on the verge of an "Internet of things," according to the International Telecommunications Union (ITU) (2005).

Consequently, the digital economy can be described as *a pyramid* (Figure 1.2). The top section—the "spearhead"—includes the products of the silicon foundry and semiconductor industries. Itself not a large sector, it is critical, because its products are at the core of computers and electronic components, which are embedded in an increasingly far-reaching range of products, including consumer electronics, automobiles, machines and industrial equipment, heating, ventilation, and air conditioning (HVAC) systems, and household appliances. Major distribution chains such as Wal-Mart have started to attach "smart tags" (RFID or radio frequency identification devices) on packages of basic goods sold in supermarkets.

The second level of the pyramid, below the spearhead, is comprised of the computer and telecommunications industries, both manufacturing and services. These industries may be regarded at the "core sector" of the digital economy, because this sector enables the working of the lower parts.

The third level represents the "main body" of the digital economy, including both manufacturing and services activities which rely heavily—sometimes almost exclusively—on digital technologies. A large fraction of people in these industries spend most of their working time on computers and telecommunications devices. In the service sector, we must mention: electronic commerce, media and entertainment, and IT-enabled business services (such as call centers and shared service centers), and financial services. In the manufacturing sector, we can identify the IT-intensive sectors where "product lifecycles" (a concept explained in Chapter 4) are fully or partially IT driven, such as the automobile or aircraft industries. However, even in the most advanced industries, digitalization varies among departments and regions. Companies have departments that are fully computerized (R&D, design, data treatments), but others which

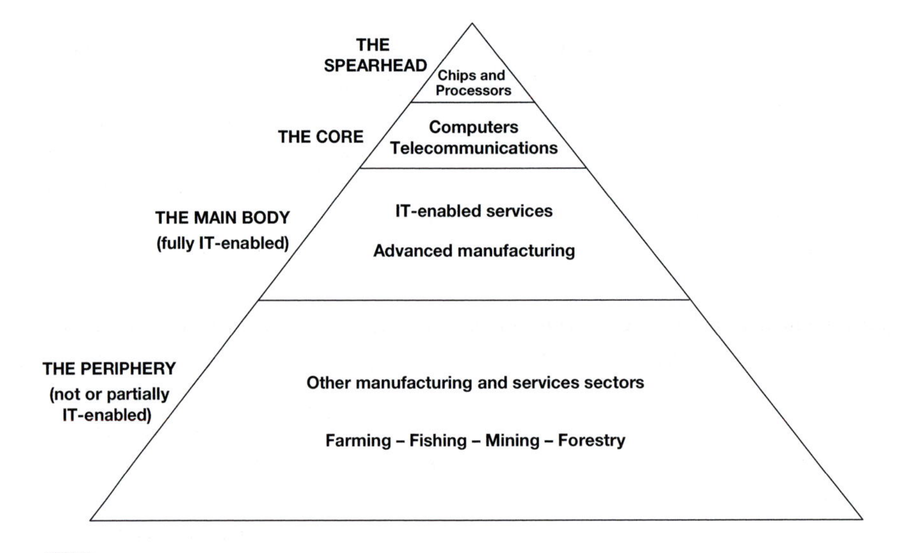

Figure 1.2 ***The pyramid of the digital economy***

Source: Authors' own research

work using more "traditional" means. In a typical automobile company such as Renault, 50 percent of the whole workforce is in front of a computer either full-time or part-time. In retail banking (where 100 percent of the workforce is computerized, at least in developed countries), employees in back-office functions perform fully IT-enabled tasks (which are increasingly subjected to offshoring), while front-office people still have face-to-face meetings with clients. The logistics industry and the medical sector also belong to this category.

The base of the pyramid is made of sectors whose digitalization remains almost nonexistent, or at most partial, as they are slowly permeated by digital technology. This is the case for a large part of the farming sector, especially in developing countries. In home and consumer services, as well as in public services, a vast range of people rarely or never use a computer, such as housemaids, garbage collectors, hairdressers, and police officers. From kindergarten to university, the use of computers and the Internet in education is very diverse, varying among levels, disciplines, and regions. For example, primary schools in Finland are all IT-equipped, but some university professors in arts and humanities, in almost every country, have never used a computer. Because face-to-face contact between students and teachers remains the base of teaching, the main body of the educational sector is not in the core of the digital economy, even if it is increasingly permeated by IT uses. What we are facing, typically, is the very difference between knowledge workers (or sectors or economy) and digital workers (or sectors or economy). This important issue is more thoroughly addressed in Chapter 2.

For companies, the effect of technological convergence has increased speed of access to, and processing of, information, allowing them increased control over decentralized systems, "not so much abolishing geography as assisting its more efficient exploitation" (Charles 1996: 101). Rather than shrinking space, telecommunications enables "human extensibility" to distant locations (Abler *et al.* 1975). Information and communication technology (IT), as a set of enabling technologies, has permitted markets to be extended to the global level (Dicken 2007; Melody 1991).

Thanks to broadband networks, computer-aided design (CAD) software and secure, virtual, collaborative workspaces, an engineer can work from abroad on the latest Airbus or Boeing project or monitor drilling in an offshore oil field thousands of kilometers away. Such an IT-based combination of reach and richness allows firms to get control of operations within the product lifecycle, regardless of distance. Through the use of IT, functional integration does not mean geographic centralization: "Integration of production is defined as bringing together diverse industries or previously disunited production stages, into a new comprehensive resource-sharing relationship, either geographically distributed or centralized" (Amirahmadi and Wallace 1995: 1764). In the utopia of a "friction-free" capitalism (Kenney and Curry 2001) and a "weightless world" (Coyle 1998), even fully IT-driven enterprises have to deal with the "thick" aspects of geographical space, such as cultural and political barriers, idiosyncratic market features, varying regulation, and transportation bottlenecks. Reducing the friction of distance, however, makes capital more rather than less sensitive to local geographical variations. The combined effect of freer trade and reduced transport costs is not greater equality of power and wealth through the evolving territorial division of labor, but growing geographical inequalities (Harvey 2006).

The key issue is to study the two-pronged challenge which companies face: making the most of the potential of IT to overcome both geographical and corporate thickness, in a context of space-time fragmentation (even if space and time have not completely collapsed) (Brunn and Leinbach 1991). The principal response, in organizational terms, is to become a network-centric firm, or even a virtual company.

Real-world complexities intrude on location decisions. For example, large pools of educated workers are in places where the infrastructure does not meet the standards elsewhere. Large metropolitan regions have the advantage of proximity to customers, suppliers, and knowledge brokers, but they are always relatively high-cost locations. Consequently, while the infrastructure of the Internet has focused on large cities, as Chapter 6 will detail, some key pieces of that infrastructure—data centers—have begun to seek out locations where the cost of electricity is lower.

The benefits of the digital economy are not only profits for business and convenience for customers. Among the benefits are the ability to transcend distance for specialized expertise and capabilities which are not found everywhere. A long-anticipated benefit of IT is telemedicine, bringing medical expertise to remote villages and towns. Indeed, surgery can now be conducted remotely, if broadband connections are in place.

Although telemedicine can be practiced to the benefit of people on islands and other remote locations, it has not yet become routine. Nicolini (2006) suggests that this is because, as in businesses, IT in the health-care sector requires the reconfiguration of coordination mechanisms, work processes, and power relationships. Indeed, enterprise-wide IT transformations are very difficult, because they define tasks, sequences, and data formats. Because their use is mandatory, employees usually dislike enterprise IT technologies: "they don't just enable new ways of working; they dictate them" (McAfee 2006: 147).

Structure of the book

In the following chapters, we analyze how the convergence of computers, advanced telecommunications, and the Internet has made firms and workers increasingly sophisticated consumers of space. IT implementation has made the production and consumption of goods and services less dependent on locational constraints but not yet completely footloose. In particular, cities have become more—not less—important as telecommunications has linked remote and rural places to the global economy in new ways.

Through case studies of companies, such as Amazon, Dell, Li & Fung, and Volvo, we demonstrate that the geography of digitally driven production is the outcome of both dispersion and agglomeration dynamics—both centrifugal and centripetal forces. Global corporations have footprints that ignore—to some degree—distance and time. However, creative and coordinating activities remain anchored in urban innovative ecosystems such as Silicon Valley and Bangalore.

Business has become network-centric, and more flexible. Through global networks of modular production, offshoring practices, and telework implementation, companies seek to make the most of resources and talent wherever they are located. These trends have been made possible by the development of a worldwide and integrated telecommunications network, whose unequal presence dictates the capabilities of places and communities to be connected to the global economy. However, the threat of the digital divide must not be overstated. In cities, rural areas, and emerging countries, local development is wrapped up in human capital, rather than technology.

We begin in Chapter 2 by setting the digital economy in its historical and economic context. Powerful and inexpensive combinations of computer processing and communications capability are thought to have created a Second Information Revolution, characterized by computer-enhanced communication with usage costs and prices that are insensitive to distance or distance-neutral and tending towards zero. The combination of reliable inexpensive mail service and widely available telegraph and telephone service in the middle of the nineteenth century was the First Information Revolution (Brock 2003). The dot.com bust in 2000–2001 dampened some of the enthusiasm for the rise of a new economy, leading some to suggest that both "the new economy and the Internet constitute nothing more than a passing fad" (Kotkin 2000: 11). The more useful conclusion might be that of Quah (2003: 316): "Aspatiality in digital goods does not imply that space no longer matters. Instead, the correct inference is that for digital goods transportation costs do not matter, so that all other reasons why geography is important now assume heightened significance."

The economic impact of IT has been uneven from country to country and across industrial sector. In part, this is because improvements in productivity do not take place simply because workers now use computers and access data from the company and elsewhere via the Internet. Large as well as subtle changes are needed—in organization, flexibility, links with suppliers, and marketing to existing and potential customers. These changes require expertise and skills specific to IT as well as an ability to understand and predict the market. Consequently, the jobs that are growing are those that require knowledge and generally, therefore, education. A second focus of Chapter 2 is the economics of digitized information, including its historical context. Previous waves of technological innovation have provided precedents, including "technological unemployment." The coincidence and overlap between the digital economy and the knowledge economy—and the jobs that will never be digitized—must be understood even as new technologies continue to enable economic space to splinter.

Antonelli (2003) sees that the specific design of the digital economy as it emerged in the US in the last decades of the twentieth century has positioned advanced telecommunications services as the backbone of the new economy.

> The provision of an array of dedicated, advanced and low costs communication services is a fundamental condition for the digital economy to spread into the system and for adopters to implement the sequential array of specific tailored incremental innovations which make the system more and more effective.
>
> (Antonelli 2003: 197)

Chapter 3 elaborates the layers and scales of the worldwide, integrated network and the economic geography of long distance, broadband infrastructures which enable the digital economy. The nodes of the digital network, the broadband technologies, and the spread of competition comprise a complex equation of "last mile" teleaccess. Broadband technologies, enablers of the digital economy, are diffusing unevenly, in response to distinct national variations in deregulation, unbundling, convergence and competition. Wired cities and connected communities are partly reality, partly hype, yet progress is being made towards widespread, ubiquitous access to networks. At the same time, it is only too evident that infrastructures remain uneven in their availability, especially in poor regions.

In Chapter 4, we examine how companies use information technology to adapt their operations and their organization to an increasingly global and competitive environment. Several trends should be identified, including the emergence of an integrated, IT-driven production cycle, which is an answer to the requirements of a customer-centric economy. An important purpose of the chapter is to disentangle the "cook-book" expressions, which have flourished in the business literature, such as "network company," "virtual enterprise," and "players and orchestrators." Simply stated, the core issue is about outsourcing and the creation of complex, multi-firm value chains, which endeavor to maximize the benefits they derive from the spatial division of labor. These "global networks of modular production" require sophisticated supply chains and IT-enabled logistics.

We untangle the multiscale geographies of electronic commerce and electronic finance in Chapter 5. Concepts and definitions are important for understanding electronic platforms for both business-to-business (BtoB) transactions and business-to-consumer (BtoC) e-commerce, or e-retail. Perhaps, as Drucker (1999: 50) suggested, the most revolutionary impact of the information revolution has been e-commerce, which was "totally new, totally unprecedented, totally unexpected. . . . In the mental geography of e-commerce, distance has been eliminated." The takeoff of retail e-commerce has been much slower than predicted, however, because there are both benefits and costs to virtual shopping—the tyranny of distance. The discussion of electronic finance shows the same tradeoff between electronic transactions and transportation of people, between globality and place-embeddedness. Multichannel, "brick-and-click" banking is becoming standard. Branches have not disappeared, and the digitalization of the financial industry has been concomitant with the growth of international financial centers, because the production of advanced financial services requires a great deal of face-to-face contact and embeddedness in a dense business milieu.

Chapter 6 addresses a topic on the policy agenda of many countries: the offshoring of corporate services. By splintering the economic space, companies have located offshore a host of information technology-enabled services (ITES). We assess the factors and limits of offshoring as well as its costs and hazards. The strategy for offshoring decision is based on both organization and geography. Organizationally, the segmentation of business services offshoring leads to the controversial subject of measuring offshoring and its consequences. How many jobs were lost to offshoring, and how many are likely to be lost in the future? Are the global consequences for developed countries a real threat or a "win–win game" for all?

We focus on India's "exceptional development pattern" and how India has become a major player in IT services and business process outsourcing (BPO). But the Indian IT industry might be oversold. It is challenged by other countries, and we are observing some cases of "backshoring"—that is, Western companies returning these activities back to their home countries. Indian firms themselves are investing in North America and Europe ("reverseshoring"). And telework, seen as a kind of "homeshoring" is growing as a credible alternative to some offshoring practices.

In Chapter 7 we address telework and telecommuting, central concepts to the digital economy with implications for both labor and production processes. Telework is a "multifaceted" concept whose problem of definition makes it both hardly measurable, but at the same time a seemingly massive phenomenon. For both employers and employees, there are pros and cons to telework. For companies, telework in practice is one of several "flexible work arrangements." For workers, telework centers and telecottages provide a third place—neither home nor business location—where they can connect with their employer, whether government or businesses. Self-employed teleworkers can operate micro-enterprises from home or be part of a "virtual" call center. The benefits of telework for energy consumption and urban traffic have been long promised, but the complexity of people's mobility and behavior biases calculations.

In Chapter 8 we grapple with the paradox of a "double-edged geography" of the Internet age, with its tendencies toward specialization and agglomeration, on one hand, and spreading out on the other (Leamer and Storper 2001). The Silicon Valley of California has become the archetype of the high-tech business cluster. Dawson (2001) has compiled 79 names, associated with 105 locations, which evoke more or less directly the name of the original. Many of these names, perhaps invented by local boosters or visiting journalists, have "stuck" and are now commonly used, such as Mobile Valley, Telecom Corridor, and Silicon Glen for Stockholm (Sweden), Richardson (Texas, US), and Edinburgh (Scotland, UK), respectively. Attempts to reproduce the Silicon Valley model results in a mixed evaluation: some places are succeeding in creating new "silicon valleys," such as Bangalore or Grenoble, whereas others will not succeed in creating anything but hype. The growth of IT industries, and their perceived nature as a strategic sector—along with biotech, have made them the driver of a kind of "digital urbanism." Although the phenomenon has passed his prime in developed countries, science parks and IT business campuses are mushrooming in emerging countries, especially in Asia and the Middle East. In these countries, the digital city is dreamed of as a living laboratory for the latest IT applications.

Chapter 9 focuses on the digital economy in peripheral regions. The widely popular concept of *digital divide* is at the core of the discussion. We make the distinction between, on the one hand, rural and industrial areas of developed countries and, on the other hand, developing countries. Rural areas suffer from an intrinsic penalty, which is defined by low density and remoteness from business agglomerations. Old industrial regions suffer from low entrepreneurship and low skills. Could IT help to bridge the gap? The broadband divide is shrinking—to a certain degree—and digital business is emerging in rural regions, through the creation of call centers and IT-enabled small businesses, or the development of telework. But no miracle should be expected, because economic development is mainly a question of improving human capital, which requires global solutions. In the developing world, the issue is posed in similar terms. Because IT was regarded as a tool for development, many have foreseen a leapfrogging phenomenon, some regions "jumping" from agrarian production straight into a digital economy. For sure, a degree of digital development can be identified, even in the poorest countries. Mobile telephones in Africa, for example, are a surprising success. People and communities in peripheral areas have proved unexpectedly innovative in adapting IT applications to local conditions, but these changes are incremental, and will probably not lead to the collapse of well-established paradigms such as the benefits of agglomeration.

INFORMATION TECHNOLOGIES AND THE "NEW ECONOMY" DEBATE

2

In the previous chapter, we have suggested that the digital economy is embedded in a larger concept, the digital paradigm, which encompasses technological and economic, as well as social and political aspects. However, the overabundance of metaphors has generated a great deal of ambiguity and technological determinism. In this chapter, we focus on the emergence of digital technologies, and the role they play in the so-called "revolution" we have been witnessing since the beginning of the 1990s.

The digital economy can be defined as "the pervasive use of IT (hardware, software, applications and telecommunications) in all aspects of the economy" (Atkinson and McKay 2007: 7). Therefore, it must not be confused with the information economy or the knowledge economy, although all three have evolved rapidly and simultaneously since the end of the twentieth century. But the links between the evolution of technology and the rise of a global digital society should not be underestimated. There is an economics of digitized information whose basics fundamentally explain the features of the whole information-based economy. Digitalization allows tradability and scalability of business information and changes its value as an economic good, since it can be stored, retrieved, duplicated, and communicated at very low cost.

We describe and assess an economy based on digitized information. The impact of IT implementation on productivity has been subjected to diverse interpretations and discussions in academic circles. The debate also revolves around the overall effects of IT within the emergence of a "new economy." We place the digital economy in historical perspective, as but the latest in a series of "waves" of innovation. Finally, we examine the knowledge economy as distinct from the digital economy.

The economy of digitized information

Information is a multifaceted concept. From a theoretical viewpoint, information measures the reduction of uncertainty, or the enhancement of certainty (Shannon and Weaver 1949). Hence the definition of the "bit," the acronym of "binary digit" in use in computers and telecommunications, which is the elementary particle of information which indicates the choice between two occurrences: zero or one.

Most importantly, the information of mathematics, computers, and telecommunications—the data of a series of zeros and ones—is decontextualized from any social or economic context. The quantity of information (the length of the zero/one series) has no relationship with its social or economic value. Access to billions of dollars in a bank account may require a few dozen bits, while terabytes of free content that circulate on the Web often have little, if any, actual value. Such a disconnection between the mathematical and the economic value of information has enormous consequences on the whole digital economy. Rallet (2000) puts forth a fundamental distinction between the two types of information, seen from an economic viewpoint.

On the one hand, information is a tool for reducing uncertainty in economic transactions, broadly speaking (and in line with Shannon and Weaver's classic ideas). But this information has no value in itself. Its value is fully contextualized. Information is most valuable when it is known by one person or a relatively small number of people, as suggested by this advertisement:

> At 12:43 p.m. EDT on Thursday, April 5, 2007, Dow Jones News Service broke this story: "Tracinda Prepared To Offer $4.5 Billion To Acquire Chrysler"
>
> In the 10 minutes before any other news outlet ran this story, DaimlerChrysler shares gained 4% and added more than $3 billion in market value.
>
> Time *is* money.
>
> (Dow Jones 2007)

On the other hand, information is the essential ingredient of a category of tradable goods and services we call "informational." In this case, information is a good in itself, whose production and market is the core of the "information economy": for example, media products, movies, and books (Varian 2000). Broadly, digital products are of three types, all of which embody information: (1) information and entertainment products (including paper-based information, graphics, audio, and video); (2) symbols, tokens, and concepts (including tickets and financial instruments); and (3) processes and services from businesses, governments, education and others (Whinston *et al.* 1997).

However, these categories are not mutually exclusive. Intangible goods and services, in particular, such as software products, licenses, business advice, or even school and university courses, are at the intersection between the two fields, as decision-making tools. For example, executives buy and read the business press because they need to be immersed in the business culture, which is made of a flow of general information, rather than of a single, specific piece of information. Another example: the decision to launch a new product may require the purchase of informational goods and services such as consumer databases and surveys.

The increasing digitalization of both kinds of information has enormous, and sometimes fearful, consequences, because of the increasing disconnect between the information itself (a series of zeros and ones) and the body of the information, which has shifted from word-of-mouth and paper to hard disks and flash memory. The first consequence is that the tradability of much information has virtually no limits and has become distance-neutral. For example, digital motion pictures can reach instantaneously millions of TV sets, at a cost approaching zero. Digital transactions and shipments are far cheaper:

> Processing a check costs banks approximately $1.40 compared to just 8 cents for processing an electronic bill payment. Shipping plastic movie reels to

> movie theaters across the world is significantly more expensive than transmitting digital movie files to theaters through broadband connections. Selling music in stores on CDs is much more expensive than selling it online. Taking pictures with film and processing each print is more expensive than using digital cameras.
>
> (Atkinson and McKay 2007: 24)

Therefore, we face a quasi-disembodiment of the economic good. Hence the current debate on intellectual property which has risen to unprecedented levels. Digitalization and the Internet endanger entire economic sectors, such as the music industry. The economic value of goods as diverse as video games, software and drugs, whose duplication costs have plummeted, is no longer in the good itself, but in the complex apparatus of patents, licenses, firewalls, legal action, and legislative lobbying which is necessary to create and to protect the rights of inventors and producers (Wusch-Vincent and Vickery 2005).

Changes are taking place rapidly in the value chains of digital content industries, with the creation of new commercial business models to exploit new opportunities. Digital content has become important not only in all media and publishing industries but also in sectors not previously considered to be content producers or users (for example, business services) and in the public sector: weather information, archives and cultural content, education and health (Batty 2006; Currah 2006; Leyshon 2001; Vickery and Wusch-Vincent 2006).

Reach and richness

The concepts of *richness* and *reach* provide a different view of digital production and IT-driven organizational changes in business, which helps to explain the emerging spatial patterns of the digital economy (Evans and Wurster 2000). "Reach" is the capacity to do business on a large scale, with a larger number of people, both in quantitative and geographic terms. "Richness" is the key characteristic of high value-added transactions which require mutual trust or the command of complex cultural codes, which incorporate a complex body of information and knowledge, or require real-time interaction between vendors and buyers. Richness often means geographic proximity, face-to-face relations, and operations within a narrow geographic scale (see also Leamer and Storper 2001). Therefore, economic transactions constantly face the tradeoff between richness and reach.

According to Evans and Wurster, the combined history of transportation and IT shows the constant shift of the tradeoff line between richness and reach, which become more compatible. The combination of richness and reach can be seen in consumer goods and services. For example, travelers in the pre-Internet era had to queue in front of transportation companies or the offices of travel agents to buy tickets. Today, they can make complex travel arrangements on websites or through call centers which are increasingly interactive and customized. In the case of e-commerce, the Internet clearly has expanded the *reach* of businesses that have products to sell. The market for information goods is truly global for anyone who has an Internet connection and a credit card. *Richness*, the depth of the information that a business gives to or collects about its customers, such as customization and recommendations based on prior purchases, is greatly enhanced by the Internet (Evans and Wurster 1999; 2000). The combination of greater richness and longer reach is also seen in business-to-business (BtoB) transactions. It can be seen, to the greatest degree, in the emergence of tools for product lifecycle management (PLM), as we show in Chapter 4.

Network effects

The value of a network increases with the number of people using that network or connected to that network. One frequently-cited version of this network effect is Metcalfe's Law, named after the inventor of Ethernet technology. Metcalfe's Law states that the value of a network is proportional to the square of the number of people using it (Metcalfe 1995, 2006). If the number of people using a network is doubled, the value to each participant is doubled, and the total value of the network increases fourfold.

Metcalfe's Law overstates network effects for several reasons. First, some users and pairs of users will derive more value from links than do other users and, in particular, early users benefit more than later users (Rohlfs 2005). Second, the interconnection of networks shows that, if two Internet service providers (ISPs) interconnect, the smaller ISP gains more than the larger one gains; this also reduces the incentive for large ISPs to interconnect. Third, not all other users are equally valuable to be connected to; connection to local users may be more valuable to local organizations, and users with the same language will be more valuable than users of other tongues (Briscoe *et al.* 2006). While communications networks—and other networks—do increase in value as they add members, they do so "much less than proportionately to the square of the number of users" (Briscoe *et al.* 2006; Rohlfs 2005: 89). Metcalfe's Law might apply more strictly for networks internal to an organization, where such inequalities are minimized (Rohlfs 2005).

Two sets of implications result from these relationships. First, the differentiated demand means that many niche markets at the shallow end—the long tail—of the demand distribution can collectively outweigh the more popular "hits." The costs of reaching those niche markets have fallen dramatically. Thanks to digital distribution, powerful search technologies, and a critical mass of broadband penetration, online markets have made it possible to offer a massively expanded variety of products. Moreover, a variety of tools and techniques, or "filters," such as search engines, recommendations and rankings, effectively give consumers ways to find products that suit their particular needs and interests (Anderson 2006; Briscoe *et al.* 2006). Much Internet content is not truly new, but rather consists of pre-existing forms of information now made available more cheaply and conveniently—such as airline schedules, stock quotes and trades, and catalog shopping (Gordon 2000).

Second, the economics of information and knowledge are what economists call "non-rival," meaning more than one person can have and use the information at the same time without diminishing the utility to others. This contrasts with many resource-based industries. The new economy also is based on a range of "intangible assets," including copyrights, patents, trade secrets, brand names, trademarks, and reputation. Investments on these, through research and development (R&D), advertising, marketing, and creativity, typically are not considered investments in company financial or national economic accounts (Nakamura 1999).

The new economy debate and the productivity paradox

Much of the discussion about the existence of a new economy focuses on the effect of IT implementation on productivity growth. The new economy is an economy "in which IT and related investments drive higher rates of productivity growth" (US Department of Commerce 2002: v). More specifically, the new economy "involves acquisition, processing and transformation, and distribution of information. Its three major components are the hardware (primarily computers) that processes information, the communications systems that acquire and distribute information, and the software that, with human help, serves to manage the systems" (Nordhaus 2002: 221–3). Oliner and Sichel (2002) broaden the definition of the new economy to include not only sectors that produce IT (hardware, software, communications equipment, and communication services) but also the many sectors that *use* IT. Distinct from IT producers and users is e-commerce as a class of transactions. Haltiwanger and Jarmin (2003) conclude that labor productivity growth after 1995 was driven by both greater use of IT capital goods and the more rapid efficiency gains in the production of these goods. Whether this growth is sustainable will depend on the pace of technological advance in the semiconductor industry and on the extent to which products embodying these advances diffuse throughout the economy.

The bulk of the evidence about the new economy and the benefits of investment in IT comes from the US, often ignoring completely the rest of the world (e.g. Alcaly 2003). The evidence at first was largely anecdotal and reported by journalists, beginning in the mid 1990s (Shepard 1997). The discourse was further promoted by politicians (Gadrey 2003). The new economy concept was promoted in Europe by the Organisation for Economic Co-operation and Development (OECD), connected to a series of policies: high technology, national (and regional) systems of innovation, globalization, knowledge-based economy, and information economy (Godin 2004).

Measuring the digital economy

The measurable impact of IT production and utilization is quite recent. Computer specialists began to be identified and counted in national censuses only about 1960, yet since that time they have been among the most rapidly growing occupations (Wyatt and Hecker 2006). The digital economy, narrowly conceived, or DigiWorld, as defined by IDATE (2006), encompasses all sectors in the upper parts of the pyramid of the digital economy (Figure 1.2). This definition allows us to track an indicator of the growth of the digital economy. These include:

- telecommunications services (fixed and mobile telephony; data and image transmission);
- telecommunications equipment (public network gear, private systems, devices, software, and services);
- computer software and services, including data processing;
- computer hardware (mainframes, PCs and peripherals, data transmission gear).

We exclude audiovisual services (television, radio, and film) and consumer electronics (audio and video equipment), whose outputs are end products rather than tools serving the entire economy. IDATE's definition is very similar to that used by the US Department of Commerce in its estimates of the digital economy in the US, the last of which was released in 2003. The IT-producing industries in the US include: hardware and communications equipment industries, software and services, and communications services (US Department of Commerce 2002, 2003). Note that many IT-enabled businesses are not included in these data of IDATE, such as financial services and CAD.

How big is the core of the digital economy? Although computers were adopted and utilized by companies since the 1960s, two inventions sparked more widespread digitization. The first was the personal computer, which meant that many more activities (word processing, spreadsheet accounting, database activities) could be done by many more people, including away from the workplace. The second was the Internet, made consumer-friendly by the World Wide Web and Internet browser software, and the standards of Internet protocol (IP), which enables encoding into digital format phone conversations, photos, music and other audio, and video.

As Table 2.1 illustrates, the three core sectors of the digital economy continued to account for growing shares of economic activity. Prices of telecommunications equipment and computer hardware have fallen since 2000, but telecom services and computer software have grown at a rapid pace. Until the mid 1990s, the digital economy (narrowly conceived) accounted for no more than 4.1 percent of world total gross national product (GNP). This share nearly doubled in the next decade, to 8.1 percent. The DigiWorld estimates are similar to those reported by the US Department of Commerce (2003: Appendix Table A-1.2), which showed a peak of 8.8 percent of US gross domestic product GDP in 2000, which fell to 7.9 percent in 2003.

IT and the productivity paradox

As people and companies take advantage of these new capabilities and those of other new technologies, the benefits are measurable as productivity growth: more output from a unit of capital and labor input. Productivity growth is the principal means of economic growth, by which a nation (or region) can increase its income and the well-being of its population.

Therefore, most of the debate about a "new economy" has logically centered on the effect of IT on productivity. The perceived phenomenon that huge investments in IT

Table 2.1 ***The growth of the world's digital economy core sectors (in billions of euros)***

	1995	*2000*	*2005*
Telecommunications equipment	150	306	252
Telecommunications services	439	732	1,025
IT hardware	220	276	263
IT software	263	498	623
Total (€)	1,239	1,934	2,339

Source: IDATE (2001; 2006)

over the past four decades have yielded very small gains in productivity has been dubbed the "IT paradox" or the Solow paradox. The paradox that "you can see the computer age everywhere but in the productivity statistics" was highlighted by Solow (1987: 36), a Nobel laureate in economics.

Productivity—measured as output per worker, as output per unit of input of other inputs, such as energy or capital, or as "total factor productivity" (TFP)—is the principal economic measure of economic progress. However, the concept applies more readily to outputs that are tangible than to those that are intangible, such as *services*. Services account for two-thirds or more of the GDP of developed economies. Nevertheless, they are very diverse and therefore are more difficult to define than manufactured goods. "Every definition of services is slippery" (Bryson *et al.* 2003: 7).

IT has been particularly important to service industries which require substantial information handling, such as banking, finance, health, transportation, utilities, communications, and business services (Quinn *et al.* 1987; Waite 1988). This trend is significant because it has radically changed the skills necessary for such jobs (Cyert and Mowery 1987; Guile and Quinn 1988; Levy and Murnane 2004). The expected measurable rise in productivity from capital investment in IT, however, has not accompanied this increase in capital intensity.

It was once the case that all internationally-traded services—whether consumed abroad by tourists, film producers, or medical patients, or produced abroad by construction workers, teachers, artists, or investors—were embodied in some way in people or products (Grubel 1987). This distinction has been reduced by "the 'copy at will' nature of the Internet" as digital files have made many more products "intangible" (Soete 2000). As the variety of digital information and entertainment increases, a larger number of companies are able to produce electronically delivered products. The greatest impact of the Internet has been felt in six sectors: computing and electronics, telecommunications, financial services, retailing, energy, and travel. All of these benefit from the fact that the Internet "puts the customer in charge as never before" (Hof 1999).

> Until the Net, buyers faced huge obstacles to extracting the best prices and service. Research was time-consuming, and everyone from producer to retailer guarded information like the crown jewels—which it was . . . For many companies, customer ignorance was a profit center.
>
> (Hof 1999: 86)

The intangible nature of services makes the measurement of productivity (output/worker) more difficult than in manufacturing where products are standardized. In many services, technology will not produce greater output, "the labor is an end in itself" and increases in productivity are sporadic at best (Baumol 1967). Many services involve "person-processing" in addition to information-processing, changing the state of people through educational, medical, or social services (Gershuny and Miles 1983: 137–8). The output of services is hard to measure, as in, for example, an education or a surgical operation, because quality is an important dimension (Waite 1988).

Brynjolfsson (1993) clears up much of the confusion in the productivity paradox surrounding investment in information technology (IT). Most of the explanation results from poor measurement inherent in productivity accounting, especially in its inability to capture either changes in the quality of output or benefits from greater flexibility (but not greater quantity) of output. Indeed, Brynjolfsson and Hitt (1996) have found, using a detailed disaggregated data set for the US, that investment in IT is significantly related to output growth. One of the first studies to demonstrate benefits of IT investment, it also attests to the need for reorganization of work in order for firms to benefit from computer technology (Gleckman *et al.* 1993).

Information technology can be classified into three types, according to McAfee (2006), each of which provides companies with a particular level of change. *Function IT* encompasses technologies, such as spreadsheet applications, that streamline individual tasks. *Network IT* includes capabilities that help people communicate and network with each other, such as e-mail, instant messaging, and departmental blogs. *Enterprise IT* brings with it larger goals, such as customer resource management and supply chain management, and allows companies to restructure interactions with business partners and between groups of workers. In general, enterprise IT mandates many changes, which workers resist, slowing change (McAfee 2006).

The quality of management can avoid employee resistance and, generally, lead to higher firm productivity. Better management is related to setting financial and nonfinancial targets, and creating incentives to boost flexibility and motivation (Bloom and Van Reenen 2006; Hubbard 2006). US multinationals are significantly better managed than European firms, Van Reenen and Sadun (2006) suggest, and therefore are able better to manage the complementary factors, producing better productivity gains.

Is the US unique?

Perhaps the US experience is unique. Bloom and Van Reenen (2006) find that competitiveness of the business environment, which is toughest in the US, is related to superior productivity performance. Examining labor productivity in France, Germany, the UK, and the US, Matteucci *et al.* (2005) find that the UK experience with IT has been closer to the US than to France and Germany. The evidence generally suggests that the US is exceptional, as IT effects remain elusive even in Finland (Daveri and Silva 2004) and Sweden (Edquist 2005). The Solow paradox "has fled the US and come to Europe" (Daveri 2002: 345). A larger study of 42 countries by Pohjola (2002) found no significant correlation between IT investment and economic growth. He cites three possible reasons for this: first, many countries, other than the US, have not yet invested greatly in IT. Second, even if an economy has invested in IT, it may not have invested sufficiently in complementary infrastructure, education, and skills. Third, the neoclassical method may not be able to capture essential aspects of the new economy or of the IT revolution.

Van Ark *et al.* (2003) found that US productivity has grown faster than in the EU because of both a larger employment share in the IT producing sector and faster productivity growth in services industries that make intensive use of IT. Two sectors (wholesale and retail trade and the financial securities industry) account for most of the difference in aggregate productivity growth between the EU and the US. Why? They conclude that structural impediments in product and labor markets hamper the successful implementation of IT in Europe. For example, limits on shopping hours, transport regulations, and restrictive hiring and firing rules make it hard for producers to organize their organizations to reap the full benefits from IT. Other barriers to entry also limit competitive pressure. However, it is not a story "based only on excessive European regulation. The more rapid takeoff of wireless technology in Europe suggests that some regulation, for example, setting standards can be productivity enhancing as well" (van Ark *et al.* 2003: 313).

The biggest differences between the EU and the US are in the contributions from computers and software and lower IT investment levels in the EU (Timmer and van Ark 2006: 703). The actual use of IT-dependent organizational changes, such as enterprise resource planning (ERP) software systems and online procurement, are positively correlated with labor productivity growth in France, Germany, Italy and the UK (Falk 2005). In the end, IT needs to be combined with substantial investments in new ways of working and new organizational structures, and with conceptual innovations that are often entirely of a non-technological nature, to result in better performance (den Hertog *et al.* 2003).

Measurement issues plague productivity analyses, casting doubt on the glowing reports of new economy "miracles" (Edquist 2005; Oz 2005). Indeed, Godin (2004: 687) concludes that, "After nearly 50 years of studies, one still looks in vain for hard data on the links between science, technology, and productivity." Part of the difficulty is the fact that there are several innovation models: science-based innovation, horizontally organized innovation and collaboration among users and/or doers; and modular structures within an innovative system (Foray 2005). Territorial innovation models explain how innovation affects places (Moulaert and Sekia 2003).

The observed contribution of computerization is accompanied by relatively large and time-consuming investments in complementary inputs, such as organizational capital, that may be omitted in conventional calculations of productivity (Brynjolfsson and Hitt 2003). As computers become cheaper and more powerful, the value of computers to firms is derived less from their computational capability than by the ability of managers to invest in new processes, procedures, and organizational structures that leverage that capability (Brynjolfsson and Hitt 2000; Mason and Apte 2005). However, early attempts to promote "digital economy" thinking and internetworked business (e.g., Tapscott 1996) were ahead of actual company experience with the Internet and with new forms of knowledge flow. Gordon (2000) makes a criticism which, despite a trend towards multi-tasking, remains valid:

> The fundamental limitation on the contribution to the productivity of computers in general and the Internet in particular occurs because of the tension between rapid exponential growth in computer speed and memory on the one hand, and the fixed endowment of human time.
>
> (Gordon 2000: 72)

Oz suggests that, in addition to measurement problems, research has failed to find a clear impact of IT investment on productivity because of the evolution of IT. He suggests that the economics of a new technology follows a simple cycle:

- first, adoption of IT results in economic profit for the first early adopters;
- second, most adopters enjoy increased profit from the new technology, but other firms have difficulty implementing it;
- third, new IT becomes standard as the technology diffuses throughout the industry, firms can no longer survive without it, and productivity gains shrink;
- fourth, a continuous decrease in prices (enabled by the productivity gains), and eventually,
- fifth, productivity disappears as a larger quantity of the products is sold for lower prices at lower profit margins.

(Oz 2005)

At the end of the cycle, productivity gains are not measurable but all businesses must adopt the new technology to survive. Indeed, firms might choose to make investments in IT for several reasons—only one of which is enhanced productivity. The other, intangible outputs of IT include improved quality, convenience, variety or timeliness, all of which could benefit customers and thereby improve a firm's market share and competitive position (Brynjolfsson and Hitt 2003). "For most companies, just staying in business will require big outlays for IT" (Carr 2003: 48). In most industries, websites and e-commerce amount to little more than "flashy but mainly superficial change" (Schlender 1999: 138).

> The key question is not whether to deploy Internet technology—companies have no choice if they want to stay competitive—but how to deploy it. . . . Gaining a competitive advantage does not require a radically new approach to business. . . . The Internet per se will rarely be a competitive advantage.
>
> (Porter 2001: 64)

Such changes take place very unevenly in an economy, differing among industries and among firms within each industry (Cortada 2004). Just six sectors of the US economy accounted for 99 percent of labor productivity growth from 1995–1999; in 53 other sectors, there was little or no productivity improvement (Lewis *et al.* 2002). The six sectors, in order of the size of their productivity gains, were: wholesale trade, retail trade, securities, semiconductors, computer manufacturing, and telecommunications. In the sectors with improved productivity, "fundamental changes in the way companies deliver products and services" were sometimes "aided by technology, sometimes not" (Lewis *et al.* 2002: 33).

In US retailing, innovations pioneered by Wal-Mart included a larger, "big box" format and "everyday low prices," and scale economies in purchasing and warehouse logistics, none of which involved IT, as well as electronic data interchange (EDI), which centered on IT. Indeed, "more than half of the productivity acceleration in the retailing of general merchandise can be explained by only two syllables: Wal-Mart" (Johnson 2002: 40). IT explains part of Wal-Mart's success. "No set of industries did more to make information tools and artifacts so obvious or so present in American life as did retailers" (Cortada 2004: 318). Wal-Mart invested in various systems (very small aperture terminal (VSAT) satellites, bar codes, wireless scanners and, recently, RFID) earlier and more aggressively than did its competitors, and these investments were focused on the goal of low prices (Housel and Skopec 2001; Johnson 2002).

The benefits of IT come about only through innovations in business practices, implemented incrementally rather than in "big bang" initiatives (Brown and Hagel 2003). However, "knowing how to use IT effectively is still a relatively scarce skill" because it is not IT itself that produces competitive advantage, but the people who know how to use it effectively (Varian 2003: 112). This set of not just complementary innovations, but also "organizational complements," such as business processes, decision making structures, incentive systems, human capital, and corporate culture, represent complementary "organizational capital." However, these intangible assets and investments in them are not typically treated as capital assets in firm or national accounts and have therefore not been fully incorporated into economic analyses of the impact of computers (Brynjolfsson and Hitt 2005).

Lipsey *et al.* (2006) propose a wider concept called *technological complementarities*. New general purpose technologies (GPTs) rejuvenate the growth process by creating technological complementarities, which are adequately measured neither by total factor productivity (TFP) nor by externalities. Growth is driven by the technological complementarities created by successive GPTs. The new connections made possible by the Internet have permitted a vast range of new combinations and collaborations in a new digital economy (Carlsson 2004).

It is still too soon to judge the long-term effects of IT on productivity (McKenzie 2003). In the absence of clear relationships between investment and productivity, Port *et al.* (1988) suggest that the "new math of productivity" uses time as the key variable, not return on investment, especially in investments to improve flexibility. Further, measuring knowledge within firms is now an objective in order to measure company intellectual capital and intangible assets such as knowledge and their contribution to company productivity and competitiveness (Engelbrecht 2003; Stewart 1997).

From studies of the impact of IT investment in developed countries, Indjikian and Siegel (2005) conclude that there is evidence of a strong positive correlation between IT and economic performance. In addition, both IT-induced changes in workforce composition in favor of highly skilled or educated workers and organizational changes allow firms to implement IT more effectively. More than in developed economies, policy-makers in developing countries must address two key deficiencies: first, a lack of knowledge of "best practices" in IT usage and, second, IT-related skill deficiencies in the workforce.

Table 2.2 lists the most IT-intensive industries in the US, measured as IT investment as a share of total investment. Such lists and definitions vary from country to country but, generally, the industries that are highest in IT-skilled employment are both "creative" industries, such as financial services, and "high-tech" industries, which produce equipment for IT-using industries. The industries lowest in IT intensity are construction, agriculture, and various labor-intensive services, including hotels and restaurants, health and social work, and education, all of which have IT intensities below 10 percent (van Welsum and Vickery 2005).

Box 2.1 General purpose technologies

The pervasive nature of IT is captured by the inclusion of the Internet as a *pervasive* or *general purpose* technology (GPT). GPTs are a small number of drastic innovations that create innovational complementarities that increase the productivity in a downstream sector (Bresnahan and Trajtenberg 1995; Helpman 1998). A GPT is characterized by the potential for pervasive use in a wide range of sectors, by its technological dynamism, and as an "enabling technology," opening up new opportunities rather than final solutions (Bresnahan and Trajtenberg 1995). GPTs have wide impact because of their scope for improvement, wide variety of users, wide range of uses, and strong technological complementarities. Their positive externalities spread both horizontally within a sector and vertically to other sectors.

The most important GPTs from history can be grouped into six main classes:

> *Materials technology*: domesticated plants, domesticated animals, bronze, iron, biotechnology
> *Power technologies*: domesticated animals, the waterwheel, the steam engine, electricity, the internal combustion engine, the dynamo
> *Information and communications technologies*: writing, printing, the computer, the Internet
> *Tools*: the wheel
> *Transportation*: domesticated animals, the wheel, the three-masted sailing ship, the railway, the iron steamship, the motor vehicle, the airplane
> *Organization*: the factory system, mass production, lean production.
>
> (Lipsey *et al.* 2006: 133)

Atkinson and McKay (2007: 9) consider IT as a general purpose technology for three reasons. First, IT is pervasive, found in nearly every sector from farming to manufacturing to services to government. Second, price declines and performance improvements in IT have been unprecedented and show few signs of ending. Third, IT has enhanced dramatically the ability to develop new business models (such as business process outsourcing (BPO), e-businesses, and new logistics systems); new products and services (such as smart phones, the iPod, and digital media services); new processes (e.g. self check-in at airports); and fundamental new inventions, such as mapping of the genome years faster than most had predicted.

Creative activities overlap but do not coincide with the digital economy. Creative work in the arts and in education remains relatively untouched by the digital economy, yet these fields are among those whose workers are *symbolic analysts* (Reich 1991). It is notable that, although the retail sector in Europe is quite low in IT intensity, several retail sectors in the US have IT intensities greater than 30 percent (van Welsum and Vickery 2005).

Table 2.2 ***Industries with highest IT intensity (IT investment as share of total investment)***

Industry	*IT intensity (%)*
Business services	89.5
Insurance	87.6
Communication	84.6
Information services and data processing	82.3
Drugs	77.8
Household audio and video, and communication equipment	75.7
Motion pictures, including TV tape and film	72.3
Electric and electronic components and accessories	68.0
Electric and electronic components, not elsewhere classified	62.9
Printing and publishing	59.8
Finance (except depository institutions)	59.0
Transportation	56.5
Computer and office equipment	48.1

Source: Adapted from Nachum and Zaheer (2005: 767)

IT has "potentially large effects for long term economic growth" through three important channels: first, a "production effect" in producing sectors; second, a "use effect" in manufacturing and service industries that have made investments in IT; and, third, a "spillover effect" sparked by the inventions and innovations that emerge through the role of IT as a general purpose technology (van Ark 2002). Atkinson and McKay (2007) distinguish between the direct and indirect benefits of IT. The direct effects are those on productivity: IT lets workers do more things at the same time, allows routine tasks to be automated, thereby increasing economic output, enables organizations to dramatically improve the efficiency of internal operations, and lets firms restructure their supply chains.

The digital economy also has indirect effects. Larger markets and better decision-making enable goods and services to be allocated more efficiently by making it easier for consumers to get more information to make better purchasing decisions. IT enables higher quality products and services through quality monitoring and mass customization. IT gives researchers powerful new tools, enables small firms to expand their R&D, boosts innovation by giving users more of a role, and lets organizations better manage the existing knowledge of their employees (Atkinson and McKay 2007).

The latest wave: IT and the Fifth Kondratieff

Among the supporting evidence for the "revolutionary" character of IT are long waves or Kondratieff waves that describe the ups and downs of economic activity, in cycles of 50–70 years duration, over the past three centuries. Identified first by the Russian economist Kondratieff in the 1920s (Kondratieff 1935), scholars have continued to identify upturns associated with clusters of innovations and of innovative industries (Boschma 1999; Hall and Preston 1988). There is considerable agreement that a "Fifth

Kondratieff" wave has begun or is about to begin, based on a new core input: the integrated circuit or "chip" and biotechnology (Figure 2.1). Freeman (1996b) looks further towards a "green" future, based on renewable energy and teleworking.

Kondratieff waves can be thought to combine as larger "revolutions." The First Industrial Revolution is the combination of the first and second Kondratieff waves, and can be characterized as the factory-based industrial economy. The third and fourth waves comprise a Second Industrial Revolution, resulting in the corporate, mass production economy. The current Information Revolution is the Third Industrial Revolution—a digital, entrepreneurial, knowledge-based economy, based on continuous discovery (Atkinson 2004; Freeman and Louçã 2001; Suarez-Villa 2000). The current wave is underpinned by the cluster of information and communication technologies, comprised of computers, telecommunications—"the infrastructure of the new economy"—and by a wave of organizational and management changes in enterprises and in the regulatory regime (Louçã 2003).

Kondratieff wave	**Approximate timing**	**Managerial and organizational changes**	**Transport and communications infrastructure**	**Constellations of innovations**
First: Industrial revolution	1780s-1815	Factory systems, entrepreneurs, partnerships	Canals, turnpike roads, sailing ships	Water-powered mechanization of industry
Second: Age of steam power and railways	1848-1873	Joint stock companies, subcontracting to craft workers	Iron railways, telegraph, steamships	Steam-powered mechanization of industry and transport
Third: Age of electricity	1895-1918	Professional management systems, Taylorism, giant firms	Steel railways, steel ships, telephone	Electrification of industry, transport and the home; chemical products
Fourth : Age of mass production	1941-1973	Mass production and consumption, Fordism, hierarchies	Radio, motorways, airports and airlines Mass media	Petrol-powered motorization of transport, civil economy, and war
Fifth: Age of microelectronics and computer networks	1990s-	Networks: internal, local and global	Information highways, digital networks, the Internet	Computerization of the entire economy; biotechnology
Sixth: The green techno-economic paradigm	2050-	Teleworking	Telematics	Renewable energy

Figure 2.1 ***Successive waves of technological change***

Source: Based on Freeman (1996b: 131, Table 7.1) and Freeman and Louçã (2001: 141, Table II.1)

Some aspects of work, reliant on the best means of telecommunications available, have been more or less the same since the invention of the telegraph. Media hype similar to that about the Internet also surrounded the emergence of the telegraph and the telephone (Thrift 1996). Almost immediately after the invention of the telegraph, businesses—notably banks and other financial service firms—began to take advantage of the ability to exchange information without simultaneous human movement (Beniger 1986; Headrick 1991). News agencies such as Reuters also "followed the cables" to create a "global electronic space" in the mid-nineteenth century (Rantanen 1997).

Innovations, including those in IT, generally have diffused much more rapidly in the twentieth century (Mensch 1979; Suarez-Villa 2000). Figure 2.2 illustrates that it has taken less and less time for new products to spread into the US population. It took 46 years for 25 percent of American homes to be wired for electricity. Recent digital innovations have diffused much more quickly: the PC took only 16 years, the cellular phone 13 years, and the Internet seven years.

The digital economy is the product of many individual innovations. The twentieth century was remarkable for the coalescence of several distinct innovations. A list of the 20 greatest engineering achievements of the twentieth century includes several that are fundamental to the digital economy: electrification, without which none of the others could operate, electronics, the foundation of the digital economy, the telephone, the Internet, and computers. Lasers and fiber-optics have transformed telecommunications further, and applications in household appliances, imaging, and health technologies combined to make the twentieth century a "digital century" (Constable and Sommerville 2003).

Despite the apparently more rapid diffusion of individual innovations seen in Figure 2.2, technology systems take decades from their discovery and development in a laboratory

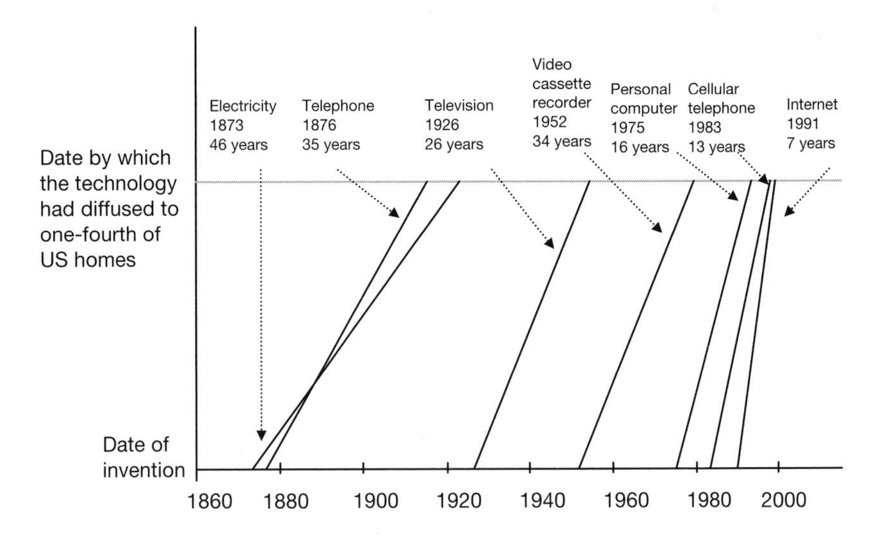

Figure 2.2 ***The increasing speed of adoption of new technologies***

Source: Adapted from data in Cox and Alm (1997: Exhibit D)

through their widespread implementation, and many (e.g. railroads) are used in several subsequent epochs. Equally important is the fact that regime of regulation must change "to accommodate not only the new technologies, but also changes in the balance of power in international relations, in the economic strength of the various contending powers, and in the culture and ideology of the dominant social groups" (Freeman and Louçã 2001: 365; Kogut 2003). The effects of IT take some time to provide general gains in productivity and welfare. The widespread utilization in industry and homes of one of those past inventions, electricity, took several decades (David 1991).

A final, large-scale shift coincided with the rise of the digital economy, and they have reinforced each other: the rise of flexible, post-Fordist production. Not reliant on IT, flexible production takes advantage of diversification of consumer tastes towards new varieties, versions, colors, and flavors. This is in sharp contrast to Henry Ford, whose Model A car in the 1930s was available in "any color you want, as long as it's black." Brynjolfsson *et al.* (1994) see the rise of smaller firms as an outcome of new, alternative forms of organizing, including outsourcing and partnerships rather than vertical integration.

The resultant product differentiation means that fewer products follow the traditional product lifecycle to become mature, high-volume, low-profit commodities. More products, aimed towards niche markets or even customized for individual consumers, remain profitable. To manufacture products, such tailored production requires IT: digital designs, computerized machinery, and inventory systems. In services, databases of customer preferences are the norm, now linked to service centers worldwide via the Internet. In several sectors, product lifecycles are extremely short. Financial advice "can become obsolete almost immediately." Advertising campaigns may last only a few weeks. New television and print broadcasting and media are produced daily or weekly. In fashion and performing arts, products change with the seasons (Schoales 2006: 167). In all these industries, uniqueness is a combination of creativity and constant awareness of ideas and imitators, which is greatly enhanced by the Internet and other digital technologies.

A knowledge economy: more than IT

The digital economy has emerged and grown in conjunction with what has become known as the *knowledge economy*. Although the term knowledge economy is most common (Drucker 1999; Foray 2004; Neef 1998; Stewart 1997), a number of other metaphors are used to describe an economy based on production of knowledge rather than production of goods:

- Cognitive capitalism (Ascher 2000)
- Knowledge-based economy (OECD 1996)
- Knowledge capitalism (Burton-Jones 1999)
- Knowledge economies (Cooke 2002)
- Net economy (Muller 2001)
- Post-industrial society (Bell 1973)
- Technocapitalism (Suarez-Villa 2000)

As the most intangible product, knowledge is more than information, just as information is more than simply data. In general, Boyer (2004) believes, the knowledge economy has replaced the new economy as a source of innovation and economic growth. Perhaps it is even more than that: knowledge as the critical production factor is bringing about "an all-encompassing realignment of the cutting edge of capitalism to emphasize information and knowledge creation" (Kenney 1997: 87). These new concepts and metaphors in the business lexicon are part of attempts to encompass the changes—"reengineering" the roles and activities of workers in new configurations of the work process as well as developing entirely new products (Boyer 2004; Kenney 1997).

The progression from economic activity based primarily on production of goods to one based on production of information and knowledge is part of the larger process of *structural change*. The shift in economic structure, among advanced economies—from primary economic activities (agriculture, fishing, forestry, and mining) to manufacturing and, more recently, to services—remains central to development economics (Syrquin 1988). The sectoral shift to services is fundamental to the "post-industrial society" (Bell 1973). More recently, services have been separated into those which involve a substantial information content and transactions, analysis, research, or decision-making (*quaternary* services) and those which do not (*tertiary* services) (Gottmann 1961; 1970). Harper (1982) has proposed a fifth "quinary" sector which represents control functions but not routine information processing activities.

Even more complex than information are knowledge, competence, expertise, and creativity (Table 2.3). While data and information can be codified and digitized, this is not the case for knowledge, competence or creativity, concepts central to a knowledge economy and, perhaps even more, to a creative economy (Florida 2002a).

Table 2.3 ***From data to creativity: concepts related to information and knowledge***

Concept	*Characteristics*
Creativity	Creativity presumes a capacity to order and reorder information with the aid of a knowledge system.
Expertise	Specialized, deep knowledge and understanding gained via experience. Expertise is personalized. An individual with expertise is able to create new knowledge is his or her area of expertise.
Competence	Embodied knowledge. There are at least three types: (1) instrument-oriented competence, (2) sector-specific competence, and (3) regional-specific competence.
Knowledge	Structurally ordered information. Includes reflection, synthesis, and context. Information laden with experience, truth, judgment, intuition and values. Concepts, ideas and patterns are subsets of knowledge. Often tacit, hard to transfer.
Information	Data endowed with relevance and purpose.
Data	Simple observations of states of the world; easily structured, easily captured on machines, easily transferred.

Source: Adapted from Andersson (1985: 13), Bender and Fish (2000), Bohn (1994), Davenport (1997), Earl (2001), Huseman and Goodman (1999), Malecki (2000), Törnqvist (1983), and Wood (2002b)

The measurement of the "knowledge-intensive" or information-based sector, was defined first by Machlup (1962) as consisting of education, research, publishing, and broadcasting; it accounted for 31 percent of the US workforce in 1958. The composition of the "information economy," developed by Porat (1977), is broader. In this conceptualization, information workers are defined as those engaged in creating or processing information, in three categories: 1) those whose final product is information; 2) workers whose main activity is informational in nature, including information creators ("knowledge workers"), information transmitters, and information processors; 3) workers who operate information technologies (Katz 1988: 5). Most recently, Drennan (2002) considers three sets of industries as the information sector: 1) financial producer services, 2) communication, business, professional and legal services, and 3) advanced consumer services (including professional sports, music, theater, motion picture production, museums, private health services, and private education. "In a nutshell, the economy is becoming more information intensive, more based on intangible assets and skills (both technological and non-technological), and progressively more dependent on new and emerging services and less on traditional manufacturing industries" (Castellacci 2006: 853). Knowledge-based competition requires, as we emphasize in Chapter 4, "new forms of collaboration and competition that the changing organizational patterns (e.g. networking and e-commerce) and the increasing degree of globalization of innovative activities are leading to" (Castellacci 2006: 855).

The economics of information, discussed above, is the basis of "new growth theory," based on knowledge and, technology and human capital or education. This theory, proposed by Romer (1990) and others, is typically taken to illustrate that knowledge and its production are the basis of the wealth of nations (Warsh 2006). However, the new growth theory still lacks several key elements of the real, working economy: first, technological advance as an essentially disequilibrium process; second, a theory of the firm in which capabilities and differences across firms are central elements; and, third, incorporation of a richer body of institutions, especially universities (Nelson 1998).

The variety of goods and services, and the knowledge on which they are based, means that work itself has become more complex and varied. Levy and Murnane (2004) distinguish among five broad kinds of tasks carried out in modern economies: expert thinking, complex communication, routine cognitive tasks, routine manual tasks, and non-routine manual tasks.

Knowledge-based tasks are the norm in the first two of these. *Expert thinking* includes solving problems for which there are no rule-based solutions. Examples might include diagnosing the illness of a patient whose symptoms seem strange, creating a good tasting meal from the ingredients that are fresh in the market that morning, or repairing a vehicle that does not run well but that has no problem according to the computer diagnostics. Computers cannot be programmed to do these tasks, but IT can complement humans in performing them by making information more readily available. *Complex communication* involves interacting with humans to acquire information, to explain it, or to persuade others of its implications for action. Examples might include a manager motivating the people whose work she supervises, a biology teacher explaining how cells divide, or an engineer describing why a new design for a machine is better than previous designs (Levy and Murnane 2004).

The number of jobs involving these non-routine, professional tasks has been growing for decades (Shapero 1985). Reich (1991) calls them *symbolic analysts*, whose work includes problem-solving, problem-identifying, and strategic-brokering activities. These jobs require face-to-face contact and collaboration as they customize their service to specific clients (Reich 1991).

Johnson *et al.* (2005) sort all jobs into three categories: *transformational* (extracting raw materials or converting them into finished goods), *transactional* (interactions that unfold in a rule-based manner and can therefore be scripted or automated), and *tacit* (more complex interactions requiring a higher level of judgment, involving ambiguity, and drawing on tacit, or experiential, knowledge). Transactional interactions include clerical and accounting work, which companies have been automating or eliminating, as well as work that IT specialists, auditors, and biochemists do. Demand for tacit workers varies among sectors. The jobs of most employees in air transportation, retailing, utilities, and recreation are transactional. Tacit jobs predominate in health care, financial services, and software. All sectors employ tacit workers, however, and demand for them is growing, perhaps most critically for "savvy frontline managers" (Johnson *et al.* 2005: 26). Florida (2002a) proposes a broader set of "creative class" workers, most of whom are high-tech, professional knowledge workers.

While creative, artistic jobs may seem to be little involved with IT, the pervasiveness of IT is having large impacts on many creative industries. Power and Jansson (2004) detail how the music industry—in particular firms that combine in innovative ways music and IT—has shifted from a focus on the production of albums and songs to include links to a varied music services industry that produces everything from remixes to music marketing strategies. The new combinations include: selling and distributing music over the Internet; web design and computerized advertising services tailored to music products; software design focused on multimedia products and virtual instruments; high-tech post-production and mixing services; and virtual communities of music industry actors.

Knowledge as both an input and as an output provides a useful framework for viewing economic activities. Arthur (1996) suggests a distinction between "Marshall's world" and neoclassical economic thinking where decreasing returns prevail, as remains the case in traditional, bulk-material manufacturing, and the world of increasing returns—based on knowledge, which generates not equilibrium but instability and cumulative causation. Hallmarks of the knowledge-based world are market instability, unpredictability, ability for lock-in, possible predominance by an inferior product, and fat profits for the winner. The knowledge-based world, including services, is managed best by constant adaptation, not optimization (Arthur 1996). Providers of "knowledge-intensive business services" or KIBS, provide *specialist knowledge* by selecting, adapting, and applying relevant knowledge to specific circumstances (Hipp and Grupp 2005; Wood 2002a).

Knowledge jobs

IT has multidimensional impacts: those of a benign information society and the job creation of the IT sector, along with automation, job destruction and the creative destruction of IT paradigm change (Freeman 1996a). IT is also changing the labor market in important ways. Computerization and use of the Internet are associated with greater hours worked as well as higher wages, and IT occupations are rapidly increasing their share of employment (Freeman 2002). Many of the new jobs that have been created over the past two decades are fundamentally different from the ones that have been lost, and the "new jobs tend to favor educated workers over those with less education and skills." Powell and Snellman propose a first-mover advantage: more educated workers are garnering the lion's share of gains from the knowledge economy, because they were "present at the founding" to "capture increasing returns from the knowledge economy" (Powell and Snellman 2004: 214).

The polarizing nature of the digital economy can be seen in projections by the US Department of Labor of occupations that are expected to gain—and to lose—the largest numbers of jobs between 2004 and 2014.

> Among the ten major occupational groups, employment in the two largest in 2004—professional and related occupations and service occupations—is projected to increase the fastest and add the most jobs from 2004 to 2014. . . . These major groups, which are on opposite ends of the educational attainment and earnings spectrum, are expected to provide about 60 percent of the total job growth from 2004 to 2014.
>
> (Hecker 2005: 71)

The US economy (and other advanced economies) will have a massive need, on the one hand, for low-skilled, non-cognitive workers (retail salespersons, janitors and cleaners, truck and light truck drivers, home care aides, food preparation and serving workers, nursing aides, and child-care workers). This portion of the economy is becoming neither cognitive nor digital. The US also is projected to need many middle-range skilled workers in non-IT centric occupations, notably nurses and teachers. The need for people to perform routine, non-cognitive tasks is not decreasing. On the contrary, a larger elderly population to care for, more people to serve at restaurants and fast food establishments, and more trucks to drive (stimulated in part by e-commerce) expand the demand for labor for these occupations (Hecker 2005).

On the other hand, however, employment in professional and related occupations, including IT-centric, knowledge economy occupations, is projected to experience the greatest job growth.

> Of the eight occupational subgroups within professional and related occupations, three—health care practitioner and technical occupations; education, training, and library occupations; and computer and mathematical science occupations—are expected to account for nearly 75 percent of the new jobs. . . . Computer and mathematical science occupations are projected to add 967,000 jobs and grow the fastest among the eight professional subgroups.
>
> (Hecker 2005: 72)

In addition, while not as ubiquitous as the occupations expected to grow fastest, *IT-enabled occupations*, including customer service representatives, accountants and auditors, office clerks, receptionists and retail clerks, and executive secretaries and administrative assistants, are among the top 30 in numbers of new jobs projected during the decade. All of these jobs involve a combination of *tacit* interactions, complex communication, and expert thinking.

By contrast, the occupations expected to see the largest declines in total numbers of jobs—including farmers and ranchers, stock clerks and order fillers, sewing-machine operators, file clerks, and order clerks—tend not to use computers. That is, they are generally non-IT occupations.

If more jobs require skills, who will do those jobs? Levy and Murnane (2004) predict that the greatest impacts of computerization will not be "mass unemployment," but a continued decline in the demand for workers in routine IT-enabled jobs (such as typists, or low-level clerks), whose tasks can be replaced by IT solutions. Freeman (2005) has examined the recent entry of China, India and the former Soviet bloc into the global economy. The large numbers of educated and skilled workers in those countries has doubled the global labor force. Levy and Murnane (2004) believe that the shift to India and China is only temporary, and will moderate as wages there continue to rise. The trend towards outsourcing work to these and other countries is part of the globalization of the digital economy, as later chapters will show.

Measurement of the new, knowledge-based economy is not easy. A respected series of reports in the US use five key economic dimensions for the new economy at the state and local scale: knowledge jobs, globalization, economic dynamism, the digital economy, and innovation capacity (Table 2.4). The dimensions of the overall "new economy index" include a mix of measures of knowledge and innovation, measures of IT connectivity and use, global links, and entrepreneurship (Atkinson and Correa 2007).

A set of benchmarks of 125 of the world's leading knowledge economy regions measures four sets of input components—human capital, knowledge capital, financial capital, and knowledge sustainability—and a set of regional economy outputs (Huggins *et al.* 2005). The combination of 19 variables yields an index that favors regions in the US; the top regions include Silicon Valley, San Jose, Boston, and San Francisco. Stockholm, ranked seventh, is the highest-ranked region in Europe; Uusimaa (Helsinki), ranked twentieth, and Île-de-France, ranked twenty-ninth, are the other European regions in the top 30. Tokyo, ranked twenty-third, is the highest-ranked region in Asia.

While these measures or benchmarks focus explicitly on knowledge and IT, the inclusion of knowledge and IT is growing in rankings of competitiveness, such as the annual *Global Competitiveness Report* of the World Economic Forum (WEF) and the *World Competitiveness Yearbook*, produced by the International Institute for Management Development (IMD). In general, knowledge-based priorities, including learning and external (even global) scanning are what "intelligent" cities and regions do, as well as continual monitoring and periodic benchmarking of what "the competition" is doing (Malecki 2004b).

Table 2.4 ***Indicators in the New Economy Index***

Dimension	*Indicator*
Knowledge jobs	Information technology jobs (as percent of total jobs) Managerial, professional and technical jobs (as percent of total jobs) Workforce education level (educational attainment of the entire workforce) Immigration of knowledge workers Manufacturing value-added High-wage traded services
Globalization	Export focus of manufacturing and services (as percent of jobs dependent on exports) Foreign direct investment (as percent employed by foreign companies) Package exports
Economic dynamism	"Gazelle" jobs (as percent of jobs in fast-growing companies—those with sales revenue that has grown 20% or more for four straight years) Job churning (business start-ups and failures as percent of all firms) Fastest growing firms (the number of Deloitte Technology Fast 500 and Inc. 500 firms as a share of total firms) Initial Public Offerings (IPOs) (value of IPOs as percent of gross state product) Entrepreneurial activity Inventor patents
Transformation to a digital economy	Online population (as percent of adults with Internet access) Commercial Internet domains (number per firm) Technology in schools E-government Online agriculture Broadband telecommunications
Technological innovation capacity	High-tech jobs (jobs in electronics, software and computer-related services, and telecommunications as percent of total employment) Scientists and engineers (as percent of workforce) Number of patents issued (per 1,000 workers) Industry investment in R&D (Industry-performed research and development as a percentage of total worker earnings) Venture capital invested (as a share of worker earnings)

Source: Atkinson and Correa (2007)

Conclusions

The information economy—the shift towards knowledge-intensive work—is not the same as the Internet-based "new economy." The information economy began long before and separately from the development of the technologies that comprise IT. In combination with the emergence of the knowledge economy, a wide range of impacts—both beneficial and detrimental—are seen. In addition to an (uneven) increase in productivity and complementary organizational changes, global production is widespread through offshoring, which is facilitated by global networks. At the same time, many small firms do not see the need to fully join the Internet age. In a survey of 500 small US businesses, 40 percent do not have a website (Holmes 2007). Thus, the digital economy is still characterized by a large base of firms and jobs that barely participate in the digital economy, as seen in Figure 1.2.

What this suggests is the parallel growth of two large groups of industries and occupations—those which depend on IT and those which do not—and an overlap or intersection of industries and jobs characterized by both. While we focus in this book mainly on several examples of business organization which take advantage of the new capabilities of the digital economy, we also recognize that more than IT is needed to create a viable business and regional development. As this chapter has shown, knowledge is central to awareness and successful implementation of IT, and is part of a larger process of economic development. Thus, there is a mutual dependence of the digital economy and the knowledge economy. Rather than pervasive, the digital economy is growing but does not yet—and probably never will—absorb all economic activity, as we have suggested in the previous chapter.

In the next chapter, we turn to a technological aspect of the digital economy: the telecommunications networks which are fundamental to global economic and social interactions. The digitalization and the convergence of telecommunication technology, both wired and wireless, favor the integration of the networks and the rising spatial ubiquity of connection capabilities. The geographic patterns of the digital economy and the economic development of places and regions, in turn, depend to a large degree on these networks.

3 WHERE LOCAL MEETS GLOBAL

The rise of the digital network

The digital economy is based on three interlinked, converging technological trends: first, the commoditization of fast, cheap computing capacity and data storage; second, the domination of standardized software platforms which allow digital interconnectivity—notably the Internet; and third, the building of a worldwide, integrated, digital network. "In short, IT has consistently gotten 'faster, better, cheaper, and easier'" (Atkinson and McKay 2007: 9).

This third element—the network—is critical in the way it shapes the ubiquity of the digital economy across time and space. From virtually everywhere on earth, at any time, every economic agent can have access to digitalized information, the essential raw material of the "new economy." However, the map of information accessibility is uneven. While there seems to be no real limit to broadband availability, every marketed service, based on a given technology, always includes a spatial component. The continuing fall in telecommunication costs and the rise—at constant prices—in speed, quality, mobility and reliability are key factors in the growth of the digital economy. The traffic of digital data—whether production orders, e-mailed documents, video, voice, or Web page—travels over some portion of the global network.

Depending on where you work or live, Internet communications vary greatly in their data rate or bandwidth, which affects the speed at which data can be transmitted and received (Goldsmith and Wu 2006). This variability is a result of the real-space location of both data and the Internet hardware through which the data travel, including routers and exchange points, fiber-optic cables, phone and cable lines, and microwave and satellite transmitters and receptors that interconnect the communications system.

The global network—including the public Internet, private intranets of corporations and governments, and other virtual private networks—is actually the aggregation of many networks at various scales. The networks include the public switched telecommunication networks (PSTN) and various private networks of governments, businesses and other organizations. The long-distance backbones, particularly the intercontinental submarine cables, are fewest in number, because the traffic of several telecommunications carriers is aggregated to "transit" these cables. Since most telecommunications firms target the same corporate customers and clusters of high-income broadband users, urban areas attract both higher levels of infrastructure investment and a larger number of competitors in metropolitan area networks (MANs) and fiber-optic loops that target business and institutional customers. These hubs of the digital economy are the locations where the interlinked networks, which today comprise the Internet, interconnect.

According to Greenstein (2005: 291–2), Internet infrastructure is the accumulation of "durable investments in software, communication and computing equipment, and related activities associated with operating information technology," including capital equipment such as mainframes, minicomputers, PCs, servers, local area networks (LANs), wide area networks (WANs), local and long-distance telephone equipment, switching equipment, wireless networks, software (both packaged and customized), as well as human capital.

In this chapter, we examine the providers of telecommunications, their technologies, and their geographies. We address first the growth of the worldwide, integrated network, and the technologies on which that network is based. Next we describe the economic geography of the "information highways"—that is, long distance, broadband infrastructures. We turn next to the hubs—cyberlocations—within cities, which have added a new dimension to the urban market for land, and changed the focus of real estate from "location, location, location" to "location, bandwidth, location" (Graham 2004: 140). Finally, we discuss the uneven local accessibility of broadband.

> The 'last mile' is a serious challenge in cities, where networks must be threaded under the congested roads and pavements of the urban fabric, to the smart buildings, dealer floors, headquarters, media complexes and stock exchanges that are the most lucrative target users. . . . Fully 80 per cent of the costs of a network are associated with this traditional, messy business of getting it into the ground in highly congested, and contested, urban areas.
>
> (Graham 2004: 139)

Prior to the telegraph (1844), messages travelled only as fast as a horse and rider or a ship at sea—or, later, a steamboat or a railway locomotive. As the global telecommunications network spread, corporate needs have shaped the geography of telecommunications infrastructure and continue to influence the Internet (Baran 1998; Hugill 1999; Schiller 1999). Financial services and other producer services have taken advantage of global telecommunications to coordinate activities, and to form a "digital archipelago" of world cities (or global cities) where capitalist investment concentrates (Graham 2004; Sassen 2006; Taylor 2004).

The original Internet network connected only four university nodes in the US (Abbate 1999). As computing and communications technology converged, private networks grew to serve corporate clients (Bakis 1987; Langdale 1989). The individual networks of new telecommunications carriers, as well as the old telecom monopolies—many of which have become global players through acquisitions, mergers, consortia, and other arrangements—make up the present Internet. Together, the networks of the Internet form a global hub-and-spoke network, focused on major Internet exchange points (IXPs), similar to the major airport hubs found in air transport networks (Langdale 2001; Townsend 2001a). Like airlines, major telecom carriers operate a small number of main hubs, as shown in the global network of US-based AT&T (Figure 3.1). Taking advantage of such networks, long-distance activities such as telesurgery have become intercontinental.

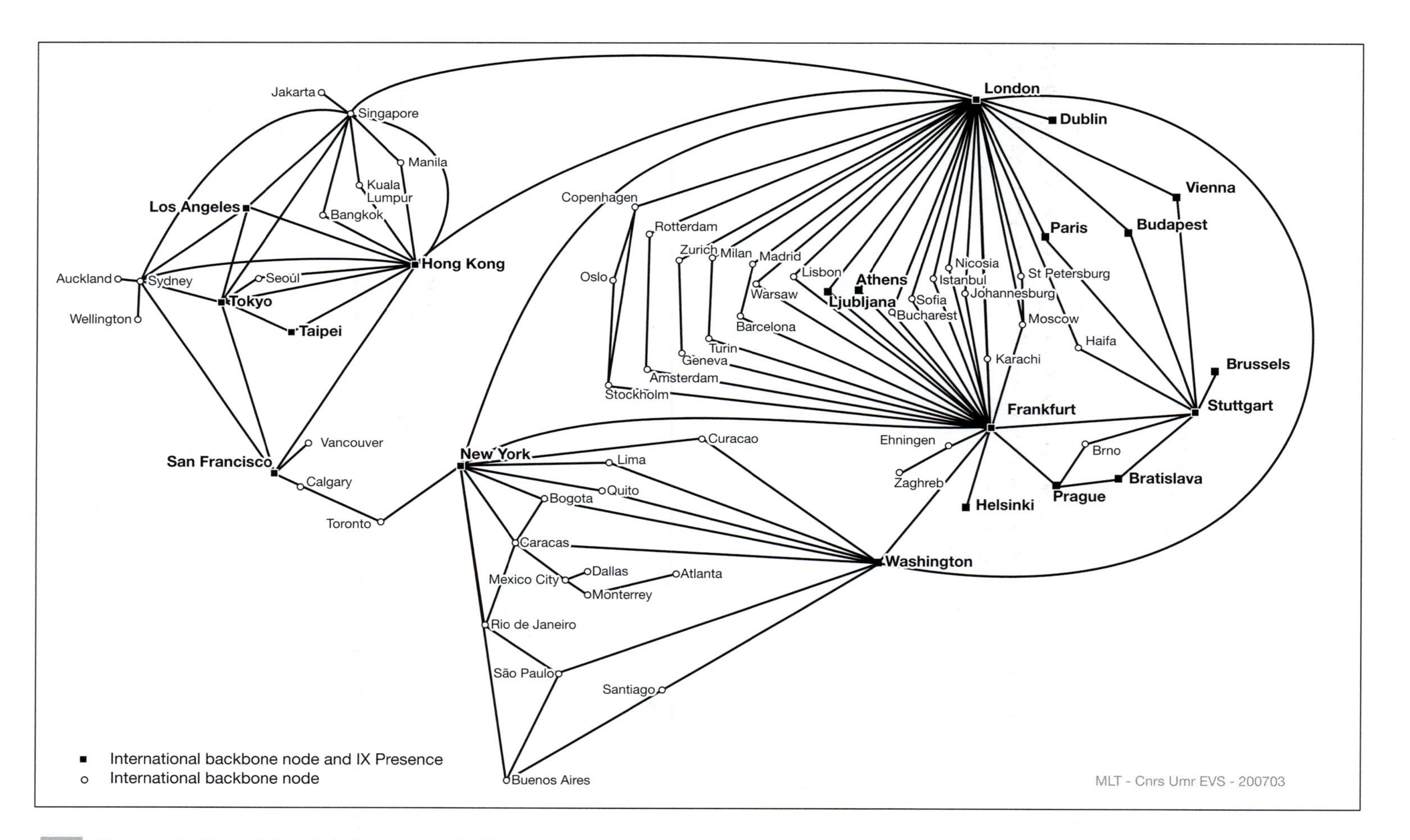

Figure 3.1 ***Map of the global network of AT&T***

Source: Based on TeleGeography Research (2006a)

Emergence of a worldwide, integrated network

Different systems of layers

The *layers* metaphor has become common when referring to telecommunications, the Internet, and the digital economy. However, analysts and pundits have conceptualized different architectures, which embrace more or less broad spectrums of infrastructures and services.

The telecommunications system was easy to describe in the days of plain old telephone service, or POTS: in each country, a monopoly carrier provided voice service to homes and businesses. Through the mid 1980s, the telecoms industry operated with three layers: (1) an equipment layer, comprised of switches, transmission systems, and customer premises equipment (CPE); (2) a network layer, the circuit-switched network; and (3) a services layer, comprised of voice, fax, and toll-free calling services.

Things became much more complex as technological change brought data, different devices, and multiple computers and their software to the formerly voice-only network. The need for standards arose in the 1970s as several networks were being developed simultaneously. Among them were ARPANET, developed at several universities sponsored by the Advanced Research Projects Agency (ARPA) of the US Department of Defense, a network developed at the National Physical Laboratory in the UK, and the CYCLADES network developed in France (Abbate 1999; Shahin 2006).

Critical to the development of the Internet was the development in the late 1970s of standards of an architecture for open systems interconnection (OSI), utilizing seven layers (Zimmermann 1980). The seven layers of the OSI model can be divided into two sets of layers. The lower layers (physical, data link, network, and transport) involve both hardware and software and are involved in *data transport*. The upper layers (session, presentation, and application) deal with *application* issues and are implemented through software (Cisco Systems 2002). The concept of the Internet was to allow interconnection of all sorts of networks and all sorts of equipment and equipment platforms. The OSI model allowed these diverse networks to interconnect, and opened the door to innovations by many new competitors to the incumbent telecoms providers.

The Internet transformed the telecommunications industry into an *infocommunications* industry, which spreads beyond the traditional boundaries of the telecommunications sector, to embrace Internet-based services. To understand this new system, Fransman (2002: 18) suggests an alternative model with six layers (Table 3.1). The old telecoms industry operated only at Layers I and II of Fransman's model—the equipment and network layers. Layers III through VI, all needed in an open systems model, are bridged to the lower layers by the IP interface.

A third system of "layers" describes the interconnection of the various networks that comprise the Internet (Gorman 2005). In effect, it is a disaggregation of the transport layer in the OSI model and of the Network layer (Layer II) in Table 3.1. The first layer consists of network access points, or NAPs, where networks exchange data between each other in a process called *peering*. The second layer is comprised of long-haul transit or backbone providers, including national carriers and global Internet firms. The third level of connection is provided by regional network providers, including municipal

Table 3.1 The six layers of the infocommunications industry

	Layer	*Contents of the layer*
VI	Customers	
V	Applications layer, including contents packaging	Web design, on-line information services, broadcasting services
IV	Navigation and middleware layer	Browsers, portals, search engines, directory assistance, security, electronic payment
III	Connectivity layer	Internet access, Web hosting, ISPs
IP interface		
II	Network layer	Optical fiber network, DSL local network, radio access network, Ethernet, frame relay, etc.
I	Equipment and software layer	Switches, transmission equipment, routers, servers, customer premises equipment (CPE)

Source: Based on Fransman (2002: 18, Table 1.1)

networks, which connect smaller cities. Internet service providers (ISPs) comprise the fourth layer, connected by (dial-up or broadband) modem to a user's home or business. Each of the networks at any of the four levels is an *autonomous system*, or AS, in the Internet. The number of ASs has increased dramatically since the late 1990s, as Internet content and service providers have acquired AS status (Paltridge 2006).

This architecture describes the *hubs* (IXPs, co-location centers, and telecom hotels) and *spokes* (the links of the individual networks) that make up the Internet. The interconnection of these networks is an enormously important influence on the interconnectedness of any given location.

The convergence of networks and applications

Technological convergence based on digitalization has permitted all forms of transmission—voice, images, and data of all kinds—to use the global Web, as we have discussed in Chapter 1. Video can be "streamed" over an Internet connection, which is perhaps also a telephone line or a television cable. Voice over Internet Protocol (VoIP) converts voice to digital signals that are, to the transmission medium, merely data. An array of competitive forces arrived, beginning in the 1960s with satellites, and more rapidly in the late 1980s with fiber-optic cable. Four very distinct forces of competition have emerged: between firms, of course, but also between technologies, between networks, and between products and services (Fransman 2002).

Predicting demand and revenue for each of the various technologies, networks, and products and services is impossible. Famously, in the mid 1980s, AT&T asked the consultancy company McKinsey to predict the number of mobile phones that would exist in 2000. McKinsey's answer was 900,000. The actual number turned out to be 40 million (Fransman 2006a: 3). As Green (2006: 23) says:

> One striking thing about wireless has been the imagination that has built such a useful portfolio of services on such an unpromising medium: telephony, fax,

text, voice mailboxes, test messaging, stock quotes, global positioning system (GPS) position location, rudimentary video, even speech recognition for hands-free usage.

The wireless phone has become the technology of choice for all sorts of communication, including new technologies for video, multimedia, and Internet access (Edwards and Ihlwan 2006; *The Economist* 2006e). Around 2.8 billion mobile phones are already in use, and 1.6 million are added each day (Cukier 2007).

Improvements in phones spill over into other areas of wireless communications, which are now being used to link machines, sensors and objects. In 2007, nearly 10 billion microprocessors will be sold, embedded in devices ranging from computers to coffee-makers. The cost, size and power requirements of wireless functions are falling rapidly, allowing bridges and buildings to be monitored for structural integrity by small sensors, and for farmland and weather conditions to be watched, switching irrigation systems on and off remotely (Cukier 2007). The phenomenon of networked systems of embedded computers "could well dwarf previous milestones in the information revolution" (National Research Council 2001a: 2). The International Telecommunications Union (ITU) (2005) calls it "The Internet of Things."

An array of new wireless technologies have expanded greatly the potential uses of mobile phones, permitting wireless Internet access and, increasingly, embedded chips that monitor and communicate, prompting concerns about privacy (Cukier 2007). Promising business uses include RFID, which is beginning to see widespread use by retailers and other businesses, as we note in Chapter 4.

Deregulation of telecommunications and competition

During the past decade, the exploitation of technology improvements in telecommunications has been dramatically driven by competition. If left entirely to markets, there would be limited competition in telecommunications, but rather monopolies or oligopolies for two reasons. First, economies of scale, resulting from the very high fixed (investment) costs of building networks, reinforce a trend towards larger firms. The second reason is network externalities, in which the benefit of membership in a network increases with the number of members of the network (Fransman 2006b).

Until 1970, in most countries, it was a government-sanctioned monopoly or part of national Post, Telegraph and Telephone (PTT) ministry. At that time, new technologies permitted competitors to nibble at the edges of PTT monopolies and, even earlier, of AT&T in the US (Brock 2003). As deregulation occurred, competition provided lower prices, more networks, and more choices for households and businesses as well as for governments.

In general, "competition is a key factor in lowering prices and competition between alternative platforms is especially important," concludes the European Commission (2005a: 13). As Figure 3.2 shows, the average cost of international telephone calls has fallen nearly 80 percent since 1988. Box 3.1 compares the experiences of France and the UK with regard to competition.

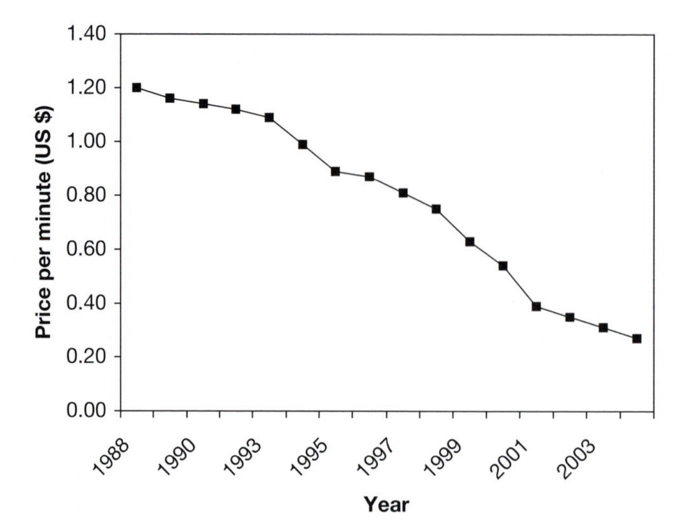

Figure 3.2 ***World average revenue per minute for international telephone calls, 1988–2004***

Source: Adapted from TeleGeography Research (2005b: 85)

Network competition and technological competition are strongest in the local access market, which now includes Internet access in addition to voice telephony and television. The providers who can offer customers the "triple play"—voice, television, and Internet—as a single bundle tend to be the winners over single-solution providers. A "quadruple play" or "grand slam" service bundle includes mobile telephony as well (Bauer 2006).

Many incumbent network operators, such as BT, FT, NTT (Japan), and Deutsche Telekom, have managed to survive the shrinking demand for voice traffic over their fixed networks. These providers have diversified into the fast-growing markets of mobile, Internet, and international services. In the past decade, the shakeout of some new entrants—or their consolidation into incumbent firms—has reduced both the number of competitors and the intensity of competition (Fransman 2002). Less competition has resulted in lower rates of adoption of broadband, however, particularly fiber-optic technologies (Fransman 2006b).

In 1990, only Japan, New Zealand, and the US had liberalized their PSTNs, usually only in long-distance and international services, rather than local service (OECD 1999). By 2004, all 30 members of the Organisation for Economic Co-operation and Development (OECD) had competition in telephone service. Competition also is evident in mobile communications, where 27 countries had three or more operators in 2004 (OECD 2005b). In many developing countries, public policies that favor regulation and incumbent monopoly providers have so far hindered Internet growth. Countries that regulate ISPs have fewer Internet hosts and users, and countries that regulate Internet pricing have higher prices for Internet access (Wallsten 2005a).

The European Union mandated full competition and pro-competitive regulation by 1998, influencing liberalization and dismantling of the monopolies of incumbent "historical" carriers (Cawley 2003). However, considerable variation remains from country to country—notably the large-scale difference between "Anglo-American

Box 3.1 A comparison of telecommunications competition in France and the UK

Both countries had their monopoly telephone provider: France Télécom (FT) and British Telecom (BT). Telecom liberalization in the UK began in 1981, when BT was separated from the Post Office and Cable & Wireless was privatized. Mercury Communications was given long-distance and international licenses. More importantly, Mercury decided to construct a digital fiber-optic network, focusing on large business users in the City of London. This led BT to upgrade its network to fiber in 1987 (Rutherford 2004). New competitive entrants followed in the early 1990s, with the effect that London had more telecom providers (six) in 1998 than any other city in Europe. By 2001, 12 operators offered services in London (Rutherford 2004: 188).

In France, by contrast, FT was sheltered from competitors until 1998, the EU deadline for opening up markets. Paris had three networks in 1998, the same number as Amsterdam and Frankfurt (Finnie 1998). Although Paris had 12 operators with 100 km or more of infrastructure in 2000, London, with the same number of operators, had lower costs and a greater concentration of data centers and other infrastructure (Rutherford 2004: 187–9).

In both cities, the new operators did not build widespread networks, but engaged in "cream skimming" or "cherry picking"—concentrating investment in zones with agglomerations of large companies (Graham 1999; Rutherford 2004). In France, liberalization was slower, took place later, and was more gradual than in the UK. It also was in response to external pressures, such as the European Commission, as well as large multinational companies which demanded international data transmission for Internet use (Rutherford 2004: 91). The Téléport Paris–Île-de-France was the principal response to this demand, so Paris lagged by a few years in boosting its fiber-optic connections to Internet backbones. A prohibition in France on the leasing of "dark fiber" also served to maintain both higher price levels for telecommunications and an overall "deficiency" of private telecomm infrastructures in the Paris region.

Ironically, it was France's early lead in electronic commerce via Minitel (launched in the 1980s) which slowed the Internet's development in France. Also, through the mid 1990s, French civil servants and entrepreneurs "were blind to the possibilities presented by the Internet." Nevertheless, the Minitel experience promoted national learning. When deregulation was enforced, the Internet spread rapidly (Benghozi and Licoppe 2003). FT has moved decisively since 1996, when it established Wanadoo as its Internet subsidiary (Fransman 2002: 177). Now, France is emerging as one of the leading countries in broadband (Abboud 2006), and is projected to pass South Korea and the United Kingdom to rank fifth in broadband by 2007 (TeleGeography Research 2006b).

liberalism" and "continental European mercantilism." The result is a number of distinct strategies on the part of former monopoly providers (Hulsink and Davies 2003: 417–18).

"Unbundling" the local loop of the fixed network has emerged as the common strategy to encourage competition. Unbundling means that incumbent telephone companies must give competitors access to the "last mile" loop. In the case of DSL technologies, newcomers are allowed to connect their own line—at its cost or at wholesale price—to the digital subscriber line access multiplexer (DSLAM), which is located within a short distance of customers' homes. This distance dictates the maximum bandwidth of the line, hence the problem faced by rural communities where homes are often great distances from the DSLAM.

Unbundling has increased competition mainly in DSL technology, allowing new providers to compete directly with incumbents in broadband technology in many countries, such as France and Japan (Crandall 2005; Fransman 2006a). However, broadband competition has not depended on unbundling in Korea (Crandall 2005).

Regulation remains in wireless, because the radio spectrum, on which mobile telephones rely, is a scarce natural resource allocated by governments to specific uses (Gans *et al.* 2005). Regulatory agencies have to follow the rapid pace of technological changes (Steinbock 2003). Wireless competition has increased along with improvements in technology such as W-CDMA (wideband code division multiple access) or third-generation (3G) mobile telephony. Overall, mobile communications remains uneven, still marked by different national and supranational standards and equipment (Hess and Coe 2006). Wireless technologies are converging on certain standards, which is beginning to permit developing countries to leapfrog past wired or fixed-line infrastructures, and to design networks to serve urban, suburban, and rural areas with different technologies (Gunasekaran and Harmantzis 2007).

The economic geography of long distance, broadband infrastructures

Many networks, based on many different technologies, make up the global digital network. Most of these networks are landline, either overhead wires or underground cables; others connect continents and islands by submarine cables; and others bounce off satellites to connect distant parts of the world.

Satellites and cables

Submarine cables have been a fixture of global telecommunications since 1850, connecting colonies and markets, and steadily networking the globe (Headrick and Griset 2001; Hugill 1999). In the 1960s, a new technology—telecommunications satellites—became an alternative to submarine cables for intercontinental communication. Television and radio quickly switched to satellite, beginning in the late 1970s, followed by news gathering and remote printing and publishing. Corporate demand was great enough that teleports were built in many large cities in Europe, Japan, and North America. One motivation was the ability of large users to bypass the monopoly telecom provider. Large TNCs, such as Ford and Hewlett Packard, followed the example of a satellite link installed by Texas Instruments in Bangalore, India in the mid 1980s (Hudson 1990; Singhal and Rogers 1989). Corporate decentralization and the creation of global firms was impossible before global telecommunication networks enabled their operations to be seamless from place to place.

By the late 1980s, 3,300 satellites had been launched, used primarily for point-to-multipoint services such as broadcast television (to cable systems or directly to viewers) and broadcast data (e.g., wire services, financial news services, and weather reports). After broadcasting, the biggest users of communications satellites are the US Department of Defense (DoD) and news organizations, for access to war zones and other remote areas (GAO 2003). Satellites can also carry data at high transmission data rates over continuous long-distance paths, utilized by many transnational corporations (TNCs) (Hudson 1990).

At first, satellite communications offered substantial cost savings over monopoly telecom providers for corporations that wanted to bypass the public switched network. From 1965 until 1988, satellites provided almost ten times the capacity of submarine telephone cables for about one-tenth the price (Whalen undated). Satellites forced a shift from analog to digital transmission, a shift that soon worked to the advantage of fiber-optic cables. TAT-8 (for transatlantic telephone), the first fiber-optic cable, placed in service in 1988, provided a tenfold increase in capacity over its copper predecessor, TAT-7. Telephone demand grew, aided by facsimile (fax) technology, which could be used to send images as well as text. The cost per circuit on submarine cables dropped dramatically as fiber-optic technology improved, from $59,000 per circuit in 1988 to $155 per circuit in 2000 (Hsu 2004).

The capacities on the new cables, utilizing dense wave lengh division multiplexing (DWDM), are many times greater than on earlier cables, and they can be upgraded to increase their capacity while in place.[1] Three transatlantic cables (Yellow/Atlantic Crossing-2 [AC-2], TAT-14, and Apollo) have current capacities of 640 gigabits per second (Gbps), far more than the 30 Gbps of TAT-12/13. AC-2 is upgradeable to 1,280 Gbps and Apollo is upgradeable to 3,200 Gbps. The C2C Cable Network in Southeast Asia and VSNL Transpacific Cable have upgradeable capacities of 7,680 Gbps.

DWDM technology and new consortia for installation of submarine or undersea cables brought about a global explosion of fiber-optic capacity during the 1990s. Competition and overcapacity led to price declines that satellites could not match, added to fiber's advantages of real-time transmission (near-zero latency) and very low error rates. Satellite communications add a delay to communications, owing to the distance from the earth to satellites in orbit, which makes interactive data transmission difficult. Increasingly, customers of satellite providers are shifting to fiber-optic cables (Bax 2005).

A severe "bandwidth glut" beginning in 2001, coinciding with the dot.com bust, precipitated fierce price competition. It resulted in spectacular bankruptcies among telecommunications carriers such as Global Crossing, KPNQwest, and WorldCom, and a dramatic shift in the geography of global telecommunications. Prices for Internet transmission, measured in megabits per second (Mbps) for a 155-Mbps link in London, have dropped dramatically (Figure 3.3). Monthly prices in 2006 were only 6 percent of their level in 2000 at the height of the bubble. Only by early 2006 was there evidence, seen in stabilizing prices, that the glut was at an end (Heinzl and Young 2006).

During the past decade, cables have far surpassed satellites as the preferred means of communication, in part because they can be owned by telecommunications carriers whereas satellite capacity is merely leased (Wheelon 1988). "Fiber serves specific fixed locations with enormous capacity and reach, while wireless can go everywhere but with

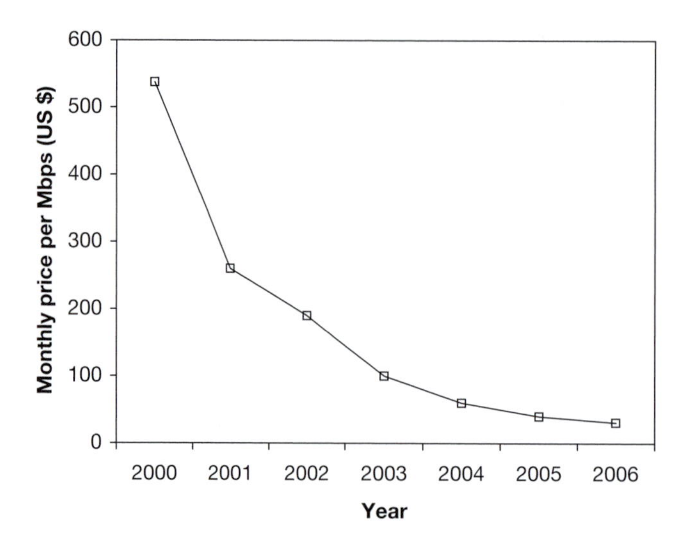

Figure 3.3 ***Prices per megabit per second (Mbps) on 155 Mbps circuits in London, 2000–2006***

Source: Adapted from TeleGeography Research (2006a: 90)

very limited capacity, variable and uncertain quality of service, and high density of base stations" (Green 2006: 23). As broadband technology has increased, cables also can now be used for video transmission, eliminating this advantage of satellites as well.

The generalization that, in general, "rich countries use fiber, poor countries use satellites" (TeleGeography 2001: 98) is changing. Mobile operators, such as Reliance in India, Globacom in Nigeria, Orascom in Algeria and Pakistan, and Indosat in Indonesia, increasingly need fiber-optic backhaul to international hubs. Islands previously not connected by submarine cables and reliant on satellites have begun to be connected or are planning connections via submarine cables. Examples include Jamaica, New Caledonia, Tahiti, and Madagascar. "Soon, every island with a significant population will be connected via submarine cable(s) in addition to satellite links" (Krebs 2006: 19).

The unique ability of satellites to serve isolated and hard-to-serve areas, however, would seem to guarantee them a segment of the market (Hatfield *et al.* 2005: 60; Warf 2006). For example, global monitoring systems for nuclear explosions and for natural disasters such as tsunamis rely on satellites to transmit data from numerous remote locations, many at sea, to central sites where they can be analyzed (Morrissey 2005; Padma 2004; Preparatory Commission 2006).

A second exception to the shift from satellites is the US retail industry. Wal-Mart, the world's largest retailer, has relied on its Wal-Mart Satellite Network (which the firm calls the largest private satellite communication system in the US) since 1987. The network links all operating units of the company and the headquarters in Bentonville, Arkansas, with two-way voice and data and one-way video communication, including identical music in all stores (Wal-Mart Stores 2006c). New applications of the satellite network continue to be found. In mid 2006, Wal-Mart and the Trailer Fleet Services division of GE Equipment Services announced an agreement whereby GE will supply

Wal-Mart with trailer tracking technology for its fleet of 46,000 over-the-road trailers (GE 2006). Indeed, Wal-Mart received praise for the firm's ability to coordinate assistance and logistics, when the US government could not, after Hurricane Katrina hit New Orleans in August 2005.

City hubs of the global network

The economic geography of the digital network includes both its hubs or nodes and its links. The hubs are where individual network providers aggregate their links from users, and interconnect to each other. Each network has its own topology, and maps differently onto actual geographic space (Gorman and Malecki 2000; Hepworth 1990). Each individual network user (such as a large firm or organization) has a set of locations of offices and major facilities, often spread among many countries, and the geographic structure of each firm's network is unique. Despite that uniqueness, the concentration of headquarters and of producer service firms means that telecommunications demand—and therefore network infrastructure—has agglomerated in a limited number of world cities (Hepworth 1990; Sassen 2006; Warf 2000). To a certain degree, this concentration also means vulnerability to attacks, failures, and accidents (Gorman 2005; Grubesic and Murray 2005).

"Far from destroying cities by making place irrelevant, the production and consumption of Internet content, and the infrastructure to support it, are concentrated in cities" (Goldsmith and Wu 2006: 56). Moreover, the level of interconnection of the networks that comprise the Internet—typically higher in cities and, in particular, in the telecom hotels found in them—increases the resiliency of the overall network (Gorman 2005).

This agglomeration of both telecommunications infrastructure and business services is seen in the case of first-tier world cities, such as London, Paris, Frankfurt, and Amsterdam, and second-tier cities, such as Brussels, Zürich, Stockholm, and Vienna. Some European cities, such as Barcelona, Dublin, Milan, Madrid, and Prague, rank higher for their economic connectivity than for their telecommunications connectivity. Others, such as Berlin, Copenhagen, Düsseldorf, and Hamburg, are technological nodes or gateways, ranking higher for telecommunications connectivity than economic connectivity (Rutherford *et al.* 2005). The largest city in a country is not necessarily first in Internet access or penetration levels, perhaps a result of population and economic diversity common in large cities (Kellerman 2004; Zook 2005).

The corporate networks that already criss-crossed the globe prior to the Internet typically used leased lines, proprietary software and various network configurations (Hepworth 1990). As fiber-optic technology became widespread, firms leased high-bandwidth lines to form their private networks, which later became in most cases their links to the Internet. Today, most data transmission capacity is in secure virtual private networks (VPNs), which are replacing leased lines for many firms. Corporations prefer to have an amount of bandwidth for times of peak use even if the bandwidth is under-utilized at other times (Odlyzko 2000; 2003). As corporate demand for bandwidth has grown, private or leased lines are "absolutely essential for low transaction latency" (Odlyzko 1999).

The need for low latency is clear in experiments with telesurgery. High-bandwidth links are needed to ensure that video images of a patient are seen instantaneously by the surgeons. Fiber-optic lines, connecting medical teams separated by thousands of kilometers, display video at speeds faster than 330 ms, the maximum safe lag time (BBC News 2001; Marescaux *et al.* 2001). Satellite signals are not always dependable and their communication delays jeopardize surgical outcomes. Telesurgery does not mean the "end of geography" in medicine. In addition to high-bandwidth connections, telesurgery also needs critical medical facilities at the physical operation point.

The high-energy physics community requires even lower latency—100 ms or less—for scientists to do interactive visualization, simulation steering, and collaboration. This speed is impossible to obtain on commercial Internet links. Data-intensive e-science demands lambdas: high-bandwidth, low-latency connectivity "with known and knowable characteristics." A lambda, "is a fully dedicated wavelength of light in an optical network, capable of greater than 10Gbps bandwidth" (DeFanti *et al.* 2003: 36). A second emerging scientific application for regional submarine networks is undersea oceanographic research.

The growing use of digital technologies in oil and gas exploration—three-dimensional (3D) and four-dimensional (4D, or 3D over time) seismic visualization—has greatly increased the success rate of drilling and exploration. Beginning in the North Sea, the Gulf of Mexico and the Persian Gulf, fiber-optic telecommunications cables connecting to the shore are now standard for offshore oil and gas fields. Indeed, the oil and gas industry represents a major demand for new submarine cable systems, as offshore fields are linked to onshore operations centers and thence to global networks (Bax 2005; Wallace *et al.* 2003). Massive data streams emerging from 4D seismic technology are transmitted to centers of the oil and gas industry in Aberdeen, Houston, London, New York, and Stavanger where visualization experts analyze them.

As the "global grid" expanded during the past decade, the Triad regions of East Asia, Europe, and North America were the first to be provided with ample bandwidth. This was the outcome of an explicit strategy to serve the concentrations of demand represented by TNCs and financial services in world cities. New competitive providers, such as COLT (City of London Telecom) and WorldCom (now part of Verizon), bypassed PSTNs and offered their customers "route diversity" in the archipelago of world cities (Graham 1999; Rutherford 2005). Thirty companies currently provide Internet connectivity to 12 or more countries. AT&T connects 52 countries; Verizon Business (formerly MCI and WorldCom) and Cable & Wireless connect 30 or more countries; and six firms connect more than 20: Sprint, REACH, Global Crossing, KPN, NTT Communications, and VSNL. Although the two largest are US-based, Asian companies such as REACH, NTT, and VSNL have joined the list of major Internet providers (TeleGeography 2006a).

Over 20 networks have been built in Europe by telecommunications providers whose customers demand seamless global communications. Rutherford *et al.* (2004) conclude that backbone network deployment across Europe is based on a broader set of cities than in the US. The "Golden Square"—London, Paris, Frankfurt, and Amsterdam—has been the focus of greatest investment.

Although the early Internet connections were mainly within and between North America and Europe, the global Internet is now filling in. A great deal of new fiber-optic capacity has been installed throughout the world. Much of it is "dark" fiber in anticipation of future demand, not yet "lit" by the optoelectronic equipment that facilitates transmission of data. Indeed, many companies in the electricity, pipeline, and railroad sectors install fiber along their rights-of-way. Russian networks, joined with those in China, now provide a low-latency route from Western Europe to Hong Kong (Stubbings 2004).

Existing infrastructures (railways, highways, bridges, sewers) are often used to support fiber-optic backbones, nationally, regionally, and locally (Gorman 2005). It is not unusual for the signal transmission of several different networks and services to use the same conduit, the same fiber bundle, or even the same single fiber (National Research Council 2001b). Similarly, submarine cables often utilize the same beaches, beach manholes, and duct routes for their backhaul to urban markets (Bishop and Walker 2006). There is therefore a cumulative or self-reinforcing aspect to the economic geography of telecommunications: the more densely populated and wealthy a country or region, the more transport and sanitation infrastructure in place, and the less costly it is to implement broadband.

The combination of demand for voice, video and data communication over the Internet, and the ability of fiber-optic technology to meet that demand thanks, notably, to DWDM technologies, has resulted in an enormous expansion in global telecommunications capacity. During the 1990s, spurred by Internet-related demand, 209 new submarine cables were installed, tripling the world total in 1989 and increasing total submarine bandwidth 100-fold. As they replaced copper cables, the new fiber-optic technologies also permitted future expansion; the new cables had "unlit" capacity awaiting demand. The full capacity of the submarine cables in place in 2005 was over nine times the operating capacity. This is a normal situation for terrestrial or landline networks as well. In major European cities, no more than 32 percent (Helsinki) and as little as 9 percent of fiber pairs are actually lit (TeleGeography Research 2005a).

The full capacity of the world's undersea cables in 2005 was 387.6 terabits per second, or 387.6 million Mbps. The top ten countries in total "lit" bandwidth now included Japan and Singapore as well as the US and several European countries (Table 3.2). Perhaps the most important shift is the most recent one: the "fiber glut" in developed countries that has all but halted investment in new cables connecting the North Atlantic, but new cables have continued to be installed to Asia and the Middle East. The right side of Table 3.2 illustrates the future of global telecommunications: eight of the top ten countries in submarine cable bandwidth are in Asia, and the UK and the US hold only a very slim lead over third-ranked Japan. Cable links to and between Asian countries have taken advantage of DWDM technology, allowing them to leapfrog ahead of many other countries in the infrastructure of the Internet.

Organized by continent, Table 3.3 highlights the great disparities that remain among the world's regions. The largest city-to-city Internet backbone links connect London with New York and the major cities in Europe with each other. Capacities on the major Latin American routes now approach those of trans-Pacific routes, the largest connecting to Miami, which is the hub for airline networks as well as Internet links. New intra-regional links connect São Paulo, Buenos Aires, Lima, and Santiago (TeleGeography 2006b).

Table 3.2 ***Leading countries connected by submarine cables, 2005 (ranked by total bandwidth in service and at full capacity)***

Country	*"Lit" bandwidth 2005 (Mbps)*		*Total bandwidth at full capacity 2005 (Mbps)*
United Kingdom	8,224,040	United Kingdom	36,929,040
United States	6,337,032	United States	34,742,032
France	4,022,484	Japan	34,495,588
Denmark	2,366,485	Singapore	26,693,365
Japan	1,875,588	South Korea	19,741,005
Netherlands	1,688,000	China	18,441,685
Belgium	1,471,770	Taiwan	16,775,185
Sweden	1,431,544	Hong Kong	16,773,505
Germany	1,340,666	India	13,771,139
Singapore	943,365	Philippines	12,890,853

Source: Based on Malecki and Wei (2006, Tables 8 and 9)

Table 3.3 ***Largest international Internet backbone links, by continent (Mbps), 2006***

	Route or link	*Internet capacity 2006 (Mbps)*
Europe		
	London—New York	386,601
	London—Paris	257,845
	Amsterdam—London	209,433
	Frankfurt—Paris	196,926
	Frankfurt—London	161,953
Asia		
	San Francisco—Tokyo	78,451
	Taipei—San Francisco	40,223
	Sydney—Los Angeles	35,581
	Taipei—Tokyo	33,550
	Seoul—Tokyo	32,174
Latin America		
	Miami—São Paulo	54,500
	Lima—Miami	40,600
	Bogata—Miami	25,196
	São Paulo—Lima	20,000
	Mexico City—Dallas	18,396
Africa		
	Cairo—New York	5,378
	Alexandria—London	5,000
	Palermo—Rabat	3,110
	Paris—Rabat	2,488
	Madrid—Rabat	1,866

Source: TeleGeography Research (2006a)

Asian countries have moved to leading positions. Tokyo, the leading Internet hub of Asia, is followed by a second tier of cities: Taipei, Hong Kong, and Seoul. A third tier includes Singapore, Beijing, Shanghai, Guangzhou, and Osaka. Cities in the Middle East were only recently connected by high-bandwidth links to other parts of the world. The Emirates Internet Exchange (EMIX) in Dubai now has direct connections to New York and San Francisco.

Africa remains by far the most poorly linked continent. Several cables were placed in service since 1999, but only a small portion of these are utilized as Internet backbones. Africa's connections remain about 2 percent of the capacity of those of European cities and one-tenth the capacity of routes to Latin America. Africa's major Internet hub cities are in North Africa (Rabat, Cairo, Alexandria, and Tunis) or in South Africa (Johannesburg) (TeleGeography 2006b).

The Internet is evolving from the centralized (and US-centric) hub-and-spoke system of the late 1990s, when communications between Asian countries and between European countries were typically routed through Internet exchange points in the US (Cukier 1999). Four of the five cities with the highest international Internet bandwidth are in Europe (Figure 3.4). More bandwidth now links key cities in Western Europe to each other than to the US, and intra-regional links in Asia are growing faster than anywhere in the world (Paltridge 2002; TeleGeography 2004: 18).

Cyberlocations: critical nodes of the global network

The new telecommunications carriers that have emerged must interconnect with both existing carriers and with each other to provide global service to their corporate customers. "The Internet cannot bypass mega-cities: it depends on the telecommunications and on the 'telecommunicators' located in those centers" (Castells 2000: 440; Sassen 2006). Among the key nodes in the infrastructure of global networks are telehouses, telecom hotels, and carrier-neutral colocation facilities. This interconnection—peering between equal large networks or "transit" purchased from a larger network—requires colocation. Increasingly, peering and transit relationships take place not only among ISPs but also between ISPs and content providers such as Google (Paltridge 2006).

An industry has arisen to provide these facilities, where interconnection to numerous networks can take place in highly secure and equipped buildings, with power and generators for uninterrupted power and, in many cases, managed services such as maintenance (Moriset 2003b). Interconnection, usually without managed services, is also possible at IXPs (PriMetrica 2003). Some of the IXPs are truly critical hubs in the Internet. Three cities—Amsterdam, London, and Tokyo—have Internet exchanges with traffic volume over 100 Gbps (Packet Clearing House 2006).

Within cities, interconnection takes place at three types of facilities: carrier hotels, colocation specialists, and bandwidth providers. *Carrier hotels* provide basic space in which operators can establish interconnection. *Colocation specialists* such as 60 Hudson Street in New York City, provide not only space but also power systems. *Bandwidth providers* include metropolitan and long-haul network operators.

> 60 Hudson Street houses over 100 different domestic and international networks. The Meet-Me-Room offers users a carrier-grade colocation facility allowing

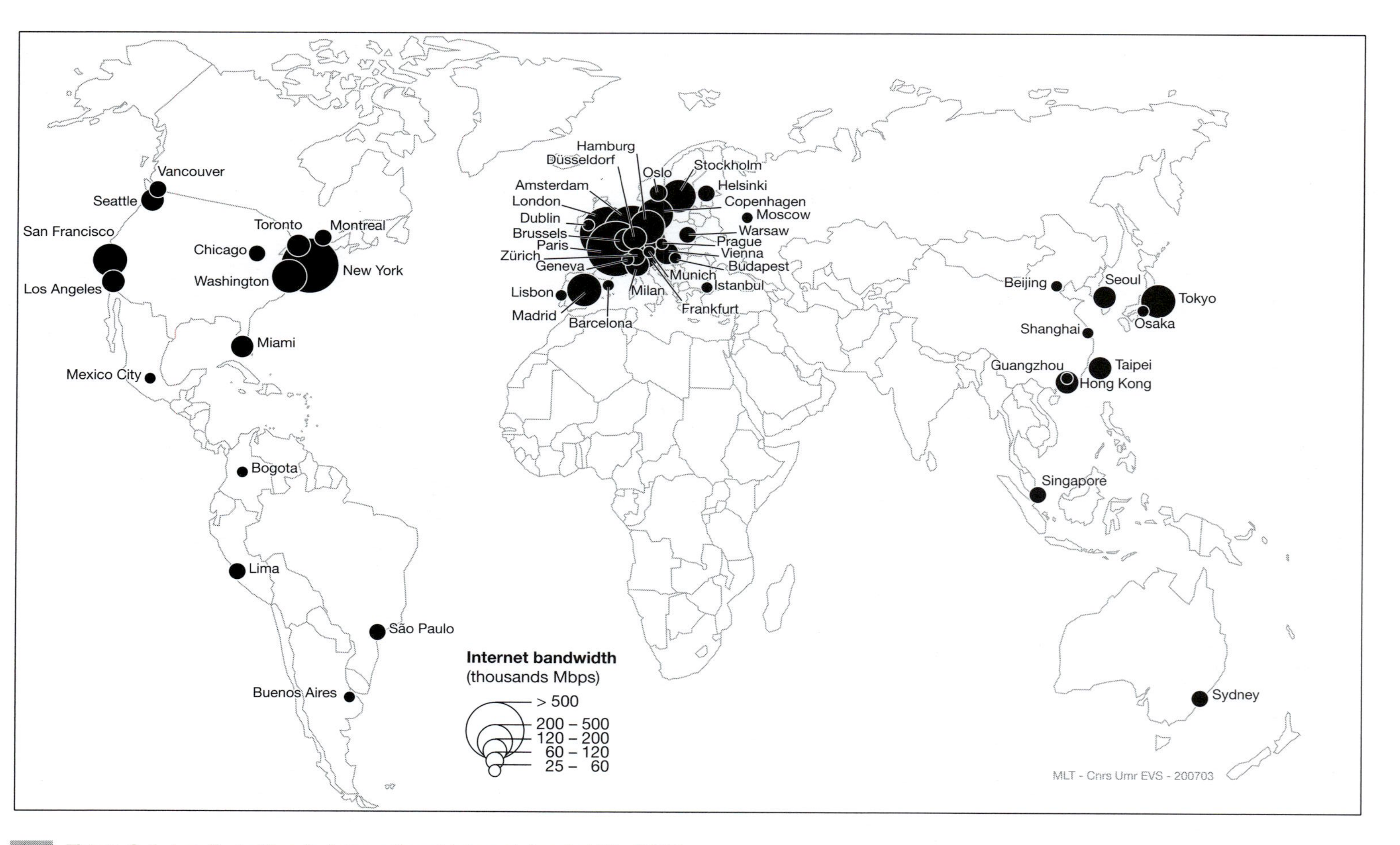

Figure 3.4 ***Leading cities in international Internet bandwidth, 2006***

Source: based on TeleGeography Research (2006a: 49)

> for immediate interconnection access to these networks. . . . A dense conduit system has been constructed to allow tenants access to locations within the building as well as networks located within the New York City conduit system.
> (FiberNet Telecom Group 2005)

These facilities permit private peering, which has long been practiced by the largest and oldest backbone providers, including Cable & Wireless, Sprint, and UUNet (part of MCI). Smaller networks can connect to the larger backbones, paying for a transit link to make the connection (Kende 2000). Private peering emerged to prevent the problems of uneven quality at public peering points, and now, according to some accounts, private peering accounts for 50 percent, and as much as 90 percent, of all interconnections (Malecki 2002).

Certain types of network traffic, such as streaming multimedia, IP telephony or VoIP, video teleconferencing, and video-dependent applications such as remote telesurgery, may require a defined (high level of) quality of service (QoS). QoS requires fixed routing and flow-aware routing "to ensure that all packets of a given flow follow the same path" (Roberts and Oueslati-Boulahia 2000: 2204). Such routing is possible only if the data travel on a single network, or via private peering with specific partners. Various end-to-end networks for "data intensive science" have been built or allocated from existing networks so scientists worldwide can interact with facilities at network nodes such as CERN in Geneva, Switzerland and Fermilab near Chicago (Newman 2006).

A key characteristic of colocation is the possibility to connect with more than one network. A number of large firms, such as Global Switch, Telehouse, Redbus Interhouse, and Interxion, provide colocation and carrier-neutral interconnection in major cities. Indeed, major cities have dozens of colocation facilities, connected to as many as 20 or more bandwidth providers. Facilities in Europe tend to be larger than in the US (TeleGeography Research 2006c). Many of the buildings are no longer prime real estate; most are aging and in declining neighborhoods or in the center or edge of city centers. Although a few new, custom-built buildings are being built, many cyber-buildings are "recycled" factories, office buildings, and department stores (Moriset 2003b; Strom 2000).

Proximity to colocation sites and other point-of-presence (POP) locations is a priority for office clusters and their tenants in order to minimize the cost of last mile connection. Walcott and Wheeler (2001) estimate that the cost of laying fiber-optic cable underground in the Atlanta area ranges from $100,000 per mile ($62,500 per km) in the dense central business district (CBD) core to $20,000 per mile ($12,500 per km) in suburban areas. Large office buildings and clusters of technology-intensive companies like to locate near, if not directly on top of, a POP (Graham 2004; Walcott and Wheeler 2001).

Sometimes in the same locations are *data centers*, which provide data storage for large companies such as banks and serve as destinations where Web pages of Internet content providers are stored. Storage needs are growing not only from business demand for security and multiple back-up copies of critical data, but also from Web logs, or *blogs*, of individuals, which frequently contain videos—vidblogs or vlogs—audio files and podcasts. Websites such as YouTube (now part of Google), MySpace, and FaceBook have become enormously popular among young people in the US—as are similar sites in other countries—adding to traffic on Internet routes and their cables and to storage needs in data centers (Boutin 2006; Hempel 2005). While not technically colocation, server and Web hosting sites may well be connected to more than one backbone network provider. Data storage and "server farms" are found in both urban and rural locations (Box 3.2).

Box 3.2 The new geography of data centers

In addition to interconnected telecommunications networks, the digital economy needs to store the files—websites, photographs, video clips, audio files, financial and inventory data—that now can be accessed and swapped so easily. This means that data storage and "server farms" add to the real estate needs of the digital economy. Both natural and political hazards, such as power shortages and terrorist attacks, have demonstrated the need for data storage facilities in multiple locations (Moriset 2003b). Both risk avoidance and power costs have contributed to a new trend: many now are being built outside major cities, in remote places.

A new state-of-the-art data center can cost up to approximately $10,000 per sq m ($1,000 per square foot), five times the cost of conventional office space. More importantly, as larger numbers of smaller "blade servers" are packed into a data center rack, they generate a great deal of heat and therefore demand for cooling systems. Power costs have risen dramatically. Equinix, whose clients include Google, Yahoo!, and MySpace, needs more storage space, and its new data centers being built near Chicago and New York will consume 20–30 megawatts of electricity (Mehta 2006).

Consequently, the best places for data centers are not first-tier cities, but places where the combination of land, electricity and labor costs is low, and which offer access to high-bandwidth fiber-optic networks. A new Google data center will be in The Dalles, Oregon, two hours from Portland, Oregon (Spellman 2005). Both Yahoo! and Microsoft are building new data centers in Quincy, Washington. The locations of data centers are strongly attracted to cheap power which in Quincy, for example, is one-eighth the cost in Manhattan and Silicon Valley (Burrows 2006b). This new development contradicts somewhat the urban bias in most analyses of the geography of the digital economy: "the reliance of contemporary communications on optic fiber creates a new centrality, a concentration of strategic resources in giant metropolitan areas" (Graham 2004: 141). Remote locations are not an option for firms such as Equinix and AT&T, which have multiple customers using their facilities, so they tend to remain in urban regions (Mehta 2006).

Google has between 150,000 and 1 million servers on which it stores data, including a cached version of every page it finds on the Internet. Copies are stored or replicated on three to six servers. Before January 2004, Google's searches took place in ten data centers in one location. The number of data centers grew to 61 by early 2006, several in northern California's Silicon Valley (Mountain View, Santa Clara, and Palo Alto), in Dublin, Ireland, and in the Washington, DC area (Sterling and Herndon, Virginia), and new ones near Atlanta and Beijing (Arnold 2005; Aubuchon 2006).

The number of "hops"—the number of routers through which data must travel from a user to a website—imposes both a cost of time and potential data loss from dropped packets (Dupuy 2002). "Mirror" sites in several locations can speed results of searches and downloads to users (Gorman and Malecki 2001). One of the earliest firms to provide "mirror" sites for Web pages of other firms as its business model was Akamai, whose major customers include CNN and Yahoo! (Staple 1999). Akamai's server farms are housed in secure building complexes located in the major global cities of the world, offering close proximity to users (Graham 2004).

Too many data centers can be a disadvantage, however, as Hewlett Packard (HP) and Intel have learned. HP is in the process of consolidating its 85 data centers into six newly built data warehouses—two each in Austin, Atlanta, and Houston—each connected to the firm's global network via a newly built fiber-optic network (Burrows 2006a). Intel is consolidating nearly 100 data centers into a small number of regional hubs, including Albuquerque (New Mexico) and Hillsboro (Oregon), near Portland (Dunn 2006).

In summary, the reliance of contemporary communications on optic fiber creates a new centrality, concentrating strategic resources in large metropolitan areas (Graham 2004). In large part, this is because the infrastructure of earlier network technologies—sewers, bridges, poles—can be used once again with the new. However, it is too simplistic to say that only large cities are privileged by the new technologies. Early in the growth of the Internet, Townsend (2001b) identified a number of "new network cities." Bandwidth investment in the US was attracted not only to cities with larger populations but also to cities prominent as "knowledge economies," indicated by doctoral degree-granting institutions and economic dynamism, an indicator of recent job growth and firm formation (Malecki 2004a). Overall, the complexity of the Internet network results in an increasing variety of business location, on both intercity and intra-city scales (Moriset 2003c).

Local broadband: the enabler of the digital economy

The presence of the global network does not mean uniform access to it in all places. The hub-and-spoke networks are, in many instances, more like tunnels that bypass places on their way to hubs in cities. Homes and businesses need a *last mile* connection—either wireless or wired—to be connected to the Internet.

Bandwidth is the term commonly used to designate the transmission data rate, measured in bits per second. Simple access to the Internet for e-mail and text messages requires low bandwidth, no more than a dial-up modem and an analogue telephone line. However, the Web increasingly contains images, video and audio files. A simple "rule of thumb is that good video requires about a thousand times as much bandwidth as speech. A picture is truly worth a thousand words" (Mitchell 1995: 180, note 28). Broadband is commonly used to describe Internet connections that are faster than dial-up, but there is no common or consistent definition. Broadband technologies have driven a new set of user-driven "Web 2.0" interactive applications, that demand symmetrical bandwidth, such as online games, 3D modelling, concurrent engineering, and video conferencing.

Thanks to unrelenting technological improvements, our understanding of broadband is continually evolving. Official definitions vary, but generally broadband refers to a range of technologies that provide always-on functionality and allow the simultaneous use of both voice and data services (European Commission 2005b: 9). The International Telecommunications Union (ITU) defines broadband "as transmission capacity with sufficient bandwidth to permit combined provision of voice, data and video" (ITU 2003: VII), usually as a data rate greater than 1.5 Mbps.

A choice of technologies

The most common technologies worldwide are DSL, which boosts the bandwidth of traditional copper-wire telephony networks, and coaxial cables, which have a higher bandwidth than copper wires but lower than optical fiber. DSL is the predominant access technology, accounting for 65 percent of all broadband lines at the end of 2005, and 80 percent in Western Europe. The penetration rate of cable modem is at about 25 percent worldwide. It is much more common in North America, Russia, and South Korea. Other technologies such as satellite, fiber, and wireless systems account for just under 10 percent of all broadband users (TeleGeography Research 2006b).

Today, bandwidths of 10 Mbps, 20 Mbps, and even 100 Mbps are commercialized, as broadband infrastructure takes advantage of fiber-optic technology for long-distance as well as local, "metro" links between central offices, hubs and head ends. These backbone and metro links carry 2 to 8 Gbps. Personal computers also can process large amounts of data—at least 1 Gbps. The *bandwidth bottleneck* is in the "last mile" that connects individual homes and offices. These last mile links are predominantly copper, whether DSL enhancements to the existing copper loop of telephone connections or the coaxial cable used by cable television systems (Green 2006). Table 3.4 shows the data rates and ranges of the various broadband technologies. While the various DSL technologies continue to improve, the clear superiority of fiber-optic technologies remains important for a number of video and interactive applications.

Although many broadband applications are tolerable with copper technology, high-definition television requires approximately five times the bandwidth of conventional television and finds the bandwidth bottleneck unacceptable. "Video content, whether downloaded from the Web, accessed through a file-sharing application, or streamed live has become the largest consumer of network capacity" (Mauldin 2006: 28). Traffic from Web-based video content providers such as YouTube and Apple's iTunes store has grown dramatically. In early 2006, YouTube's outbound traffic was 20 Gbps and growing nearly 20 percent a month. Apple reported that 15 million videos had been downloaded through its iTunes Store as of mid 2006. Although music files were responsible for much bandwidth traffic a few years ago, video files accounted for the 61 percent of peer-to-peer (P2P) traffic in 2005, according to CacheLogic (2005). Streaming video, primarily of live events, continues to grow, as does online gaming traffic (Mauldin 2006). The utilization of telecommunications networks for video—uploads as well as downloads—makes the broadband bottleneck even more pronounced (George 2005). The bottleneck is perhaps even more critical for users with cable modem Internet access, since their bandwidth is shared with many other users.

Table 3.4 *Data rates and ranges of various broadband access technologies*

End-user access technology	*Typical data rates*	*Typical distance limits*
ADSL (current)	1–10 Mbps, Asymmetric, dedicated	6,000 m
XDSL (emerging)	10–100 Mbps, Some symmetric	1,000 m
Cable modem	1–10 Mbps Asymmetric, shared	10 km with amplifiers
Fiber to the home (FTTH)	10 Gbps per wavelength Symmetric, dedicated	100 km without regeneration
Wi-Fi	11 or 54 Mbps Asymmetric, shared	100 m Proprietary up to 10 km
WiMAX	75 Mbps Symmetric optional, shared	50 km
Microwave	1 Mbps–1 Gbps	50 km
Millimeter wave (current)	155 Mbps–1.25 Gbps	2 to 5 km
Millimeter wave (emerging)	1.25–10 Gbps	1 to 2 km
Free-space optical	100 Mbps–2.5 Gbps	1 to 2 km
Powerlines (experimental)	200 Mbps	

Source: Based on IEEE-US (2005), Table 1

The solution to the bandwidth bottleneck is to replace copper with fiber-optic (Green 2006). These links are called fiber to the node (FTTN), fiber to the curb (FTTC), fiber to the premises (FTTP), fiber to the building (FTTB), fiber to the home (FTTH); and generically, these are referred to as FTTx. Fiber-optic networks are costly to install—two to six times as costly per user as a satellite earth station and eight to twenty-five times as costly as cable or DSL (Crandall 2005). In the US, most FTTH has been installed in new, "greenfield" subdivisions. In Japan and Korea, dense populations have helped those countries to have widespread FTTB to apartment buildings and FTTH to other neighborhoods.

Demand for more downstream bandwidth is increasing along with the number of digital streams delivered to the subscriber, such as multiple TVs, TVs with "picture-inpicture" capabilities, multiple computers with broadband Internet connection, and digital video recorders (DVRs), which allow one to watch one channel while recording another one at the same time, thus requiring two digital streams of video at the same time. Kozischek (2006a, 2006b) calculates that it will be difficult, if not impossible, for asymmetric DSL and cable systems to meet this growing downstream demand, and even more the growing upstream bandwidth demand for sharing digital photographs and videos.

The demand for bandwidth and video is not only for leisure. In the US, those "with FTTH say they work at home an average of 1.3 days per month, versus 0.3 day before they got fiber" (Ross 2006: 22). Hence the role of broadband in telework implementation, which we examine in greater detail in Chapter 7. Current fiber technologies are unlikely to provide enough bandwidth for the needs of video. Consequently, several groups are advocating "ubiquitous gigabit networks" that are neither shared among users nor asymmetric, thereby increasing bandwidth to acceptable levels for likely demand (George 2005; IEEE-US 2005).

Wireless broadband technologies are improving, including wireless fidelity (WiFi), WiMax, and mobile Internet on 3G mobile or cellular networks. All are based on radio, however, and the quality of a radio signal degrades (or attenuates) as the distance between an access point and a user's wireless adaptor increases. Attenuation also is caused by obstructions, such as tunnels or, within buildings, walls or metal. New software products such as IBM's WebSphere Everyplace work to automatically connect a device, such as a notebook computer or mobile telephone, to the fastest available wireless or wireline network (Ehrle 2006).

The widespread acceptance of wireless broadband has raised a second bandwidth bottleneck: the need for landline "backhaul" from wireless network servers to the Internet (Ferguson 2004; Sharma and Searcey 2006). Wireless service, including cellular mobile telephone service, also needs a terrestrial backhaul connection from each cell to a central office (CO) or point of presence (POP) and these are typically fiber-optic cable. This "middle mile" connects the user's last mile connection to the backbone network that links one to the Internet. Mobile networks, which transmit signals over radio waves, inherently have far less capacity than terrestrial networks. Consequently, many mobile operators charge more for heavier usage or cancel service to heavy users (Sharma and Searcey 2006).

Currently, the most widespread technologies for broadband, DSL and cable, have asymmetric data rates. That is, the downstream bandwidth is typically many times greater than the upstream bandwidth. To maintain the viability of leased lines, incumbent telecommunications companies are thought to be reluctant to make symmetric broadband widely available (Ferguson 2004). The importance of symmetric broadband networks has become clear as the necessary infrastructure for full-scale participatory media (Kluth 2006).

Broadband diffusion policies

> A common thread among leading broadband nations is a national broadband policy. . . . National initiatives in Ireland, Canada, Korea, Netherlands, France, Japan, Australia, China and other nations have at least one thing in common: The idea that broadband is now a necessity to keep pace in this global economy.
>
> (Whitman 2005: 44–5)

Broadband appears to require a combination of public sector policy and private sector investment. In general, broadband has diffused faster in countries with a higher economic level, competition between alternative broadband technologies, and greater density of households. Of these, "the existence of strong competition among broadband technologies seems to be the main driver for broadband deployment," aided by "medium-intervention" policies that use public funding for infrastructure supply in underserved (rural and scarcely populated) areas, and initiatives to foster technological competition (Cava-Ferreruela and Alabau-Muñoz 2006: 456).

The ITU (2005) has compiled data on broadband subscribers in 123 countries, ranging from 37.9 million in the US to ten in Tajikistan. The top ten countries together have 81 percent of all broadband users in the world, and the top 40 countries account for 98.9 percent of all broadband users. China will soon supplant the US as the country with the largest number of broadband subscribers, according to a forecast by TeleGeography (2006b). India, which had no broadband subscribers at all in 2001, will rank number 11 by 2010.

Generally, small, dense countries are more able to connect a greater percentage of households and businesses. Thus, Denmark, the Netherlands, South Korea, and Switzerland are at the top of the world ranking for broadband per 100 inhabitants, according to the OECD in June 2006 (Figure 3.5). Several countries with very low population density but concentrated population clusters, such as Finland, Norway, Sweden, and Canada are also high in broadband use (OECD 2006).

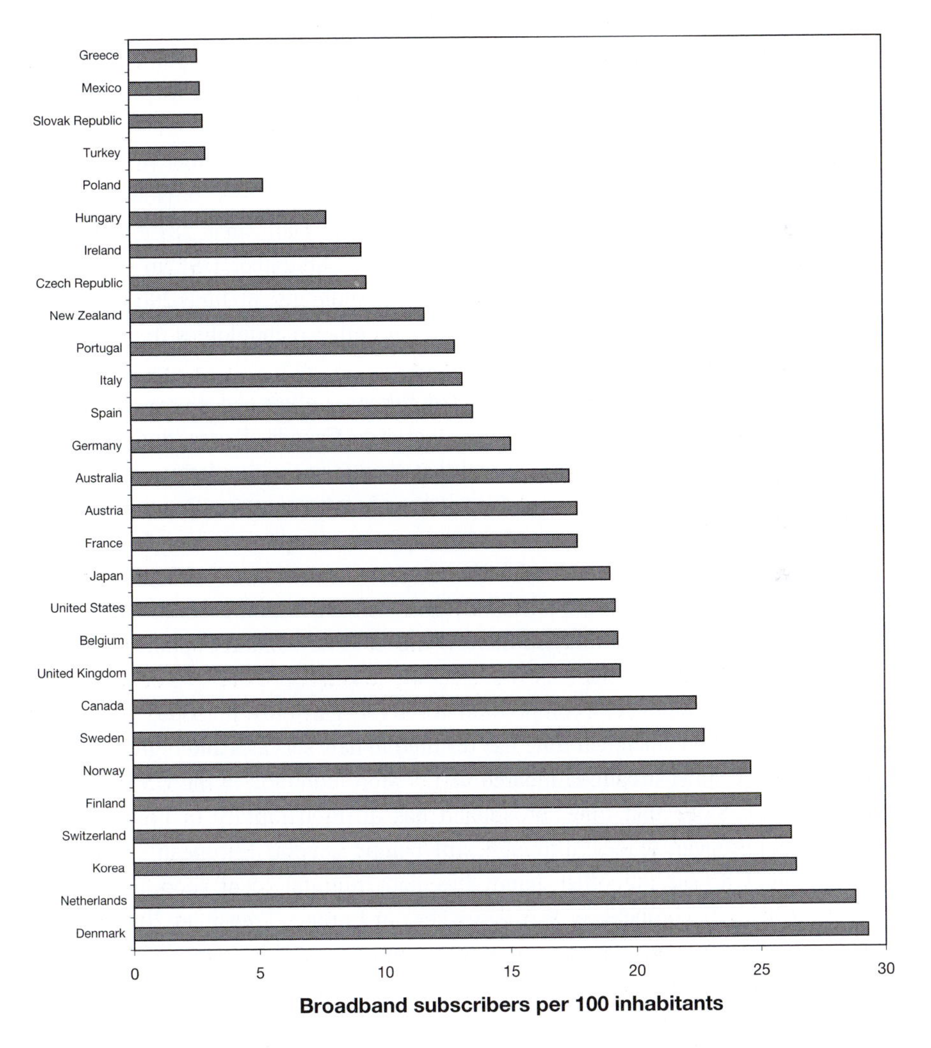

Figure 3.5 ***Broadband subscribers per 100 inhabitants (OECD countries with population 1 million or more)***

Source: Based on OECD (2006)

The Korean government's "cyber vision plan" has provided an open market that encourages both competition between service providers and technological competition. The dramatic growth of the broadband market in Korea is the culmination of government policy, growing demand for entertainment and e-commerce, and fierce market competition (ITU 2003: Box 9; Lau *et al.* 2005; Park and Yoon 2005).

Japan also has a combination of intense competition, widespread investment in optical fiber (in urban areas), and a policy of thoroughly open access to incumbent NTT's copper and optical fiber networks.[2] "It became possible for newcomers to offer ADSL and FTTH services without having their own infrastructure. As a result, communications networks have become more open in Japan than in any other country in the world," with very low charges for line sharing and dark fiber (Ida 2006: 69). Repeated price reductions have led to Japan having the cheapest and fastest broadband connections in the world. FTTH (100 Mbps) connections are very cheap: monthly charges of $64 to $74 (Hayashi and Brown 2005). Electric power companies have become leaders, installing their own optical fiber. They are not required to provide universal service or to share "dark fiber" with other users (Ida 2006).

Nowadays, Japan has more widespread availability of FTTH than any other country. The number of fiber subscribers alone in Japan outnumbers total broadband subscribers in 22 of the 30 OECD countries. DSL is still the leading platform in 28 OECD countries—that is all but Japan and Korea. Cable subscribers outnumber DSL in Canada and the United States. The breakdown of broadband technologies in June 2006 is as follows: DSL: 63 percent, cable modem: 29 percent, and other technologies (e.g. satellite, fiber and fixed wireless): 8 percent (OECD 2006).

Although the US saw the emergence of Silicon Valley and pioneered the Internet, it has "largely failed" in broadband, in contrast to Canada, Korea, and Japan (Frieden 2005: 610). This lag stems from the fact that change in the regulatory climate triggered litigation, rather than competition and investment in infrastructure (Bauer 2006; Fransman 2006a; Frieden 2005). US incumbent telephone companies complained about the unfairness of having to unbundle their local networks, and state-level policies also have generally not been effective (Wallsten 2005b). Carriers in best-practice nations accepted the regulatory mandate and turned their attention to new market opportunities (Frieden 2005).

In France, although real competition for FT began only in 2000, unbundling and intense competition since that time has resulted in new services and higher bandwidth—up to 24 Mbps downstream, compared to 1.5 Mbps common in the US (Abboud 2006; Krafft 2006). Since that time, broadband has diffused rapidly in France, which has become a European leader, although still behind several European countries to the north. The UK, on the other hand, has had minimal local loop unbundling and consequently less competition than elsewhere in Europe (Fransman 2006b).

FTTH is spreading—slowly—in Europe, mainly in five countries (Sweden, Italy, Denmark, the Netherlands and Norway). Of 166 FTTx projects in Europe, 72 percent were initiated by municipalities or power utilities (Montagne 2006). However, major telecom carriers and Internet providers have plans for FTTH. On December 15, 2006, FT announced it would launch a 100 Mbps fiber-optic service in 2007, first in Paris, then in regional capitals (Brafman 2006).

To sum up: a small number of Asian and European countries lead the world in broadband, and particularly in FTTH. In Korea, a leader in broadband, "near FTTH" is being deployed, where fiber is run to a pole and CAT5 wire is run to the home or apartment (Whitman 2005). Most importantly, a handful of European and Asian countries are consistently the most advanced in networked readiness, broadband, and other markers of the digital economy. Although many of the broadband leaders are small and densely populated, the explanation goes beyond population density, which is correlated at only 0.265 with broadband penetration. As suggested earlier, much more influential is income per capita GDP ($r = 0.598$) (OECD 2006). Indeed, the "geographical broadband digital divide is only one aspect of a wider social and economic development issue" (European Commission 2006). This important subject is addressed more thoroughly in Chapter 9.

Urban areas are central hubs in the digital economy for several reasons. Many knowledge-economy companies must be located in a city for the people, ideas, and "buzz" that create unique knowledge. Indeed, the largest cities have the greatest "buzz" for creative and competitive projects despite their high real estate costs. We will return to the issue of "buzz" in Chapter 8.

Conclusion: does broadband matter?

The smooth operation of the Internet, using high-bandwidth links, user-friendly Web browser software, and ubiquitous hyperlinks, opens up a limitless ocean of information. Some of the "pressure points" caused by the transition to a commercial Internet have been more or less solved, such as interconnection among the networks that comprise the Internet. Others are still evolving sources of conflict, including Internet telephony, quality-of-service pricing, control over domain names, intellectual property rights, and international standard for encryption, taxation, and censorship (Goldsmith and Wu 2006; Thomas and Wyatt 1999). Other issues include privacy, since digital "data shadows" allow us to be tracked and our actions monitored (Zook *et al.* 2004).

It is certain that both FTTH and wireless technologies for broadband will continue to advance and diffuse, and that they will do so unevenly. Some question whether broadband—its availability, speed, or price—makes any difference in economic performance or in the quality of personal or social life (Fransman 2006b). Most research assumes that e-commerce is beneficial, often without clear evidence. However, no really new broadband-specific "killer applications" have emerged. VoIP is growing worldwide, and interactive games typically take advantage of broadband but, overall, "most users spend most of their time using their broadband Internet connections to do much the same as they did with their narrowband" (Fransman 2006b: 25). Despite the lack of precise impacts, it is clear that people work and use their leisure time differently—particularly those who use broadband—relying less on travel to shops and doing at least some portion of their work at home (Roberts 2005). We discuss the phenomenon of telework in greater detail in Chapter 7.

Wider social benefits within a community are beginning to be documented (Gillett *et al.* 2006). Communities in the US in which mass-market broadband was available by December 1999 experienced more rapid growth in employment, in the number of businesses overall, and in businesses in IT-intensive sectors, relative to comparable communities without broadband at that time. This analysis did not find a statistically significant impact of broadband on the average level of wages.

Positive effects of IT and broadband include their role as infrastructures, similar to paved roads and telephones in the past, which gave peripheral communities better and cheaper access to goods and services, and airports, freeways, and high-speed rail links today. Broadband now provides local firms, institutions, and consumers with better and cheaper access to information, undoubtedly narrowing the information gap between central and peripheral regions. In coming years, important positive impacts of IT will perhaps be in the public sector, specifically health and education (Ida 2006; Polèse and Shearmur 2002).

The unevenness of infrastructure worldwide has direct impacts on the investment decisions of global companies. For example, US firms, which have been steadily dispersing their research and development (R&D) for the past several decades, eliminate two sets of locations from consideration: those with inadequate communications infrastructure and those with insufficient access to scientific and engineering talent (Council on Competitiveness 2005). However, determinism must be avoided. The lack of broadband may indeed have a deterrent effect, but broadband availability has almost never been in itself the propellant of economic development. On the contrary, broadband goes where business and wealth already have flourished. That is not to say that digital telecommunication, and IT in general, do not have effects on the economy. While the availability of a well-educated workforce, business-friendly regulation, good transportation infrastructures, and all the essential constituents of place and location continue to play a key role, the remote access to a virtually unlimited amount of data, as we explain in the following chapters, has deeply affected business processes and organizations, with real consequences for regions and communities.

4 DIGITAL PRODUCTION AND BUSINESS ORGANIZATIONS

The functional nature of the firm is to be a system which processes and coordinates economic transactions (Coase 1937). Since the beginning of the telegraph and telephone era, information technology has relentlessly increased its role in all kinds of transactions. IT and the Internet have permeated the *value chains* of every business sector (Porter and Millar 1985; Porter 2001) and have deeply transformed the organization of firms (Dutton *et al.* 2005). Schiller (1999) evokes the emergence of a "digital capitalism."

In the first section of this chapter, we explain how contemporary production results from the creation and management of complex, IT-driven *product lifecycles*. The cycle approach makes irrelevant the traditional distinction between material goods and services. We face the emergence of a systemic arrangement of both material and informational transactions which constitute modern *value chains*, the series of interlinked tasks and transactions which generate value-added within the firm (Porter 1985). In the same section, we present the causal loop between IT uses and the requirements of *customer-centricity* in business. IT implementation has proven a driver of customer empowerment. In turn, IT allows mass customization and helps the company to meet *time-to-market* requirements which characterizes the digital economy.

The second section surveys the emergence of *flexible*, *network-centric* or even *virtual* business organizations, which epitomizes the transition from simple value chains towards more complex *value networks* which are unstable in their temporal, spatial and organizational dimensions. This trend materializes through the practice of *outsourcing*—that is, the purchase of capabilities outside the company.

The third section places emphasis on the *supply chain* issue. According to Lüthje (2005: 338), "the acceleration of technological changes and shortening of product life cycles" lead to "commoditization, disintegration, and increasing instability across the value chain." "Coordinating the pieces of the value chain," that is the role of supply chain management, "is becoming only more challenging" (Deloitte Research 2003). *Global networks of modular production*, based on *Wintelist* methods, rely on the efficiency of the *logistics industry*—actually the alliance between IT and modern transportation.

Mastering complexity with information technology

> Organizational complexity is a key condition necessary for a firm to maximize the benefits it can derive from the new information technologies.
>
> (Sassen 2001a: 412)

Complexity is a key word of modern business understanding (Rycroft and Kash 1999). The pursuit of interlinked, basic goals such as technological innovation, new product development, new market targeting, and the widening of the geographical scope of both sales and purchasing operations, make a company and its management more complex. Gottfredson and Aspinall (2005) regard as a key issue of business strategy the existence of a tradeoff between complexity level and overall efficiency.

A major feature of contemporary business is that the most successful companies do not rely on technological or marketing innovativeness only, but increasingly on *organizational innovativeness* (Chesbrough and Teece 1996), which makes it possible to push away the complexity vs. efficiency tradeoff frontier. Therefore, it is a basic axiom of the digital economy that state-of-the-art IT implementation is a key tool for mastering complexity and developing sophisticated, but nevertheless agile, creative, and profitable organizations.

Today, companies have to deal with a series of intertwined problems that cannot be solved separately and sequentially: the rise of customer power, the shortening of product lifecycles, the globalization of sales and purchases, and the splintering of the supply chain. Contemporary management faces causal loops and paradoxes. Customers want lower prices, which would lead to low-cost offshoring. But they also want "customized solutions" and short delays, which would lead production to remain local. Large retailers and manufacturers generate a rising low-cost pressure on suppliers, but innovation and time-to-market requires an increasing level of cooperation and trust between purchasers and suppliers, which results in the concept of *extended enterprise* (Post *et al.* 2002).

In this context of tremendous organizational changes, IT implementation does not act as a cause—at least not as a primary cause—but rather as the great *facilitator*. However, we can identify a notable exception, the "bandwagon effect," which is a bad management tendency to adopt new technologies without clear, rational grounds, but only because rivals have adopted it. It must be added that if technology is usually not a primary cause of business changes, it can become a secondary cause. IT implementation, answering to well-identified commercial needs, may in turn lead to subsequent changes and, finally, to a complete restructuring or even "unbundling" of the firm's organization (Hagel and Singer 1999). Tippins and Sohi (2003) suggest that heavy IT investments result in effective business performance especially when "IT competency is mediated by organizational learning." All in all, IT implementation usually proves to be decisive in the way a company becomes a complexity master facing the emergence of an integrated production cycle (Deloitte Research 2003; Hobday *et al.* 2005).

The emergence of an integrated, IT-driven production cycle

> We believe that in the future, all manufactured goods in the world will be digitally defined, engineered, simulated, produced and managed throughout their lifecycle.
>
> (Charlès 2006)

The seminal works of Vernon (1966) and Day (1981) have made popular the concept of *product lifecycle* which encompasses the whole life of the product, from the earliest design research to the market end. The cycle approach is ideally suited to deal with complexity and systemic functioning which characterize advanced ways of doing business. It leads to conceiving production in a global way, as a system of dense and frequent interrelations which integrates closely people, places, and firms of all sizes. However, the lifecycle analogy is a convenience which fits many situations, but to which few expect strict adherence (Malecki 1997: 65).

Product lifecycles include increasingly complex combinations of time and space arrangements, which encompass a series of activities ranging from the very beginning of the project to its death:

- market analysis
- R&D and product design
- advertising
- procurement and supply chain management
- production engineering
- manufacturing, streamlining, quality control
- administrative and legal work (accounting, IT administration)
- labor and human resource management (training, recruitment)
- sales (and billing, shipment, delivery)
- sales analysis (data mining)
- after-sales service, CRM.

The product lifecycle must preferably be thought of in a systemic way, like a loop (Figure 4.1). Several of the tasks must be performed simultaneously rather than sequentially. Market analyses and feasibility studies must be done at the same time. Legal

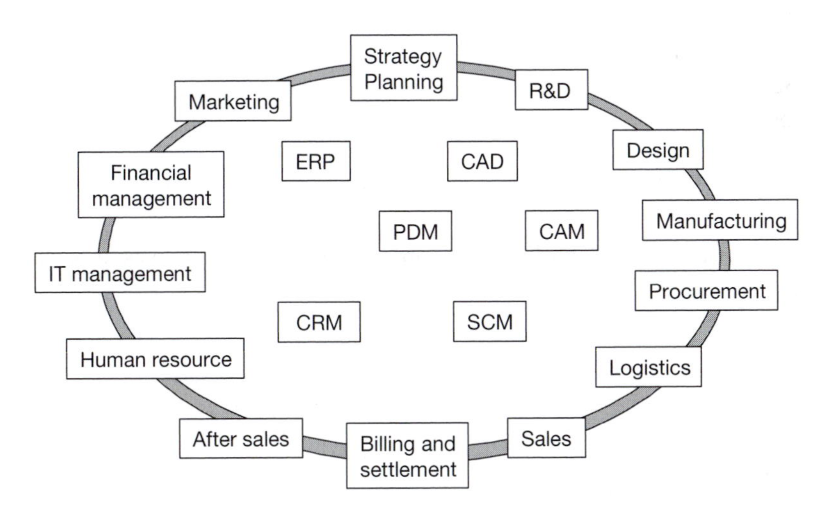

Figure 4.1 ***The sphere of product lifecycle management***

Source: Adapted from Wikipedia, http://en.wikipedia.org/wiki/Image:Plm1.png, accessed 4 November 2007

work, human resource and financial management, computer and telecom administration are increasingly parallel to core production processes. And "tasks can be separated in both time and space" (Bryson *et al.* 2003: 36–7).

Moreover, many activities which have long been described as "downstream" have actual and rapid consequences on common "upstream" tasks. Modern production is an iterative process. For example, data from customer enquiries must result in short notice in production improvements or even in design alterations. IT-enabled *lifecycle engineering* results in permanent, virtually real-time feedback processes and adjustments in the whole production cycle (Wanyama *et al.* 2003).

Advanced methods of product lifecycle management (PLM) are driven by integrated e-business software suites. In February 2003, Oracle's advertising campaign for its e-business suite boasted that it covered from "a single database" the entire product lifecycle: marketing, sales, supply chain, manufacturing, finance, and human resources. The world leading company in PLM software is Dassault Systèmes, the software subsidiary of Groupe Dassault, a French aircraft and media company. Dassault Systèmes' flagship application is CATIA (Computer Aided Three dimensional Interactive Application), first developed in 1977 by Avions Marcel Dassault, and now marketed worldwide by IBM. CATIA integrates 3D digital mockup in computer-aided design (CAD), engineering (CAE), and manufacturing (CAM). According to the company, 7,812 CATIA licenses were in use worldwide by early 2007, mostly in the automotive and aircraft industry. CATIA's most notorious users are Airbus and Boeing, whose 777—the program was launched in October 1990—was the first jetliner to be 100 percent digitally designed and prototyped.

The use of CATIA is pivotal in the new production organization Boeing has set up for its recent 787 Dreamliner program, whose configuration was completed in September 2005. CATIA, and Dassault's collaborative software ENOVIA and SMARTEAM, have allowed concurrent design and engineering on the 3D model by Boeing's 43 partners, connected from 135 sites dispersed in 14 US states and in Canada, France, Germany, Italy, the UK, Sweden, Australia, Korea, and Japan (Boeing 2007). This example illustrates the key role IT plays in *value chain unbundling*, *global sourcing*, and the emergence of value networks, presented in the following section.

Empirica (2005) delivers interesting data about the implementation of various IT applications by 5,200 companies of ten business sectors in seven European countries: Czech Republic, France, Germany, Italy, Poland, Spain, and the UK (Table 4.1). It shows that capital-intensive, innovative sectors such as aeronautics, IT services, automotive, and pharmaceuticals are IT-intensive. Companies in these sectors usually have an intranet, and often use enterprise resource planning (ERP) systems.[1] Some sectors are intensive users of a given application. For example, the automotive industry, which relies on parts purchases from thousands of suppliers, is the leading user of supply chain management (SCM) systems and electronic data interchange (EDI). Tourism and publishing, which are not IT-intensive, sell well online, a fact we explain in Chapter 5. Data based on the employment breakdown shows that IT implementation rates unsurprisingly increase with the size of the firms.

A popular thought, today, is to consider the emergence of a "service economy," which would tend to overcome the old, "industrial" economy (Walker 1985). *Product*

Table 4.1 ***IT implementation rates by European companies in 10 manufacturing and service sectors***

	Using an intranet	*Using an ERP system*	*Buying online > 5% of purchases*	*Using a SCM system*	*Using EDI standards*	*Selling goods online*
Aeronautics	98	52	38	35	16	8
IT services	87	39	60	17	10	25
Automotive	85	71	34	48	61	7
Pharma	75	71	29	35	42	18
Machinery	54	58	22	14	23	11
Publishing	52	21	32	9	13	37
Food	46	37	16	21	37	12
Textile	45	34	15	13	21	14
Tourism	41	12	36	10	–	36
Construction	29	13	18	7	4	4
Total (EU-7)	47	28	27	15	19	17
250+ empl.	79	59	29	27	43	21
50-249 empl.	53	33	26	14	14	16
10-49 empl.	29	9	20	8	4	14
1-9 empl.	18	8	26	5	2	15

Source: Selected from Empirica (2005)

lifecycle thinking requires us to abandon this traditional distinction. Zysman (2005: 268) suggests that "the distinction between service and product is blurring." Phillips *et al.* (1999) proposes the emergence of "product-services."

This tendency is illustrated by firms in the automobile industry, which no longer sell cars only, but rather a service which includes insurance, emergency repairs, and credit. Similar tendencies are seen in the travel and tourism sector, where the holiday experience, which is the actual product sold, requires a mix of pure services, real commodity (food), and brick-and-mortar equipment (hotels). The relative decline of goods in favor of services is epitomized by the growth of rental practices, especially in the BtoB sector (cars and trucks, computers, real estate and, increasingly, software). "Product-service" is also illustrated by the wireless telephone industry, whose customers often buy simultaneously an object (the handset) and a subscription.

In the end, what we are witnessing is the decline of the "object-oriented" product, for the concept of service-oriented or, better, *experience-oriented* product (Pine and Gilmore 1998, 1999). Indeed, customer experience has become another catchword in contemporary business milieus (Cisco Systems 2003a), and even in best-selling literature (Rifkin 2001).

IT runs the fast, customer-centric firm

A remarkable feature of contemporary business organization is the advent of the customer-centric enterprise (Tseng and Piller 2003). O'Callaghan (2005: 7) suggests the expression of *customer-centric ecosystems*. Market deregulation and free trade have spurred competition. The rising value of customer loyalty (Gitomer 1998) has increased the importance of marketing functions, which are no longer subordinated to production, but rather command it. Thanks to the implementation of data mining software, call center operations feed customer databases which are used to target telesales, improve design and, in the end, allow companies to adapt swiftly to customer tastes or fashion tendencies (Elmer 2004).

Another illustration of the role of IT in customer empowerment is the rise of Internet uses for the purpose of comparing products, services and prices. An example, found in greater detail in Chapter 5, is seen in consumer-to-business (CtoB) platforms such as Priceline.com. Customer experiences with firms and their products, both good and bad, can easily be posted and consulted on dedicated websites. Business failures which harm customer interests may quickly endanger the firm and its brand reputation.

The primacy of client needs has generated two major business trends: mass customization (Gilmore and Pine 2000), and the tremendous increase of time-to-market requirements. Thanks to advanced, IT-enabled marketing methods, such as extensive telephone surveys and computerized analysis of call center inputs, companies are able to identify a larger number of customer segments which are smaller in volume, and require more versions to be developed, produced and marketed (Sanderson and Uzumeri 1995).

Mass customization must be assessed at both production and marketing levels which are, as we noted above, increasingly intertwined (in modern business models, a product is often sold first, then manufactured). The website of Dell illustrates the outcome of mass customization in the personal computer industry, and the help adequate IT can provide at the sales level: as of the end of 2006, the home products portal was designed to help visitors to review and choose between eight basic desktops, twelve processors, seven monitors, plus a choice of hard drives, optical drives, video cards, memories, modem cards, embedded software, and warranty options.

At the production level, the challenge for manufacturers is to assemble thousands of parts, from hundreds of suppliers. Depending on frame, power, color, and various equipment and accessories, car makers face billions of permutations (Holstein 2001). CAD software and computer-driven machine tools allow firms to deliver shorter series of increasingly complex products.

In the service sector, travel companies could not tailor trips (including air travel, accommodation, car rental, etc.) without the integration of complex software, huge databases and advanced telecommunication networks. Mass customization has reached a high degree of sophistication in information-based services which trade fully digital outputs with marginal costs virtually tending to zero. Shapiro and Varian (1998) demonstrate that the "endless malleability of digital data" allows "versioning," which is the capability of keeping closer to customer needs through a great number of customized offers. This phenomenon is clearly visible in the home media, TV and telecommunication markets, where the digital convergence, notably through the "triple play" capability of cable or DSL technologies, gave birth to complex and flexible, "bundled" subscriptions, and to "customer-centric pricing" (Cross and Dixit 2005).

As Sassen (2001b: 97) writes, "greater product differentiation expands the marketing and selling functions of the firm." The evolution we describe above has made CRM a critical function within the production cycle. It materializes in the emergence of the call center industry, an enormous business worldwide whose regional implications we survey in Chapter 6. Mass customization is concomitant with the increase of time pressure in business. "The phrase 'time-to-market' has become a catchword in business" (Amirahmadi and Wallace 1995: 1759).

The major time-related tendency is the shortening of the product lifecycle. Given the acceleration of technological changes and growing competition, the periods when producers might expect trading in a new market with almost monopoly profits are narrowing, and firms endeavor to reduce the time between first ideas and new product launchings. A survey of 600 manufacturers in North America and Europe by Deloitte Research (2003) found that they had cut their product development cycle by 12 percent to 16 months in 2003, and were planning to shorten cycle times to 13 months by 2006. Meanwhile, the number of new products tends to increase: new products introduced within the last three years generated 29 percent of total revenue in 2003, up from 21 percent in 1998, and producers were expecting the number to reach 35 percent by 2006.

The tyranny of time results in an all-out search for more flexibility, which is recognized as a prominent feature of the contemporary economy (Beveridge and Perks 2000; Coffey and Bailly 1992; Scott 1988). And providing flexibility is the very purpose of IT implementation within the company's entire organization.

Changes in business organization and management

The digital economy is both a technological transformation and a radical upheaval in business organization and practice. The technology aspect itself is driven by steady advances in computing capability and by improvements in the capacity of the telecommunications networks to transmit digital data. This "E-conomy" is bringing about, in addition to changes in business organization, changes in market structures, government regulations, and human experience (Boyer 2004; Cohen *et al.* 2000).

An economy based on digitized information is different from economies in the past. The origins of the new paradigm date back to the 1970s, with semiconductor chips, pervasive computers, and telecommunications networks facilitating new, networked organizational forms (Castells 2000). The benefits of information technology (IT) did not occur all at once. Early uses of IT by businesses were mainly "to control and standardize their operations" (Burton-Jones 1999: 9). Centralized file handling and distributed computing were internal to the firm during the 1960s and 1970s, and focused on data management. As personal computing and local networking took hold during the 1980s to mid 1990s, information management prevailed. Only since 1995, when the Internet and global networking became standard, has it been possible to speak of managing knowledge, defined as the cumulative stock of information and skills derived from the use of information, rather than merely managing data or information (organized data) (Burton-Jones 1999).

The transformation has not been smooth, as firms and their management had to learn how to adapt to the knowledge-based information economy (Eliasson 2005). Firms now must master four sets of information and communication activities—not only internally, but also with other firms and organizations:

- Innovation—Creating business opportunities by exploring, experimenting, and expanding
- Selection—Entrepreneurship and identifying and choosing business opportunities
- Coordination—Coordinating competition in markets and coordinating management in networks
- Learning—Knowledge transfer.

An example is found in Procter & Gamble (P&G), one of the world's largest consumer-products manufacturers, which has transformed its old model of internal or closed R&D to one based on external networks. The company identifies promising ideas throughout the world and applies its internal capabilities to them to create better and cheaper products, faster. According to a P&G's senior executive, research and development (R&D) has been replaced by "connect" and "develop" (C&D) (Chesbrough 2003; Huston and Sakkab 2006):

1) connect: P&G collaborates with suppliers, competitors, scientists, and entrepreneurs, systematically scanning the world for proven technologies, packages, and products that P&G can improve, scale up, and market.
2) develop: either alone or in partnership with other companies, P&G has implemented a suite of new technologies for data mining, simulation, prototyping and visual representation—P&G calls them "innovation technology"—which support open innovation (Dodgson *et al.* 2006). To a large degree, these links with other sources of innovation have grown out of research pioneered by von Hippel (1988, 2005).

Harnessing the power of IT throughout the firm's organization

> The secret to success is not just to unbundle, but to unbundle and to rebundle, creating a new organization.
>
> (Hagel and Singer 1999: 140)

The seemingly limitless capacity of storage, retrieval, and exchange of information that digital technology provides to companies, and the global scale, speed, and complexity of operations are a rising challenge to business organization analysts and managers. Hindle (2006) suggests that business re-engineering often lags behind individual changes in the way people work.

The recent turmoil at Airbus, the arch-rival of Boeing in civil aircraft manufacturing, is revealing. Airbus has pioneered modularity and networking in the aircraft industry (a model of organization now adopted by Boeing, as mentioned above). In Toulouse (France) Airbus assembles modules built at 16 production sites in France, Germany, Spain, and the UK. Production and sales of the new A380 model have been delayed for almost two years, and the company is likely to lose billions of dollars, because the cabling of some parts of the body made in Hamburg, Germany, did not fit the other parts in the final assembling process. The problem has been identified: while the French were using the latest version of CATIA, Dassault Systèmes' CAD/CAM software, the Germans were still using a version dating from the 1980s. Airbus' failure is explained by a lack of organization and governance at the highest level, rather than a lack of technical skills. Airbus is a subsidiary of EADS, the world's second company in aerospace industries. The organization has been hampered by the dual structure at the top of EADS, and poisoned by rampant Franco–German rivalry. Rumors internal to the company suggest that the German management of Airbus did not give the priority to CATIA updating because Dassault Systèmes, its designer, is controlled by Groupe Dassault, a rival of EADS in aircraft and defense industries (Matlack 2006).

The lesson of the A380 story is that governance problems have so far prevented the company's leaders: 1) from implementing state-of-the-art IT in the whole structure of the firm, and 2) from rethinking its whole organization in order to make the most from new technologies. Perhaps a cause of Airbus' failure, from an organizational point of view (the governance and currency problem set apart), has been the lack of an IT-coordinated matrix structure.

Matrix organization has been the subject of an important literature in management studies. Large, diversified corporations tend to be organized in impenetrable *silos*, which prevent valuable information and knowledge from permeating across the divisions. In a matrix structure, dedicated transversal business units are in charge of these information exchanges. This kind of structure requires a great deal of coordination effort—and accounting prowess—which are intrinsically IT-intensive. The example of Volvo Group and Volvo IT shows that information technology plays in a company's structure the role of steel in reinforced concrete. However, the matrix is not a panacea because matrix management is not only a question of business organization, but more a question of "frame of mind," which requires building "a matrix in the minds of our managers" (Bartlett and Ghoshal 1990: 145).

Coordinating a matrix structure with IT: example of Volvo Group and Volvo IT

Headquartered in Gothenburg (Sweden), Volvo Group (AB Volvo) is a leading manufacturer of trucks, buses, and construction equipment. It has about 81,000 employees and production operations in 25 countries. Volvo's net sales in 2005 amounted to about $31 billion.[2]

The company's organization can be understood as a 3D matrix (Figure 4.2):

1) The first axis gives entry to the company's brands and industries, organized in eight subsidiaries named *business areas*: Volvo Trucks, Mack, Renault Trucks, Volvo Buses, etc.
2) The second axis describes transversal functions which line across, and supply or support the operations of the business areas. These functions are performed by eight *business units*, which are fully-owned, autonomous subsidiaries: Volvo Powertrain (engines and transmission systems), Volvo 3P (planning, product development and purchasing), Volvo Parts, Volvo Logistics, Volvo Information Technology (Volvo IT).
3) The geographic dimension is represented by administrative, R&D, manufacturing and commercial units operated worldwide by Volvo's affiliated companies.

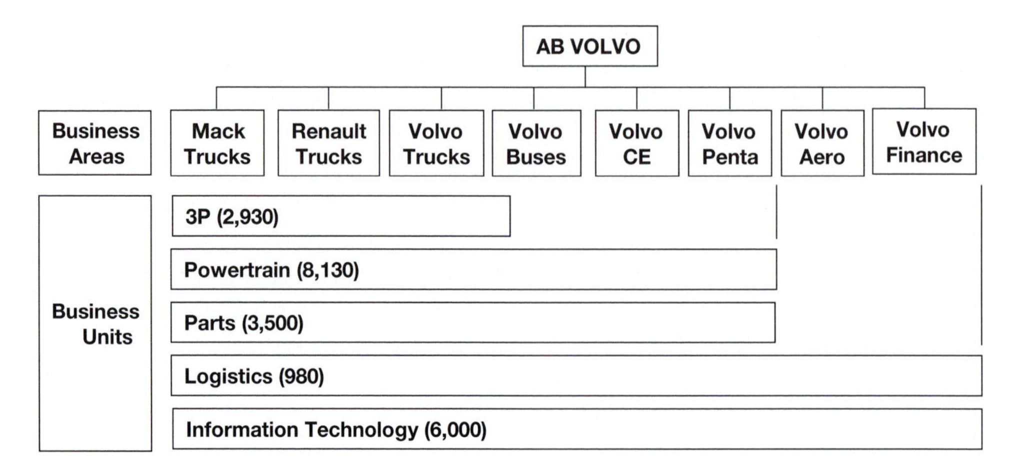

Figure 4.2 ***The matrix structure of Volvo AB***

Source: Adapted from Volvo Group (2006: 2)

Note: Figures in brackets: number of employees in the Business Unit

The mission of Volvo IT's 4,660 employees is dual: 1) serving the Group's core industries which are increasingly IT-enabled, and 2) keeping coherent and efficient the matrix structure. The "PLM Solution Center" signed in 2002 an agreement with Dassault Systèmes to develop and commercialize industrial applications around the software suites CATIA, ENOVIA, and SMARTEAM. Other services include:

- business process consulting and project management
- system development and integration
- support, operation and maintenance
- education and training.

The Collaborative Work Solution Center develops collaborative tools used by 50,000 people inside the group: mail and messaging, audioconferencing, online meetings, videoconferencing, collaborative work platform, project and time reporting. In addition, Volvo IT provides sales and after-sales IT-based services: e-commerce solutions for dealers, CRM solutions, and internal technical support for parts sales and logistics.

After the acquisition of Mack Trucks and Renault Trucks in 2000 and 2001, the new IT personnel were integrated with Volvo IT. This organizational re-engineering has led to the geographic dispersion of teams working for the same business unit. Volvo IT has six sites in Sweden, five in France, two in the UK, one each in Belgium and Poland, 11 in North America, and is present in Brazil, India, China, and Malaysia.

Volvo could have adopted the "lean and mean" model, outsourcing IT and making people redundant. Instead, maintaining and integrating the IT function within the group provides several benefits:

- It avoids costs and damage to the group's image which would have resulted from redundancies.

- It avoids knowledge leakage.
- It keeps support functions in the industrial culture of the truck industry, and makes it possible to earn profit from the cultural diversity given by geographic dispersion.
- Volvo's business units expand the group's market by selling their know-how to customers outside the group. Volvo IT's customers include Ford-owned Volvo Car, Nobel Biocare, SKF Group, and YIT Building Systems (Volvo IT 2006).

Volvo IT illustrates the spatial and organizational patterns of "unbundling and rebundling" (Hagel and Singer 1999). It is a case of a spatial disintegration process, followed by IT-enabled, functional re-integration, thanks to the transversal *business unit* concept. In this example, the discussion is restricted within the boundaries of a single company. The next section places emphasis on organizations made of a network of several, interlinked firms.

Network-centric firms or virtual enterprises?

> The firms that are typical of the emerging technocapitalist era are networked firms.
>
> (Suarez-Villa 2003: 392)

> The traditional business paradigm revolved around the firm. The new paradigm regards the firm as a node in an ecosystem—a network of partners that collaborate to create customer experiences and intelligent products that can adapt themselves to evolving customers' needs.
>
> (O'Callaghan 2005: 10)

Since the mid 1990s, a series of catchwords has emerged from both the academic literature and business circles: virtual corporation (Byrne *et al.* 1993; Davidow and Malone 1993), virtual organization (Franke 2002; Mowshowitz 1994, 1999; Stratigea and Giaoutzi 2000), virtual department (Symons 1997), virtual capitalism (Thrift 1998), agile firm (Goldman *et al.* 1994), networked firm (Suarez-Villa 2003), fractal company (Warnecke 1997), and network-centric organization (Roitz *et al.* 2004). These expressions are used to describe more or less similar concepts. Their definitions are context-dependent and uses vary. The present section endeavors to disentangle these concepts, and to separate actual business organizational patterns, supported by theory and strong empirical evidence, from the media hype where metaphors dominate.

To make things simple, we make a distinction between two main concepts: the *networked enterprise* and the *virtual enterprise*, even if the main body of the business literature rarely distinguishes between the two expressions.

The networked company

Because *network company* often describes businesses in the media, telecom and utility industries, it seems well advised to follow Häcki and Lighton (2001), Suarez-Villa

(2003), and Vatne and Taylor (2000), in preferring *networked* or *network-centric company* to describe organizations built upon, centered around, or embedded in networks.

The concept of *networked company* is useful to interpret both a single legal entity and a complex arrangement of several firms. A firm is said to be *networked* when it abandons the old centralized, core-periphery model, and adopts a decentralized organization, or a matrix model. According to Suarez-Villa (2003), decentralization, devolution, and the dismantling of a corporate hierarchy are archetypal features of the networked enterprise. Inside the firm, the network organization model favors the emergence of nearly independent profit-oriented business units, or the birth of autonomous R&D task forces. These decentralized structures may trade both inside and outside the firm as shows the example of Volvo presented above. Suarez-Villa (2003) regards this trend as leading ultimately to the "deconstruction" of the firm, and Hagel and Singer (1999) consider similarly a process of "unbundling" and "rebundling" of the company.

At the inter-company level, the networked organization may encompass a complex, geographically dispersed web of suppliers, manufacturers, retailers, and related service providers, linked together by commercial networks (Birkinshaw 2000). Peppard and Rylander (2006) consider a shift from value chains to *value networks*. Networked organizations and value networks usually materialize through the process of *outsourcing*.

The rise of outsourcing: "orchestrators" and "players"

Bahrami and Evans (1995) have described the networked enterprise as an *ecosystem of firms*, a constellation of specialized enterprises whose operations are coordinated by a company central to the system, acting as an *orchestrator*.

The completion of this pattern is usually achieved by outsourcing (or "capability sourcing") practices, whose generalization is the logical outcome of most contemporary economic trends, themselves enabled by a very diverse range of events, including "the fall of the Berlin Wall, China's embrace of capitalism, the advent of worldwide tariff reduction agreements, and the spread of cheap, accessible telecom infrastructure" (Gottfredson *et al.* 2005: 134).

Outsourcing is particularly well-developed in large companies: Gottfredson *et al.* (2005) report that 82 percent of large firms in Europe, Asia, and America have outsourcing arrangements. Nevertheless, thanks to the commoditization of e-business tools, and to the high degree of portability and standardization inherent to digital business information, small and medium-sized enterprises (SMEs) have increasing access to the outsourcing market.

The current surge in outsourcing practices is spurred by the interlinked characteristics of contemporary business we presented above: increasing competition under pressure from both customers and shareholders, the overall search for decreasing costs, and the incessant need for more flexibility. Basically, outsourcing is a response to rigidities in labor markets and labor laws. Firms that subcontract non-core production seek to reduce their well-paid, permanent staff, while suppliers' employees often do not benefit from such favorable collective agreements. When demand falls, social hazards and costs of downsizing are therefore transferred onto subcontractors. Chesbrough and Teece (1996) suggest that the outsourcing-based organization delivers higher incentives, because it is coordinated by the market place, rather than by internal transactions. Indeed, a firm can more easily put pressure on its subcontractors, in terms of costs or deadlines, than on its own employees (Veltz 2000). These are basic principles of "lean and mean" management (Harrison 1994).

Another general characteristic of outsourcing is the search for "external economies of scale." Subcontractors who are specialized in a certain task can get return on investment (ROI) through intensive use of efficient, costly equipment. Economies of scale are also observed when subcontractors serve companies operating in sectors subjected to seasonal market fluctuations, such as travel agencies and airlines. Outsourcing makes it possible to reduce costs associated with fixed capital assets such as production infrastructures and real estate. Outsourcing of a part of production also makes it possible to benchmark the activities the firm keeps in-house, and to tap into new labor pools that are unknown to the foreign principal company, but are easier to exploit by local contractors.

> We have seasonality in the business, so outsourcing gives us the ability to ramp up or down without having to change our employee count. The second reason is that outsourcing helps us benchmark our service quality and costs. You know exactly how much you are paying a third-party vendor, and that gives you a good standard to compare your own costs against. A third thing is that these vendors work for multiple clients, so working with them gives us an opportunity to learn what could be done better. Perhaps a fourth reason is risk mitigation. By spreading our centers out, we may gain a bit of control over something like a natural disaster that impacts one area. And finally, working with partners lets us tap into new labor markets without setting up our own centers in those locations. Take India, for example: there are at least eight to ten cities where one could set up centers of reasonable size.
>
> (R. Malhotra, director of Dell's Indian operations, in Kaka 2006: 24–5)

In the end, what is of premium importance for a company is not *ownership*, but *control* of the value chain (Veltz 2000), and the ability "to make the most of critical capabilities" (Gottfredson *et al.* 2005: 134). Hence the key role of IT for overcoming the tradeoff between incentives and control (Chesbrough and Teece 1996). Rottman and Lacity (2006), like Maskell *et al.* (2005), suggest that the development of outsourcing practices follows an incremental learning process. After an initial stage of "hype and fear," early adopters focus essentially on costs. Then, after best and worst practices have been identified, and as the market matures, richer practices turn to focus on quality. The final, "institutionalized" phase sees companies focusing on high value-added-oriented transformation.

Cisco Systems (2003a: 10) suggests the two-pronged distinction between core and context activities on the one hand, and mission-critical and non-mission critical on the other. "Any activity that is not 'core' to the organization is considered 'context,' an activity that is important to the overall process, but one that could be better performed by another organization for whom that activity is core." While activities regarded as both core and mission-critical (such as the design of new processors in Intel's R&D centers) must remain in-house, activities that are neither core nor mission-critical (such as janitorial services) must be outsourced (Figure 4.3).

	Core Contributes to competitive advantage	**Context** Does not contribute to competitive advantage
Mission critical Poorly performed, the organization is imperiled	Engage and control Keep in-house	Disengage and entrust Out-task and control
Non-mission critical Cannot pose an immediate risk to the organization	Disengage and entrust Out-task	Disengage and entrust Outsource and give up control

Figure 4.3 *Choice factors for capability outsourcing*

Source: Adapted from Cisco Systems (2003a: 11)

Cisco (2003a) also conceptualizes a halfway concept between in-house and outsourcing: *out-tasking* (see Figure 4.3). Out-tasking particularly fits non-core, mission-critical activities which require very tight control. For example, Dell has out-tasked warehousing and inventory management, but has kept in-house assembly lines, electronic systems and most high-grade call centers (dedicated to the business market). In Dell's build-to-order and e-commerce business model which we present more thoroughly below, these functions are both core- and mission-critical.

In such an ecosystem of networked firms, the leading firm acts as an *orchestrator*. The orchestrator owns the brand, establishes the industry's technical standards, manages the relations with customers, and controls or at least coordinates the firms involved in the supply chain. In the 787 Dreamliner program presented above, Boeing is the orchestrator. Dell and Li & Fung, presented in detail on pp. 78 and 84, are typical orchestrators. Outsourcers and out-taskers which revolve around the orchestrator are "the players" (Cisco Systems 2003a: 14).

In the manufacturing sectors, outsourcing materializes primarily through the emergence of *contract manufacturers*, such as Solectron, Flextronics, or Sanmina, often located in low-cost areas (Lüthje 2002). Therefore, former original equipment manufacturers (OEM) tend ultimately to become "fabless" (zero factory), "hollow firms" whose single asset is a brand and a few top-level R&D centers. While completed to the greatest degree in a networked context, these processes may lead the company to become a *virtual enterprise*.

The virtual enterprise

The *virtual corporation* entered the public arena in 1993 with a cover story of *Business Week* (Byrne *et al.* 1993), a short time after the book by Davidow and Malone (1992, reprinted in 1993).

> A Virtual Corporation is a temporary network of independent companies—suppliers, customers, and even rivals—linked by information technology to share skills, costs, and access to one another's markets. This corporate model

> is fluid and flexible—a group of collaborators that quickly unite to exploit a specific opportunity. Once the opportunity is met, the venture will, more often than not, disband. Technology plays a central role in the development of the virtual corporation. Teams of people in different companies work together, concurrently rather than sequentially, via computer networks in real time.
>
> (Byrne *et al.* 1993: 36)

"Virtual" characterizes something which lacks some elements that would make it "real" or "actual." *Virtual reality* describes a digital environment which creates a sensorial experience as similar as possible to those generated by a material environment. Most companies in high-tech manufacturing sectors have implemented *virtual design*, through the implementation of 3D digital mock-up. *Virtual office* describes a working environment which provides people with informational accessibility outside the office—basically, the combination of a notebook computer, *ad hoc* software and mobile or *virtual workspace* technologies (Malhotra and Majchrzak 2005).

In all cases, business requires walls and legal bodies. Companies often depicted as virtual, such as Dell (Kraemer and Dedrick 2001), JetBlue Airways (Young and Jude 2004), or Li & Fung, have real employees, real equipment, and real offices, in real locations. And these are not temporary arrangements: in 2006, Li & Fung celebrated its centennial. The consensus has emerged in academic circles that *virtual enterprise* should preferably refer to a temporary or flexible, goal-oriented arrangement between several firms, or to a unit within a company that temporarily assembles dispersed capabilities and competencies (Mowshowitz 1999). Therefore, there is room for doubts about the value of the concept. Chesbrough and Teece (1996) even regard it as "oversold." They warn that "virtuality" is not a panacea, because "the reliance of virtual firms on partners, suppliers and other outside companies exposes them to strategic hazards," and that the "virtuous virtuals" have carefully kept in-house the key capabilities that provide competitive advantage (Chesbrough and Teece 1996: 70). "Virtual" remains a metaphor. Nevertheless, beyond theoretical flaws and a tendency for hype, analyses and interpretations of the so-called *virtualness* convey interesting concepts which are useful for the understanding of contemporary IT-enabled business organizations and operations.

The switching principle

Mowshowitz (1999: 7) suggests that the switching principle "lies at the heart of the virtual organization." IT is pivotal in this interpretation, particularly relevant in the present Internet era, because a unique feature of the Internet is precisely its unmatched switching capacity. The power of IT for business is to give firms (as well as customers) the possibility of switching easily from one supplier to another one, notably through BtoB electronic platforms (see Chapter 5). The rise of online reverse auctions in the purchase process is a meaningful application of the switching principle. The rising mobility or spatial flexibility of business is a geographic application of the switching principle. According to Mowshowitz, *hollow firms* or *agile companies* are virtual in the way that, thanks to outsourcing, they can move production steps from place to place more easily than "thick and stuck" integrated firms.

> The pervasiveness of commerce on global, ubiquitous, and standardized networks is rapidly changing the fundamental nature of market structures and the manner in which buyers and suppliers interact. The most apparent observable effect is the reduction in coordination costs. This includes improved efficiency in the buyer's ability to search products, compare features, negotiate prices, and conduct transactions.
>
> (Grover and Ramanlal 2004: 71–2)

A virtual company in the "fast" garment industry: Li & Fung Ltd.

> Li & Fung "does not own a stitch when it comes to making garments. No factories, no machines, no fabrics. Instead, the Fungs deal only with information."
>
> (Lee-Young and Barnett 2001: 77)

Li & Fung, the world's largest outsourcer in the garment industry, was founded in Guangzhou in 1906, as a trading company of traditional Chinese goods. It is headquartered in Hong Kong and has 7,000 employees worldwide. From 1996 to 2005, Li & Fung's annual turnover rose nearly fivefold, to US$7.2 billion, and its profit nearly sixfold, to US$230 million (Li & Fung 2006a).

Li & Fung is a fine example of a virtual company, acting as a value chain orchestrator, in that it does not own any manufacturing capability, but rather coordinates a network of nearly 10,000 suppliers. The core business of the firm is to be a "one-stop shop" for Western retailers, by delivering a "global value-added package," including "product design and development, raw material and factory sourcing, production planning and management, quality assurance, shipping consolidation" (Li & Fung 2006b). Li & Fung illustrates the contemporary trend of the textile and garment industry, which has become increasingly fast and globalized while remaining embedded in local milieus from the viewpoint of both customers (rapid changes in fashion tendencies) and suppliers (manufacturing capabilities) (Magretta 1998).

> In May 2006, Li & Fung has substantially increased its European presence and market share, with the buying of KarstadtQuelle International Services AG. Li & Fung will handle all global import for the group. Thanks to a dramatic increase in purchasing power, the alliance with Li & Fung is designed to allow the German group to reduce prices, extend terms of payment, and to react more flexibly than before to fast-changing customer behavior, notably by making it possible for the Group to offer its customers up to twelve collections a year.
>
> (KarstadtQuelle AG 2006)

Li & Fung's worldwide presence in terms of sales and purchases is achieved by a network of 71 offices in 40 countries (McFarlan and Young 2000). The world geography of the textile and apparel industry is apparent in Figure 4.4. China, which represents nearly a half of the firm's total purchases, is over-represented, but Li & Fung has offices in nearly all the regions significant in textile and apparel manufacturing. After having located in

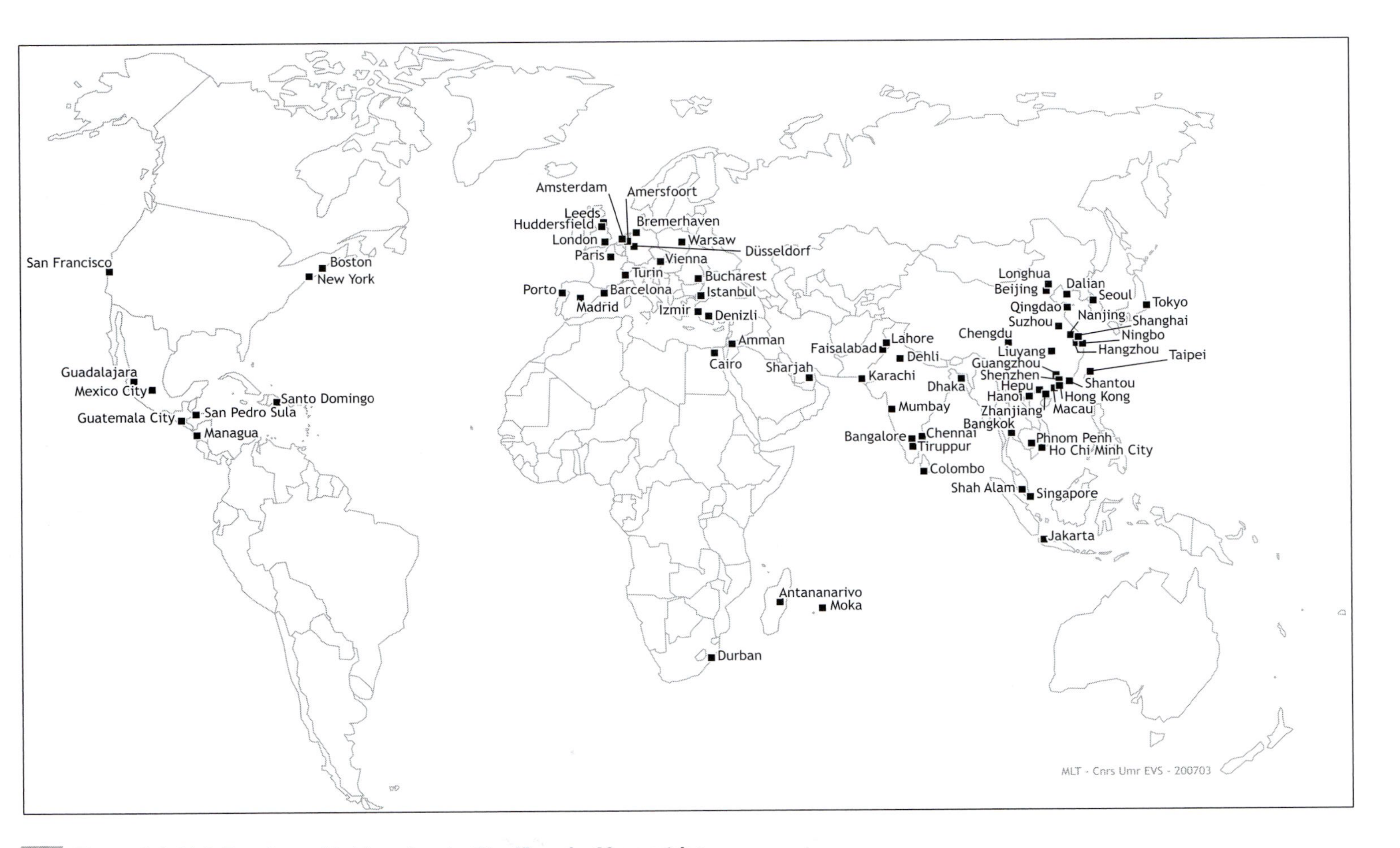

Figure 4.4 ***Li & Fung's worldwide network: 71 offices in 40 countries***

Source: Based on Li and Fung Ltd (2006a)

well-known producing regions, large and small, such as Tunisia, Morocco, Mauritius, Bangladesh, and India, the company was led by its permanent search for low costs and flexibility to explore promising, recently opened spots such as Vietnam and Cambodia.

Li & Fung-led manufacturing operations illustrate the much finer spatial division of labor that characterizes the "new economy." The fabric woven in China, and the fasteners made in Korea, will be sewn in Guatemala to become pants sold in the US market. The dispersion, density and diversity of the network of suppliers allow Li & Fung to switch easily from one manufacturer to another (an application of the switching principle we developed above). This gives agility to the company and to the retailers it serves. If a part of the supply chain—manufacturing or shipping—collapses for technical, social or political reasons, Li & Fung can easily switch to another supplier elsewhere in the world.

Li & Fung's large customers, such as Avon, Disney, Levi Strauss, Reebok, and KarstadtQuelle AG (Germany), have access to the company's dedicated extranet sites where they can place orders and track the production and shipping process. The virtue of e-commerce in terms of flexibility and time-to-market capabilities is obvious: until the fabric is dyed, the client can change the color; until it is cut, the size can be changed (McFarlan and Young 2000). This is a case of agile capacity, or the "power of postponement" harnessed for mass customization requirements (Feitzinger and Lee 1997).

However, it would be a false understanding of Li & Fung's business to think that it relies entirely on telecommunications. The main purpose of the network of offices scattered throughout the world is to maintain physical contact with manufacturers, which are often located in developing regions without reliable telecommunication infrastructures and services. Li & Fung trades with small and medium manufacturing firms located in under-equipped areas, notably in countries peripheral to the global economy, such as Jordan, Vietnam, and Cambodia. Therefore, it has to rely on telephone, fax, and frequent visits of its representatives to check that suppliers effectively fulfill orders and comply with the customers' quality standards[3] and labor management policy—especially those regarding child labor (Holstein 2002).

The company is organized in 90 autonomous subsidiaries, located close to major markets, a critical need in a fashion-oriented industry. The Hong Kong-based headquarters provides the centralized IT system and financial and administrative support (McFarlan and Young 2000). An important factor in the company's success is that it allows small and medium-sized retailers in Western countries, and small and medium-sized manufacturers in developing countries, to meet together for doing business, while benefiting from scale economies which derive from its enormous purchasing and sales power. Li & Fung illustrates the permanent need for informational intermediaries—or "infomediaries"—in globalized supply chain organizations (Hagel and Singer 1999).

Global networks of modular production and the supply chain issue

In this chapter, we have presented the emergence of IT-driven and customer-driven product lifecycles. We have suggested that product lifecycles are operated by networked organizations, complex and often unstable arrangements of parts producers and service suppliers, often coordinated by an "orchestrator" company such as Li & Fung. These organizations rely critically on the efficiency of reliable and fast supply chains. The key role of IT in this issue should be addressed at two levels: the emergence of IT-based standardization in the emergence of modular production, which some authors describe as "Wintelism," and the massive use of IT in the logistics industry, whose growth is central to the emergence of networked, modular production.

Modular production and Wintelism

> Competition in the Wintelist era is a struggle over setting and evolving de facto product market standards, with the possibility of exerting market power lodged anywhere in the value chain . . . As these fundamentals of Wintelism have evolved, the constituent elements of the product became modules.
>
> (Zysman 2005: 265)

Modularity appeared in the 1990s as a seminal conceptual tool for the understanding of contemporary, IT-enabled business organizations and their spatial features (Richard and Devinney 2005; Sanchez and Mahoney 1996; Schilling 2000; Sturgeon 2002). "A new system of industrial organization, the modular production network, has recently come into view . . . entwined with the spatial process of geographic dispersion, relocation and regionalization" (Sturgeon 2003: 199).

Richard and Devinney (2005: 90) present the modular organization as a system with three elements:

1) modules, the discrete functional activities that work on inputs to create distinct outputs;
2) interfaces which set out strict requirements that the inputs and outputs of each module must adhere to, therefore standardizing the boundaries of the module; and
3) interactions that pass over the interfaces, including communications or the exchange of material goods.

Modularity creates flexibility within the organization because the internal elements of the module may be changed whereas the interface is not. An illustration is given by the PC industry: different hard disks or video cards, from different suppliers, may be assembled at short notice on desktops, because the connections as well as the operating software are standardized. Modularity in advanced manufacturing is made possible by IT-enabled PLM methods, such as concurrent engineering and 3D digital mock-up. However, a module must not be conceived narrowly as a part to be assembled by a final manufacturer, but rather as a package, which may include complex services, such as logistics.

Modular production is often synonymous with vertical disintegration processes, followed by horizontal specialization and consolidation. In the PC industry, for example, only four makers of microprocessors are significant: Intel, AMD, Cyrix, and Freescale (formerly part of Motorola).

Sturgeon (2003) regards large contract manufacturers, such as Solectron or Flextronics, or global supply chain outsourcers such as Li & Fung, as *turn-key suppliers*. Turn-key capacity is very close to the switching capacity discussed above. The large footprint of global suppliers allows them to serve a large number of customers, from a large number of suppliers, and to redeploy production capacity easily in terms of quantity, quality, and geography. Horizontal specialization also helps customers benefit from external economies of scale, given the enormous purchasing power of major contract manufacturers.

Modular production is associated with *Wintelism*. This word was coined by Zysman and Borrus (1997) to address the fundamental role of standards in contemporary business organization, notably in modular production networks which have emerged within advanced manufacturing sectors, such as PCs and consumer electronics. Sturgeon (2002, 2003), Hart and Kim (2002), and Zysman (2005) suggest that the strong revival of the US economy in the 1990s, and the huge increase of productivity, are in part the result of the adoption by the largest and most innovative firms of the Wintelist model.

The association of Microsoft's operating system and Intel's microprocessors has indeed established two of the most powerful standards in use in contemporary production. The digital economy as a whole could not function without the numerous standards which dictate the interface between modules of production, and guarantee their tradability. In the IT industries, PDF, MP3, JPEG, JavaScript, USB, Ethernet, WiFi, GSM, and, last but not least, the Internet, are examples of dominant standards.

Capturing the geography of modular production: value chain unbundling and global sourcing

Sturgeon (2003: 201) suggests that the nexus of outsourcing practices is represented by "distinct breaks in the value chain where information regarding product and process specifications can be highly formalized." These breaks characterize *de facto* the places and times where modules are designed.

Value chain unbundling and *global sourcing* have become mantras of business literature (Ge *et al.* 2004; Wrigley 2000). Authors have placed particular emphasis on integration/disintegration tendencies throughout the production process (Hepworth 1990; Amirahmadi and Wallace 1995; Li 1995). "By resolving the tension between spatial separation and function integration, corporate networks have given large firms enormous flexibility in locational choices" (Li 1995: 1642). Global sourcing is intrinsically IT-driven in the way that Internet-based tools such as e-mail, intranets and extranets, collaborative work solutions, voice and video conferencing, and mobile telephones with roaming capabilities represent a dramatic increase in terms of data treatment and coordination/communication capabilities. "Using a corporate Intranet, for example, information can be made available to a firm's employees no matter where they are located in the organizational structure and in the world" (Afuah and Tucci 2003: 397).

Stated simply, global sourcing means that firms seek to obtain the greatest value from dispersed production factors throughout the world, in both qualitative and quantitative terms: "global sourcing involves integrating and coordinating common items, material, processes, technologies, designs and suppliers across worldwide buying, design and operating locations" (Trent and Monczka 2005: 24). Each task in the production cycle described above may be *sourced* from the place where it will be processed in the most efficient and—all things being equal—least costly way. Manufacturing operations coordinated by Li & Fung are a fine example. The global sourcing of various capabilities across the value chain has been popularized by case studies in the garment and footwear industry, Nike being perhaps the most popular example (Donaghu and Barff 1990).

A perfect accomplishment of this model would materialize the utopia of a "friction-free" capitalism (Kenney and Curry 2001). However, even fully IT-driven enterprises have to deal with the thickness of geographical space, which is made of cultural and political barriers and hazards, idiosyncratic market features, regulation disparities, transportation bottlenecks and other essential constituents of locality which support the spatial dimension of *embeddedness* (Granovetter 1985). From these elements derives the tradeoff between "the myth and realities of global sourcing" (Wrigley 2000: 308).

The example of Microsoft's Xbox

Microsoft, unchallenged in software, had planned to rival Nintendo and Sony in gaming devices by launching the Xbox by November 15, 2001. However, it had neither the capability nor the experience for such a large manufacturing project. Microsoft selected Singapore-headquartered Flextronics as principal manufacturing partner. The two companies established a collaborative work environment together with 200 component suppliers. The whole project was based on concurrent engineering and supply chain streamlining prowess. Agile Software (San Jose, CA)—already a software supplier for Flextronics—was required to implement the collaborative tools for concurrent design and supply chain operation, notably the supply of real-time testing data which encapsulate order number and serial number of all the elements inside the box.

> "We spent a significant amount of time at our suppliers' factories understanding their capabilities, their supply chains, their potential bottlenecks and hurdles, and anticipating what could potentially fail in our planning and addressing those scenarios before they transpired," said Todd Holmdahl, general manager of Xbox hardware. "We created a very open channel of communication using integrated business systems so that data could be shared in real time and problem solving was a collaborative effort." By combining its supply chain, business processes, software technology, and communication flow, Microsoft was able to ship 1.5 million Xbox consoles by the end of last year and, in the process, put itself solidly on the gaming map.
>
> (Baljko Shah and Serant 2002: 1)

The consoles for the American market are manufactured in Guadalajara, Mexico, where ten other electronic manufacturing services (EMS) companies are located (O'Brien 2001). At first, the European market was served from Hungary. In summer 2002, Flextronics relocated Hungarian operations to Doumen, China, where it benefited from massive production capacity, lower labor costs, and close proximity from suppliers (Olavsrud 2002).

In Fall 2002, Microsoft contracted for additional assembly capacities with Winstron (Taiwan), which would manufacture 25 percent of Microsoft's consoles in Zhongshan, China. In summer 2005, Microsoft hired a third EMS, Celestica (Canada), also located in China. These additional capacities did not prevent Microsoft from supply shortages. Fourth quarter 2005 sales settled at 1.5 million Xboxes, far below the initial target of 2.75 million to 3 million units (Thorsen 2006).

The story illustrates well the features of modular production and the supply chain imperatives in marketing-driven industrial sectors which have become literally "paranoid" under time-to-market pressures. From a geographical point of view, this example challenges the concept of an all-out tendency towards global sourcing. As a matter of fact, contract manufacturers increasingly demand that their suppliers be present on the same site, or at close distance. Flextronics has 16 suppliers present on its campus of Guadalajara, and manufactures a great number of components on its own (O'Brien 2001).

The Dell business model: build-to-order, modular production, and supply chain optimization

IT is pivotal in major organizational trends which some pundits have recognized in "Toyotism" (Alfasi and Portugali 2004), or more recently in "Dellism" (Sako 2003), a word which is sometimes used in business circles to describe the decrease of inventory costs through optimization of the supply chain, modular production, and build-to-order practices. Dell Inc., headquartered in Round Rock (near Austin), Texas, is perhaps the most emblematic success of the IT-driven business model in the manufacturing sector, and has justifiably received a great deal of attention from pundits and academics (Breen 2004; Dedrick and Kraemer 2005; Fields 2006; Kraemer *et al.* 2000; Kraemer and Dedrick 2001, 2002;).

Founded in 1984, Dell has become the world's leading producer of PCs and related equipment, overcoming well-established rivals such as IBM, Compaq, and HP. In 2005, every third PC sold in the US came from Dell, whose net revenue for the fiscal year ended 3 February 2006 reached $55.9 billion, with a net income of $3.572 billion (Dell Inc. 2006). Despite recent difficulties, which show that its business model might have reached a threshold (Kanellos 2006), Dell has so far demonstrated the superiority of organizational and marketing innovativeness over technical innovativeness in an industry whose core products have been commoditized. Dell's overwhelming success has been founded on a business concept which was—at the time—revolutionary in the computer industry: mail order sale, build-to-order, and home delivery.

> Dell's build-to-order model lets it receive payment from its customers immediately. It pulls the parts directly from its suppliers and builds and ships the product within four days. Yet the company doesn't pay those suppliers until 36 days after it receives payment from the customer. That means it operates with negative working capital, eliminating the need to finance its operations.
>
> (Breen 2004: 86)

Dell's "disruptive" innovation has been the complete rethinking of the whole extended production cycle, including parts supplies, assembly, delivery, and customer support.

Design and assembling are subordinated to marketing, logistics, and the management of the supply chain—all tasks fully IT-driven. Dell has managed to decrease dramatically inventory time, from 32 days in 1994 to three days in 2002 (Fields 2006). In 2003, its inventory turnover was 106.3 per year while the average in the PC industry was 88.4. As a result, Dell's inventory costs are lower than those of its competitors. The 2003 net profit margin was 6.4 percent compared to an average 4.5 percent in the PC industry, and the return on equity 44.2 percent, compared to an average 21.1 percent (Dedrick and Kraemer 2005: 126).

> Dell uncovered how to extract profit from gains of trade in two principal ways: (1) by eliminating intermediaries in the route from producer to consumer and capturing that portion of the gains from trade normally accruing to these actors and (2) by compressing time between the various adjacent steps in producing and selling PCs, primarily between the final production of the PC and final sale to the consumer, thereby cutting costs that are associated with the time expended in essentially warehousing the product in inventory as preparation for final sale.
>
> (Fields 2006: 126)

Mail order sales have freed the firm from the costs of maintaining a brick-and-mortar commercial network. The most strategic places in the firm are electronic. Dell pioneered e-commerce: it was in 1997 the first company in history to reach $100 million in sales online. On www.dell.com, clients discover, choose, configure, price, order, and pay for their systems. They track their order until the final delivery and installation. After-sale services and technical support are provided mainly by a huge network of call centers. Through digital networks and dedicated extranets, Dell shares vital information with its suppliers, with the permanent focus on shortening delays and cutting down costs.

System assembly is processed in eight sites (in September 2006, Dell announced the final decision for building a ninth factory in Łódź, Poland). Dell's assembly sites are strategically located throughout the world within regional markets (Figure 4.5). The geography of Dell's manufacturing operations reveals the importance of distance and delays in a just-in-time model.

The geographic imperative equally matters for customer relationship management (CRM), even if it is processed mostly through telephone and the Internet. Dell holds 34 operating subsidiaries worldwide. This large number meets technical as well as linguistic and cultural requirements. For example, Dell's Ireland-based call center will deal easily with Northern European customers, while facilities in Montpellier (France) and Casablanca (Morocco) serve customers in France, Italy, and Spain. Despite rampant problems with discontented customers who cannot tolerate Indian accents, Dell has decided to increase its operations in India for serving English-speaking customers. In 2005, it opened in Chandigarh a facility with 1,500 employees, bringing its total to 9,000 in the country (Chatterjee 2005).

In addition to outstanding streamlined performance in its assembly plants, and relentless pressure on suppliers, Dell's success relies on logistics. In 2002, 10,000 dockworkers went on strike and blocked the entire US West Coast for ten days—a potentially lethal

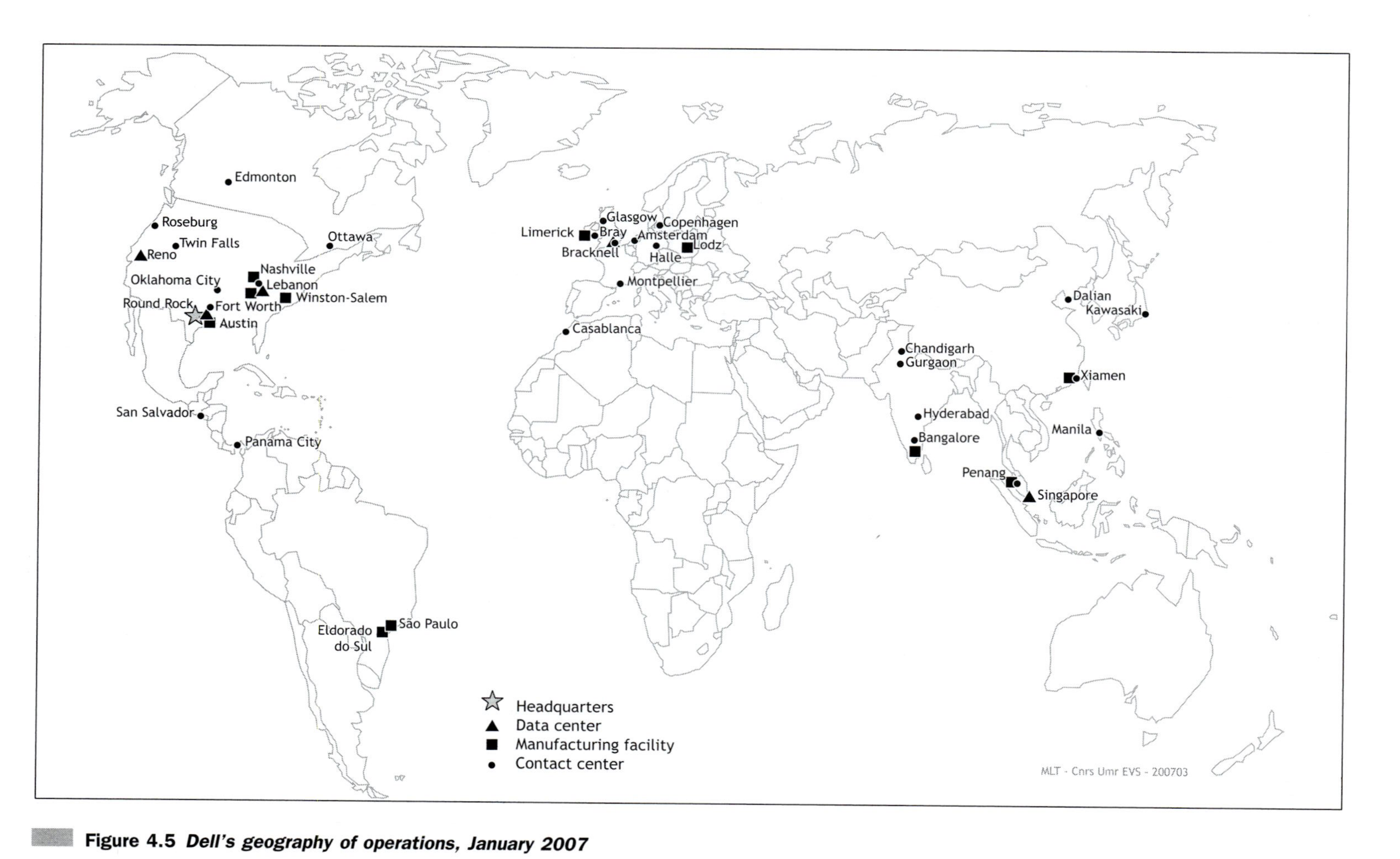

Figure 4.5 ***Dell's geography of operations, January 2007***

Source: Based on Dell Inc. (2006), Kraemer and Dedrick (2002), Fields (2006), and authors' web research

incident for build-to-order companies. Dell reacted swiftly, chartered 18 747 jumbo jets, sent a task force to Asian ports to make it possible that "Dell's parts were the last to be loaded onto each cargo ship so they'd be unloaded first when the ship hit the West Coast." In the end, Dell "survived a 10-day supply-chain blackout with roughly 72 hours of inventory, and it never delayed a customer order" (Breen 2004: 86).

The globalization of logistics

The very existence of Dell, Li & Fung, and other companies based on global production networks and supply chain optimization cannot be conceived without the existence of cost-efficient, IT-enabled logistics solutions. Standardization, tradability, continuity, and ubiquity are no longer words that describe informational flows only, but apply equally to material flows. Global networks of production require global freighters (Leinbach and Capineri 2007). Leinbach and Bowen (2003, 2004), for example, show that extensive use of air cargo services is critical to the functioning of value chains in the Asian electronic industry.

As a result of the organizational patterns we have surveyed, the contemporary economy is in great demand of long-haul, small, fast, and frequent shipments. Hence the fortune of global logistics service providers (LSPs) such as DHL, UPS, FedEx, and TNT. Table 4.2 presents a comparison between the two US arch-rivals, FedEx and UPS. The logistics sector has been subjected to a huge process of consolidation, which replicates the tendency for horizontal specialization we presented above as a key feature of Wintelism and modular production networks. DHL, the world leader, was founded as a US company in 1969. Between 1998 and 2002, it was gradually acquired by Deutsche Post World Net, and then grew through several acquisitions by the German giant, notably those of ABX Air, a US freight cargo firm, in 2003, and Exel plc, a $10 billion British company, in 2005. It is headquartered in Bonn, Germany, and had 2005 revenue worth $69 billion.

The emergence of the world-sized LSP with a global footprint is explained by the four interlinked requirements—organizational, technological, geographical, and therefore financial—of modern supply chain operations. While most large companies outsource their logistics functions (see below), they ask for "seamless" services which require both air and ground capacity. Only companies with large aircraft and vehicle flotillas can deal with major OEMs and turn-key suppliers on a large-scale basis. The digitalization of the supply chain makes the splintering of logistics functions in several independent companies increasingly uncomfortable, because IT systems of logistics service providers must be closely interconnected to client systems. The geographic application of the switching principle—a critical source of flexibility in supply chain—also leads to concentration. If a client company needs to change supply origin on short notice, it is better not to have to change logistics service provider. Hence the global footprint of each of the leading companies. The emergence of large LSPs is also explained by the specific requirement of electronic commerce (Aoyama *et al.* 2005; Rabinovich and Knemeyer 2006), and by more general tendencies in production methods such as just-in-time production, which result, first, in a decrease in the size and value of each individual package (the average revenue per piece shipped by UPS' ground service in

Table 4.2 *UPS and FedEx 2005 facts and figures*

	UPS	*FedEx*
Headquarters	Atlanta	Memphis
Revenue 2005 ($ billions)	42.5	29.4
Net Income ($ millions)	3,870	1,449
Employees	407,000	260,000
Aircraft	577	677
Vehicles	98,000	70,000
Average daily shipments	14.8 million	6 million
Average revenue per piece (US $):		
Air	20.06 (US Domestic)	20.10
Ground	6.52	6.68
Sorting and handling facilities	35 million sq.ft.	9 million sq.ft.
Information Technology		
Main data centers:	Mahwah (New Jersey) Atlanta (Georgia)	Memphis (Tennessee) Colorado Springs (CO)
Website daily usage (page views)	18.5 million	
Tracking requests per business day	10 million	3 million
Technology employees	4,700	
Mainframe and UNIX storage	471 terabits	
Servers	8,700	
LAN workstations	149,000	75,000

Source: FedEx Corporation (2005), United Parcel Service, Inc. (2006), and companies' websites

2005 was $6.50) and, second, in more demand for speed and flexibility in shipping and delivery. This explains why the fleets of LSPs are made of cargo jets and agile vans rather than 40-ton trucks.

Outsourcing logistics: the emergence of 3PLs and 4PLs

Third-party logistics service providers are the result of the global tendency for outsourcing and vertical disintegration that has been the focus of this chapter. According to Aoyama *et al.* (2006: 330), "73 percent of Fortune 100 companies outsource logistics functions."

The logistics industry spans a large range of tasks, which includes not only shipping, but line across the supply chain: shipping planning, warehousing, inventory management, customs clearing, and carrier selection. The range of tasks outsourced varies with the business segment, in application of the two-pronged categorization between core/non core and critical/non-critical activities. While it is obviously a critical activity in "Dellism" times, flying cargo jets or driving trucks is far enough from manufacturers' core capabilities that most of them outsource shipping. Warehousing offers more of a contrast.

In retail industries, warehousing is regarded as both core and critical, and most companies have kept it in-house, in both brick-and-mortar (Wal-Mart, Carrefour, Ikea), and electronic commerce (Amazon, cf. Chapter 5). Dell chose the "outsource but control" option. It hired APL Logistics (Oakland, California, 4,000 employees in 50 countries) for implementing and operating a network of "supplier logistics centers" (SLCs). Two of these centers are located in West Chester, Ohio (opened in 2004) and in Winston-Salem, North Carolina (opened in September 2005 near the brand-new manufacturing facility Dell inaugurated at the same time).

> The SLC operation, which is within three miles of the new Dell factory, will handle the inventory for suppliers that provide parts to Dell. The center will warehouse and manage computer peripherals such as monitors, speakers and computer chassis from 50 different suppliers.
>
> (APL Logistics 2005)

A phenomenon of vertical reintegration within the supply chain generated the somewhat controversial moniker "4PL" (fourth-party logistics provider). The term 4PL was created and trademarked by Andersen Consulting, now Accenture, in 1996 (Biederman 2005). Purists distinguish between 3PLs and 4PLs through the fact that 4PLs are not asset-based (no warehouses, no trucks, no 747 Jumbos), but rather rely on their informational capabilities to assemble and coordinate supply chain services, including those provided by asset-based 3PLs. However, global logistics companies such as UPS, FedEx, or DHL which offer fully integrated logistics solutions, including IT management and electronic procurement capabilities, are regarded by some pundits as 4PLs (Haddad and Ewing 2001; Biederman 2005). Actually, there are few "pure" 4PLs. A fine example is Vector SCM, which was founded in December of 2000 as a joint venture between General Motors (GM) and Menlo Worldwide, a US supply chain solution provider. Vector's primary function is to coordinate the inbound and outbound operations of GM's supply chain (Biederman 2005).

Logistics is digital

F.W. Smith, founder and chairman of FedEx, the world's third largest logistics company, said in 1979: "The information about a package is as important as the delivery of the package itself" (FedEx Corporation 2006). The fundamental complementarity between logistics and IT is captured in the title of a recent survey by *The Economist*: "The physical Internet: a survey of logistics" (Markillie 2006).

Transportation and logistics firms have had to reorganize themselves around IT. This might be illustrated by the joint venture created in October 2006 by CMA CGM (Marseille, France), the world's third largest shipping company, and IBM, the leading IT services firm. The new entity, CMA CGM SYSTeMS, is designed to help its founder to leverage IBM's business expertise to design innovative solutions to meet the rising complexity of the shipping market (CMA CGM 2006). Actually, the requirements of global production and trade networks have led major transportation and logistics companies to become worldwide electronic powerhouses dominated by the complex intertwining of commodity flows and information flows (Aoyama *et al.* 2005). Rather than simple transportation and delivery, LSPs offer a package of IT-enabled integrated services, which cover the entire shipping issue (Table 4.3).

A glance at FedEx and UPS shows clearly the role of IT in modern logistics. The key tool is the use of bar codes on shipments, which are scanned at each step of the

Table 4.3 ***IT-based services from LSPs used by companies in the Americas, Western Europe, and Asia-Pacific***

IT uses	*Percentage of respondents*
Export/Import/Freight forwarding/Customs clearance	74.5
Transportation management (execution)	74.0
Shipment tracking/Tracing/Event management	72.0
Warehouse/Distribution center management	66.4
Web-enabled communications	63.0
Transportation management (planning)	47.2
Customer order management	20.4
Internet-based transportation logistics markets	19.8
Supplier management systems	15.7
RFID and asset tracking	12.3
Supply chain planning	7.9

Source: selected and calculated from Langley *et al.* (2005)

Note: The original table presents data by regions, calculated from 1052 usable responses. The industries predominantly represented by the survey respondents included: Automotive, Chemical, Consumer Products, Food and Beverage, High-Tech and Electronics, Industrial Manufacturing, Life Sciences (Pharmaceutical) and Medical, Retail, and Telecommunications.

shipping process, from pickup to delivery. This practice makes possible the tracking of each package on an LSP's website. FedEx and its counterparts offer clients the possibility of calculating freight and insurance rates, duties, and taxes online. Bluetooth wireless technology allows FedEx couriers to remain connected to their vehicles, which are equipped with mobile computers, global positioning systems (GPS), and path-finding software.

The importance of IT in the LSP business is epitomized by the breakdown of their material assets. In 2005, FedEx's assets in computer and electronic equipment amounted to \$3.9 billion—twice the value of its ground vehicles (\$1.9 billion) and more than half the value of aircraft and related equipment (\$7.6 billion) (FedEx Corporation 2005). The IT system of UPS shows similarly impressive features: 18.5 million Web pages viewed daily, 10 million tracking requests per business day, 4,700 employees committed to information technology, a total mainframe and UNIX storage of 471 terabits (471,000 Gbits), 8,700 servers, and 149,000 local workstations (United Parcel Service, Inc. 2006).

The latest IT-related innovation in supply chain and logistics management is the Radio Frequency Identification (RFID) device or electronic product code (EPC), which could replace bar codes in the medium term. RFID consists of miniaturized electronic tags attached to packages or even embedded in the product itself, which are capable of distant detection and tracking, from the manufacturing site to the customer's location. The use of RFID has been pioneered by Wal-Mart, which launched a first experiment in April 2004 in North Texas. The trial started with 150 stores and Sam's Clubs, and was planned to expand to 1,000 stores and clubs, and 600 suppliers—which were given no other choice but compliance to Wal-Mart's demand—by early 2007 (Schwartz 2004). RFID implementation provides significant improvements at all the stages of the supply chain, from the initial packaging at the factory to the replenishment of stores' shelves.

However, RFID poses a series of problems—ethical and political, rather than economic or technological. The system may be regarded as an additional element in the building of an increasingly pervasive, IT-based surveillance system, which could threaten individual rights and privacy (along with software cookies, surveillance cameras, digital passports, and travel data files).

The sensitive nature of RFID implementation, and the damage it could inflict to the image of a brand, are illustrated by the 2003 Benetton affair. The Italian garment company, an early pioneer of fast, flexible supply chains, had started RFID testing, in partnership with Philips Semiconductors. Philips issued a communiqué revealing that Benetton had implanted transponders in millions of clothes sold under the brand Sisley. Later, some consumer organizations called for a boycott of Benetton's products. Indeed, the company acknowledged it was testing the process, but denied having industrialized it yet, and announced that it "decided not to take this technology any further as it provides no significant benefits to the productive and commercial organization" (Benetton Group 2003).

Conclusion

In this chapter, we have described in real terms, with examples of real companies, the dramatic evolution of IT-enabled production and business organization, notably in advanced manufacturing sectors. Most of the tasks that constitute the value chain, from design to sales and after sales, are computer-aided. In the context of an increasingly competitive, globalized, and customer-centric economy, IT implementation allows companies to embrace a holistic view of increasingly complex and shortened product lifecycles.

IT also plays a prominent role in the disintegration of the value chain, from both organizational and spatial points of view. Value chain "unbundling" and "global sourcing" are two key features of the digital economy which are recurrent in the present chapter. Widespread outsourcing practices are driven, notably, by the search for economies of scale and for lower costs in the supply of both goods and services. They materialize in vertical disintegration processes, followed by the horizontal reintegration of suppliers and producer services. We are witnessing the emergence of networked organizations, each coordinated by an "orchestrator firm," that is to say the company that markets the product and owns the brand, such as Boeing, Airbus, Dell, or Microsoft. The erosion of the boundaries within those organizations, and the existence of temporary arrangements between firms, have led several authors to identify virtual enterprises, a concept whose heuristic properties, however, are greater than its operational value.

The "pieces of value-added" created and assembled by the operation of networked or virtual organizations may be considered as modules. The management of such dispersed value chains (in both organizational and geographical terms) requires a great deal of coordination, impossible to conceive without the use of powerful computing capacity and cheap broadband telecommunications. Supply chain management is at the core of these global, IT-driven value chains (or value networks). Therefore, the logistics industry itself has had to become global and digital.

However, the causal effect between IT and contemporary business organization must be addressed carefully. In the process of spatial and organizational disintegration of value chains, IT acts as a tool, not as a primary factor. Networks of production have become more globalized, complex, and therefore IT-enabled, because of the opening of new countries to free trade and entrepreneurship, the uneven geography of talent and labor costs, and increasing competitive pressure.

This chapter has brought up several issues which are given emphasis in the following chapters. We have explained that modern value chains assemble production and transportation processes with the output of service-like activities, such as sales, procurement, and finance. In Chapter 5, we address the profound IT-driven reorganization of those activities and its effect, notably in the field of e-commerce and e-finance. Chapter 6 (the offshoring of services) and 7 (the rise of telework) focus on the much finer spatial division of labor, which is made possible by the digitalization of value chains.

5 THE MULTISCALE GEOGRAPHIES OF ELECTRONIC COMMERCE AND ELECTRONIC FINANCE

In Chapter 4, we have described the spread of digital technologies in production processes and business organizations. Through case studies, we have depicted the splintering of the economic space, and the much finer spatial division of labor which prevails in the Internet age. In most industries described above, from truck and computer manufacturing to logistics, IT plays a rather subordinate, although essential, enabling role. In the present chapter, we analyze two sectors which are fundamentally based upon *digital transactions*: electronic commerce and electronic finance. These are sectors where the myth of the end of distance—and geography—has been particularly flourishing. The rhetoric around e-commerce has been dominated by simplistic assertions such as the "fall of the shopping mall." Since money has become increasingly digital (at least in the developed world), some people have forecasted the end of the need of personal contact in the banking industry, resulting in the massive closure of branches.

However, neither shops and supermarkets, nor branches have disappeared. Utopian discourses have been swept away—albeit not entirely—by the spectacular collapse of the "dot.com bubble" in 2000–2001. And pure online banking has proved a flop. Nevertheless, the way of doing commerce and finance has been revolutionized by IT implementation (Whinston *et al.* 1997). Actually, what we are witnessing is the emergence of a complex, multiscale geography, which is shaped by the tension between, on the one hand, globalized, almost distance-free structures and processes, and, on the other hand, very localized forms of business sites and transactions. E-commerce websites, worldwide financial telecom networks, and electronic exchanges are characterized by a global reach that ignores, to a certain degree, geographic constraints. But the imperatives of fulfillment in e-commerce, the resilience of human, face-to-face banking, and agglomeration effects in advanced financial services, demonstrate the persistent importance of distance, place embeddedness, and local idiosyncrasies.

E-commerce: concepts and definitions

Defining and measuring e-commerce is a confusing task. Some authors make it almost synonymous with e-business. Leinbach (2001: 13–14) reports several large definitions which embrace all processes conducted over computers and telecommunications. Following the detailed discussion by Mesenbourg (2000), we prefer a more narrow definition, which makes e-commerce a subset of e-business, and places emphasis on the concept of *trade* (or *sales*), that is to say, on transactions which derive in "transfer of ownership or rights to use goods or services" (in accordance with most surveys and reports produced by public and private institutions worldwide).

Therefore, IT-enabled practices such as computer-aided production processes, transactions internal to the firm such as intra-firm EDI or Internet use, and salaried telework, while they clearly belong to the sphere of e-business, should not be considered as e-commerce practices given they do not see the transfer of property of goods or services. The definitions suggested by the OECD (Table 5.1), place emphasis on ordering, not on payment or delivery.

An important ambiguity is the status of electronic finance. Common sense does not consider shares, bonds, currencies, and money, as ordinary goods and we cannot find them in the statistics of the World Trade Organization. It is in some way paradoxical that trillions of dollars which "travel" daily around the world are rarely reported in e-commerce measurement, while they are truly electronic transactions. In fact, the actual service traded must not be measured by the amount of money which is transferred or "swapped," but by the fee or commission the intermediary (bank, exchange) earns on the transaction. For example, while the New York Stock Exchange (NYSE) traded a daily average of $69 billion in shares in January and February 2006, the operating sales of its parent company, NYSE Group, Inc. for 2005 amounted to only $1.72 billion, mainly from listing, data processing, market information, trading, and regulatory fees (NYSE Group 2006).[1] A similar problem is found in the evaluation of the turnover of e-commerce platforms such as eBay, which do not own the goods sold, but act only as intermediaries. Hence the recommendation of the OECD:

Table 5.1 ***The OECD definitions of e-commerce transactions and interpretation guidelines***

Broad definition	
An *electronic transaction* is the sale or purchase of goods or services, whether between businesses, households, individuals, governments, and other public or private organizations, conducted over *computer-mediated networks*. The goods and services are ordered over those networks, but the payment and the ultimate delivery of the good or service may be conducted on or offline.	*Include*: orders received or placed on any online application used in automated transactions such as Internet applications, EDI, Minitel, or interactive telephone systems.
Narrow definition	
An *Internet transaction* is the sale or purchase of goods or services . . . conducted over the *Internet*.	*Include*: orders received or placed on any Internet application used in automated transactions such as Web pages, Extranets . . . or over any other Web enabled application regardless of how the Web is accessed (e.g. through a mobile or a TV set, etc.). *Exclude*: orders received or placed by telephone, facsimile, or conventional e-mail.

Source: Roberts 2005: 41

> The advice offered in this paper is that agents should report the value of commissions or fees earned on the Internet transaction. For financial services, include only commissions, fees and premiums earned in respect of services offered over the Internet and, in respect of Internet-only accounts, net interest income.
>
> (OECD 2005a: 16)

However, ambiguity remains. The key issue is the fragmentation of the commercial process in numerous, tasks: research, testing, price negotiation, ordering, billing, tracking of shipping, delivery, installation, and after-sales services (Couclelis 2004a). The definitions quoted above put emphasis on ordering. Therefore, they fail to embrace the whole phenomenon, because numerous tasks in the trading process may be performed through digital networks. Online payment, is regarded by consumers as a critical step. Because most grocers' shops in developed countries accept credit card payment, a large share of the whole retail trade sector must be regarded in some way as digital. Some people like to browse the electronic catalogue before ordering by post-mail and paying by check, because they are reluctant to use their credit card identity on the Web.

Another ambiguity lies in the treatment of teleshopping and telephone ordering in catalogue shopping. While TV and the telephone are becoming digital and interactive, increasingly running under IP (a result of the digital convergence), there are fewer and fewer reasons for excluding these practices from e-commerce considerations. Following Currah (2002), we are therefore invited to consider a "multichannel, or 'brick-and-click', organizational paradigm," which makes tight subdivisions between traditional and digital commercial channels increasingly irrelevant.

A taxonomy of e-commerce

The most popular subdivision of electronic commerce looks like a 2x2 matrix (Peet 2000; Dixon and Marston 2002) presented in Figure 5.1.

- Business-to-Business (BtoB or B2B) refers to inter-firm exchanges performed on electronic market places such as Agentrics presented below) or on firm extranets dedicated to procurement.
- Business-to-Consumer (BtoC) includes retail trade of goods and services (Amazon.com, Dell.com, and Travelocity.com).
- Consumer-to-Consumer (CtoC) refers to transactions between individuals, the intermediary firm acting only as a meeting marketplace which never has ownership of the goods being sold. The most popular sites are eBay and XQL (PayPal is a subsidiary of eBay dedicated to online payment processing).
- Consumer-to-Business (CtoB) is a rare and peculiar category where enterprises themselves apply for household bids. For example, someone posts a travel demand on Priceline.com and then waits for a travel company's offer. CtoB commerce is rather aesthetic to conceptualize but is theoretically flawed: after completion of the real commercial transaction, the actual buyer is the customer, and the vendor is the travel company! Therefore it is a case of BtoC e-commerce, Priceline.com acting as an intermediary like eBay.

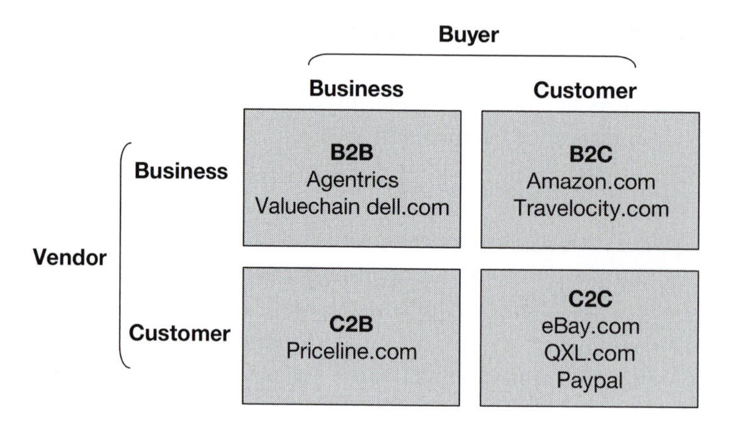

Figure 5.1 ***The classic e-commerce matrix***

Source: Adapted from Peet (2000)

Another important distinction is between incumbent, "*brick-and-mortar*" vendors and *pure players* such as E-Bay and Amazon, which conduct all sales online (Cortada 2004). In BtoB e-commerce, pure players are independent platforms, which act as intermediaries between firms. However, the distinction is blurring, since traditional firms are increasingly present in e-commerce markets. Wal-Mart and Carrefour (supermarkets) or Barnes & Noble (bookstores) conduct a small but increasing fraction of their sales online. They are becoming "brick-and-click," multichannel retailers (Steinfield *et al.* 2002). The popularity of Amazon and e-Bay must not hide the fact that pure players, acting as newcomers in their market segment, have so far captured a minority share of online retail sales. Likewise, in the BtoB area, independent e-marketplaces are of little importance by comparison with direct interfirm electronic exchanges.

Business-to-business (BtoB) e-commerce

The public attention focuses on the popular success stories of Amazon.com and eBay, but usually ignores the much greater pervasiveness of business-to-business forms of e-commerce.

BtoB e-commerce did not escape fully unscathed from the dot.com bubble bust. It has nevertheless achieved a massive growth, becoming a major—if not dominant—way of conducting business in procurement and wholesale trade. For example, electronic sales from US manufacturers in 2004 amounted to $996 billion, 23.4 percent of total shipments, and e-sales from wholesalers (including manufacturers sales branches and offices) totalled $825 billion, 17.2 percent of total sales. The same year, retail e-commerce amounted to only $71 billion, 2.0 percent of the market (US Census Bureau 2006a).

Figures vary across sectors. Table 5.2 presents data from manufacturing sectors with the highest rate of electronic sales. All in all, six industries account for 72 percent of total electronic sales by manufacturing firms. "Transportation equipment" alone represents more than a third of the total, with 52 percent of sales processed electronically, followed

Table 5.2 ***E-commerce in selected US manufacturing sectors***

NAICS	Industries	E-commerce as percentage of total shipments	Total sales ($ billions)	E-sales as percentage of total e-commerce
336	Transportation equipment	52.0	663	34.8
312	Beverage and tobacco products	47.2	111	5.3
324	Petroleum and coal products	24.8	313	7.8
335	Electrical equipment, appliance, and components	24.0	104	2.5
334	Computer and electronic products	21.0	362	7.6
325	Chemicals	19.5	528	10.3
333	Machinery	19.4	269	5.2
311	Food	12.6	509	6.4

Source: US Census Bureau 2006a: appendix, table 1

by beverage and tobacco products, petroleum and coal products, electrical equipment, appliance, and components, computer and electronic products. Other sectors, which are in some way at the periphery of the digital economy, are less involved in BtoB e-commerce: wood products, mineral products, textile mills, food.

The reign of electronic platforms

Electronic platforms acting as marketplaces have become the digital equivalent of "brick-and-mortar" logistics platforms and play a key role within the digital economy. Platforms reign supreme in the financial sector (Visa and other credit card networks, SWIFT (Society for Worldwide Interbank Financial Telecommunications) and CHIPS (Clearing House Interbank Payment System). Platforms also are critical in the travel reservation industry, which is dominated by four electronic reservation systems (ERSs) we present in greater detail below: Sabre, Galileo, Amadeus and Worldspan. Amazon and eBay have become global retail platforms. Dassault Systémes' software family CATIA-ENOVIA-SMARTEAM is a typical PLM platform.

Increasing returns (Arthur 1996), network externalities (Afuah and Tucci 2003) and the switching principle (Mowshowitz 1999) are at the core of the e-platform concept. The value of networks increases with the number of users. E-platforms are to commerce what Cisco Systems' switches are to Internet traffic routing—that is to say, places where the greatest number of buyers can meet and exchange with the greatest number of vendors of goods and services. The concept of electronic marketplaces as switching platforms is well suited to the contemporary interpretation of digital networks, often described as multi-layered systems. Related to the switching concept is the concept of e-marketplaces as *electronic hubs* (Kaplan and Sawhney 1999).

"E-hubs create value by two fundamentally different mechanisms: aggregation and matching" (Kaplan and Sawhney 2000: 100). *Aggregation* means that e-marketplaces have the capability of bringing together a large number of sellers and buyers, therefore reducing transaction costs by "providing one-stop shopping." This is typically the case of Li & Fung in the garment industry: it operates as a hub, which allows distributors to meet a large number of producers dispersed throughout the world. The augmented purchasing power of e-hubs is likely to decrease prices significantly. *Matching* results directly from the more intrinsic capability of electronic purchasing: enabling of real-time transactions. The matching concept is embodied typically by *electronic auctions*—especially reverse auctions—a tool commonly offered to vendors and purchasers by electronic marketplaces.

The efficiency of e-marketplaces results from their ability to reduce information asymmetries, which means "reduction in the cost of contract negotiation, monitoring, and enforcement" (Afuah and Tucci 2003: 398). This reduction of information asymmetries derives from both network effects and the emergence of standards, which are a series of broadly recognized norms and specifications (which, in turn, allow an increase in the level of "richness"). Transaction costs fall since a larger number of clients and vendors may be present on a platform, sharing a common system of standards, therefore reducing market uncertainties—for example, price information from various vendors becomes equally available among buyers (Figure 5.2).

During the second half of the 1990s, independent business-to-business (BtoB) e-marketplaces (or exchanges) were regarded as a promising concept, which was to revolutionize supply chain management in a broad range of manufacturing and service activities. Actually, most "pure players" failed. Day *et al.* (2003) report a peak of 1,520 existing exchanges in 2000, which was the peak year of the dot.com bubble. They were forecasting a remaining number of 180 for 2003. Zällh (2005: 2) has compiled a list of 52 significant e-marketplaces in 17 industries "based on two main criteria; the e-markets are well known globally in their industries and they have significant global traffic" (Box 5.1).

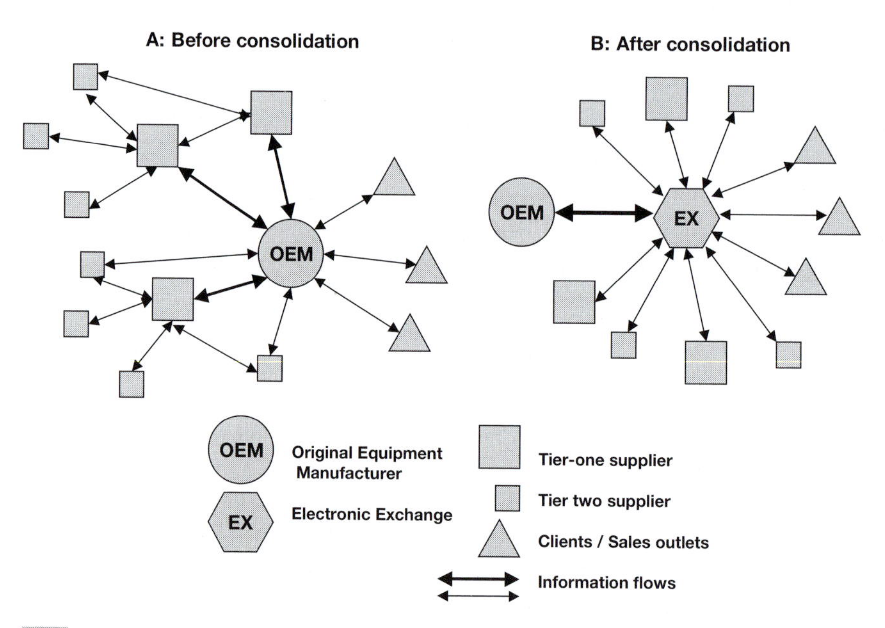

Figure 5.2 ***Electronic marketplaces and network consolidation***

Source: Authors' own research

Box 5.1 Selected major e-marketplaces

Name—Headquarters—Creation	*Industry—Selected facts (members lists, when mentioned, are not exhaustive)*
Aeroxchange www.aeroxchange.com Irving, Texas, 2000	**Aviation.** Founded by 13 airline companies. Dedicated to maximizing efficiency across the aviation supply chain.
Chemical Elemica www.elemica.com Wayne, Pennsylvania, 2000	**Chemicals.** Created by 22 of the world's largest chemical companies.
Converge www.converge.com Peabody, Massachusetts, 1980	**Electronic.** Provides semiconductors, electronic components, computer products and networking equipment. 6,500 trading partners in 139 countries. Clients include: Cisco, Apple, Intel, NEC, Philips, and IBM.
DRAMeXchange www.dramexchange.com Taiwan, 2000	**Electronic.** Founded in 2000 in Taiwan. Offers DRAM and RAM memories to major players including Motorola, Sun, NEC, Intel, and Samsung.
Intercontinental Exchange, ICE www.theice.com Atlanta, 2000	**Energy and fuels.** Major oil companies and energy trading firms. Assists companies in sourcing from suppliers. Operates as a global marketplace for the trading of energy commodity futures and OTC contracts.
Global Healthcare Exchange www.ghx.com Westminster, Colorado, 2000	**Healthcare & Pharmaceutical.** Partners include 3M Medical, Arrow International, Bayer Diagnostics, Johnson & Johnson, GE Medical Systems, etc. Serves more than 2,200 hospitals.
SourcingParts www.sourcingparts.com Switzerland, 2000	**Industrial Machinery & Equipment.** Approximately 42,000 suppliers.
Quadrem www.quadrem.com Plano, Texas, 2000	**Mining, minerals and metals.** Established by 14 of the world's largest mining, minerals, and metals companies. 47,000 suppliers and 700 buyers.
Trade-India.com www.tradeindia.com New Delhi, 1996	**Regional e-marketplace.** More than 4 million registered users.
Mercado Eletrônico www.me.com.br São Paulo, 1994	**Regional e-marketplace.** More than 180,000 registered members. Has a geographical focus on medium and large enterprises in Latin America, especially Brazil.
GlobalSources www.globalsources.com Singapore, 2000	**Regional e-marketplace.** A particular focus on the China market. Over 423,000 active members.

Source: Zällh 2005 and companies' websites, accessed 30 November 2006

Nowadays, the business is dominated by private networks or consortium-owned exchanges such as Agentrics, the world's largest online marketplace/exchange. Agentrics (for AGENT for Retail Information and Collaborative Solutions) is headquartered in Alexandria (Virginia, US) and Chicago with offices in London and Tokyo. Agentrics' 250 clients, linked with 100,000 suppliers, represent a consolidated purchasing power in excess of $1 trillion. Agentrics serves 17 of the world's top 25 retailers, including Ahold, Auchan, Best Buy, Carrefour (the second-largest retailer in the world), Casino, Karstadt Quelle, Kingfisher, Metro (Germany's largest retailer), Sears (the third-largest general retailer in the US), Tesco, and Walgreens. The absence of Wal-Mart, the world's largest distributor, is worth noting: its enormous purchasing power allows it to implement its own private exchange. Agentrics' portfolio of solutions includes: "online auctions, supply chain collaboration solutions, promotions management, operational point-of-sale data exchange, trading partner performance score-carding and management, product lifecycle management" (Agentrics 2005).

Networks and electronic exchanges: example of the automobile industry

"The automobile industry is technologically intensive and systemic" and "has extraordinary global reach." It is "arguably the world's largest coordinated production system" (King and Lyytinen 2005: 284). Indeed, car manufacturers have pioneered new production methods such as "Fordism" and "Toyotism." They are at the vanguard of IT implementation in production cycle management. Although it has long been a producer-driven business model, the sector has shifted to a buyer-driven, customer-centric model. Confronted by mass customization tendencies and inexhaustible cost-cutting needs, firms have to manage a complex workflow, which integrates thousands of suppliers.

As seen in Chapter 4, Table 4.1, the automobile industry rivals the aerospace sector in IT-intensiveness, notably through IT-enabled solutions for ERP and SCM. Car makers and tier-one suppliers have favored the emergence of value-added networks (VANs) such as American Network eXchange (ANX), and European Network eXchange (ENX) which support a great number of applications, ranging from CAD and workflow management to e-procurement.

ENX is a French Law association that was founded in June 2000 by Renault and PSA (Peugeot-Citroën). It has 17 full members, which represent almost the entire European car industry (car makers and major suppliers), including (in addition to Renault and PSA) BMW, Bosch, DaimlerChrysler, Ford, Porsche, and Volkswagen (ENX Association 2006). The basic framework of ENX is a virtual, private telecom network which was implemented by the interconnection of three telecommunications providers: France Télécom, Deutsche Telecom, and Telefonica. In addition to security and privacy, a major point of ENX is versatility. It supports:

- access to manufacturers norms and specifications
- transfer of CAD files
- digital "mock-up" applications
- access to 3D collaborative workspaces

- video conference
- supply chain and logistics management.

The migration of EDI systems towards Internet protocol and Web interfaces has favored the emergence of integrated, compatible information systems, for both vertical and horizontal transactions. In December 2000, four major car makers—Ford, General Motors, Nissan, and Renault—founded Covisint, which was designed to become a global e-marketplace for the automotive industry. However, it never captured durably a significant share of car makers' purchases. Honda, Toyota, and Volkswagen never used it. General Motors and DaimlerChrysler "developed their own private BtoB systems, even as they were participating in Covisint" (Day *et al.* 2003: 135). Finally, in February 2004, Covisint's assets were sold to Compuware, a Detroit-based software firm, and it ceased to be a single-industry exchange. The final failure of Covisint as a single industry marketplace has been thoroughly analyzed by Gerst and Bunduchi (2005). The main explanation was the reluctance of car manufacturers to share the control of standard development with suppliers. In the end, they developed their own private networks, and suppliers created their own electronic exchange, www.supplyon.com.

Covisint is now a multiple industry exchange platform, used as a supply chain management system by 30,000 companies in 96 countries. Covisint provides OEMs a single point of entry for thousands of suppliers worldwide. It allows them to use a common interface in the use of more than 160 applications, ranging from accountancy, collaborative work, procurement and supply chain management, and logistics (Covisint 2006).

The mediocre success of e-retail

The short history of BtoC e-commerce (or e-retail or e-shopping) epitomizes the evolution and the final end of cyber-utopian geographic thinking in the business sphere. During the second half of the 1990s, e-retail was a subject of hype in the media, entrepreneurial and financial spheres. Innumerable start-ups were launched, trying to sell everything, new or used: cars, food, and clothes. Most of them went bust in the early 2000s. Indergaard (2004) and Zook (2005) provide an inspired narration of the bubble's rise and bust, on both the East Coast (Silicon Alley) and the West Coast (the Bay Area). The NASDAQ index, where most e-commerce firms were listed, hit an all-time high of 5,132 in March 10, 2000. In October 2002, it had plummeted to 1,108. In April 2007, it was still far below its record height, around 2,520. At the present time, the utopian rhetoric has faded away. Pure players remain marginal, while incumbent brick-and-mortar retailers regard online shopping as a powerful tool to complement their traditional activities.

A laborious takeoff

There is evidence that BtoC e-commerce lags far behind BtoB e-commerce. Quarterly statistics from the US Census Bureau (2006b) demonstrate that, despite a vigorous growth in both absolute and relative terms, e-retail has so far captured about 2.4 percent of the whole retail sector sales, amounting to $88 billion in 2005, a 24 percent increase from 2004 (Table 5.3). These figures are consistent with those provided by the *E-commerce guide* (Maguire 2005).

It is worth noting that statistics from the US Census Bureau include sales of services, and therefore travel, but take into account only the fee that electronic intermediaries charge to customers, which represents a small fraction of total travel sales. According to *The Economist* (2005c), "one-third of America's $200 billion travel market will be booked online this year."

These facts pose an important question: why has e-retail, up to now, achieved such meager results, at a time when the Internet has become the actual standard of interfirm business exchanges? In fact, BtoB and BtoC are tremendously different. Retail trade faces the problem of ordering and shipping piecemeal quantities of low value items. It is, therefore, fully impacted by the shipping and delivery issue. Another important matter is that households, for obvious technical and financial reasons, are less efficient e-traders than enterprises. Actors involved in BtoB e-commerce—essentially wholesale trade and e-procurement—experimented with EDI networks for a long time, then shifted onto the Internet readily and with greater knowledge of the process.

However, electronic commerce, either BtoB or BtoC, must not be regarded as monolithic. On the one hand, there are success stories in particular sectors such as books, travel, consumer electronics, and computers. On the other hand, the majority of pure players in food and clothing have collapsed since 2000. The explanation of e-commerce achievements and failures must take the variety of this business into account. And geography—notably geographic scale—must be placed at the center of the analysis: the nexus of e-retail is the shipping issue (Murphy 2003).

E-retail costs and benefits

At first sight, online shopping is a fanciful business concept, because it seems to give customers access to a virtually unlimited range of goods and services, provided by a virtually unlimited list of sellers, without regard to geography. According to most

Table 5.3 ***Estimated US retail sales: total and e-commerce***

	Total retail sales ($ billion)	*Percentage change from prior year*	*E-retail sales ($ billion)*	*Percentage change from prior year*	*E-retail share of total sales*
2005	3,719.178	6.96	88.026	24.14	2.37
2004	3,477.308	6.49	70.906	25.18	2.04
2003	3,265.477	4.18	56.644	25.87	1.73
2002	3,134.322	2.17	45.001	30.37	1.44
2001	3,067.725	2.64	34.517	24.32	1.13
2000	2,988.756		27.765		0.93

Source: US Census Bureau (Scheleur *et al.* 2006)

Note: The first official US government estimate of retail e-commerce sales was released for the fourth quarter of 1999. Yearly figures have been calculated by the author on the basis of non-adjusted quarterly estimates.

economists, an electronic marketplace should deliver increasing returns, notably through "network effects." As noted in Chapter 2, Metcalfe's law says that "the value of a network increases with the square of the number of its members" (Metcalfe 1995). Since, some scholars have raised doubts about this statement (Briscoe *et al.* 2006). Even reduced to a lesser extent, network effects actually exist, notably in e-Bay-like "combinatorial networks" where customers are linked to each other. However, classical, "radial" networks such as Amazon (see below) benefit primarily from traditional scale economies.

Additional benefits for both sellers and buyers have often been expected from disintermediation processes, notably through the end of wholesalers. However, this vision hardly materialized, because fulfillment requirements have not been taken into full consideration. In fact, intermediation, disintermediation, and reintermediation are all happening (Andal-Ancion *et al.* 2003; Smith *et al.* 2000).

Actually, deterrent factors in BtoC e-commerce are numerous and strong. A 2005 survey by comScore Networks (Maguire 2005) asked customers which things would encourage them to shop more online. The five most common responses were:

- more credit card protection (70 percent)
- reasonable shipping fees (50 percent)
- less expensive prices (50 percent)
- privacy protection (50 percent)
- guarantee of effective product delivery (35 percent).

Three major drawbacks emerge from these figures: lack of trust, lack of "touch," and higher prices which derive mainly from shipping and delivery costs (a problem which equally hampers mail order sale).

Lack of trust derives from the "virtuality" and absence of personal contact of e-retail. In a brick-and-mortar shop, customers exchange their money in a real place, for a real product, with a real vendor. Indeed, the collapse of many dot.com shops at the beginning of the 2000s has been of little help for establishing trust in customer's minds. Establishing trust in a virtual brand is a long task (Schneider 1999). This explains why incumbent "brick-and-mortar" retailers, which benefit from well-established brands and reputation, have often successfully implemented commercial websites. For similar reasons, successful pure players such as eBay or Amazon.com are now a decade and more old. Lack of "touch" is strongly linked to lack of trust. In brick-and-mortar outlets, shoppers can see "real" items. They can touch them, try them on (clothes), and even taste or smell (food, wine, perfume, leather products). French customers like to squeeze the Camembert before buying it. Everybody wants to sit in the couch, or to grasp the wheel of the car, before buying it. By comparison, shopping online is a poor *experience*, a concept we develop more thoroughly below.

Shipping fees: the cost of geography

Brick-and-mortar supermarkets usually have usually lower prices than e-grocers because they do not support the cost of delivery operations. But customers of Wal-Mart, Carrefour, or Tesco do usually not take shopping travel expenses into full account. On average,

8 percent of US drivers' mileage is travelled for shopping purposes (Anderson *et al.* 2003). Nevertheless, the underestimation of the logistics issue is recognized as the primary cause of the dot.com bubble inflation and final bust (Stone 1998). The fundamental challenge of e-retail (as well as those of mail order sale) is to deliver small parcels to individual clients, without hope of scale economies in the final transportation process.

Amazon.com, which we present below, proved to be a success, at least in terms of sales, if not of profit, because books, music, and game or software packages are fully standardized items, from which customers cannot fear bad surprises, unlike in clothing or fresh food. Amazon's critical size makes it possible for the firm to capture a notable share of a category of consumers we could call "premium consumers," a minority but millions worldwide, that is those people whose time is money and who willingly pay more to be served at home. "Wal-Mart caters more to people with large families and people who aren't in much of a hurry." On the contrary, "many online shoppers fall into the 'time-stressed' category of baby boomers that like having items delivered" (Maguire 2002).

Business life and performance in e-retail often result from niche strategies and local scale of operations. This is demonstrated by SPUD, a West Coast Canadian company, which specializes in fresh and organic food home delivery and puts marketing emphasis on geographic proximity, support of local economy and environment-friendliness (Murphy 2007).

Like catalogue sale companies, e-grocers rationalize delivery routes by using drop zones, such as the dense network of convenience stores in Japan (Aoyama 2001). In the UK, Homeport has implemented safe, insulated boxes at the doors of customers' homes. This method solves the requirement of buyers' presence, and makes possible night delivery service, avoiding street congestion (Murphy 2007).

Theoretically, an optimal organization of the delivery process would overcome the cost of delivering small parcels. Figure 5.3 shows that the distance traveled on the sum

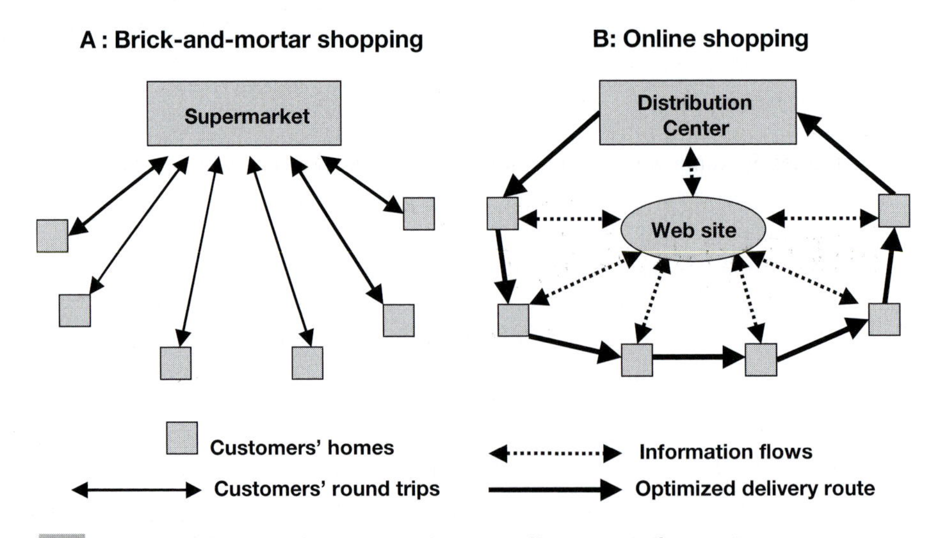

Figure 5.3 ***Home delivery may reduce overall transportation costs***

Source: Authors' own research

of individual round-trips to the shop by a given number of self-carrying customers is greater than a corresponding single delivery route, which can be optimized through GPS and path-finding devices and software. Moreover, e-shopping can benefit from some economies in the warehousing process: hyper and supermarkets are designed for marketing needs, rather than for logistics requirements. In e-retail warehouses, employee's trolleys do not have to zigzag in the crowd of shoppers carts to replenish the shelves.

Towards multichannel, "brick-and-click" shopping

A purely economic interpretation of shopping—which would consider customers as fully rational agents—inevitably misses the explanation of traditional stores' resilience. People go to the mall, rather than browsing on the Internet, because they like to do that ("people who shop at Wal-Mart like to go to the store"—Maguire 2002). In the eyes of customers, the value-added of the supermarket is not only in low prices. It derives from a bundle of services and experiences made inside or outside the shop, which may include having a meal in the nearby McDonald's or going to the hairdresser. And social contact with "flesh and blood" vendors and clerks represents an important part of the shopping experience.

Brick-and-mortar retailers have well understood that their main competitive advantage against both e-retailers and supermarket chains, in addition to finding niche markets and high quality products, is "to enhance the entertainment value of shopping" (Dixon and Marston 2002), through providing customers with living experiences that may derive only from physical proximity with shops, goods, and sale clerks (Pine and Gilmore 1999).

Take the example of FNAC, the leading French retail chain in consumer electronics. Of course, it has developed a commercial website. But it offers customers an experience which goes far beyond buying books or laptops. FNAC's shops have become lavish department stores located in the smartest neighborhoods of selected cities. Customers are invited to discover the most recent technical and cultural products of the information society through strolling in the flat-screen display area, asking shop assistants for the latest iPod specifications, leafing through a magazine while having a coffee in the cafeteria.

Barnes & Noble, which is the world's leading bookstore chain, has adopted a similar model. Since 2004, it has installed WiFi technology in all its stores. In 2005, www.barnesandnoble.com sold $440 million, which represents a tenth of its total sales ($5.103 million). Barnes & Noble Inc. (2006: 5) is explicit: "the company has a multi-channel strategy. Barnes & Noble.com serves as both the Company's direct-to-home delivery service and as a broadcast channel and advertising medium for the Barnes & Noble brand."

Traditional retailers have recorded timid debuts in e-commerce implementation. Maguire (2002) reports that Walmart.com ranked thirteenth in the US with 6.5 million visitors, lagging far behind eBay (34.4 million visitors) and Amazon (25.6 million), although it has been selling online since 1996. In the early 2000s, online buyers were mostly "affluent, tech-savvy young and male people"—not the typical customer base of Wal-Mart. Now that PCs, Internet access, and broadband are almost commoditized, the middle-class buyers are online, with notable effects on traditional distributors' e-commerce data: in mid 2006, Walmart.com and Target.com ranked third and fifth on the list of most visited US commercial websites, while eBay and Amazon remained first and second (Vara and Mangalindan 2006).

In the end, multi-channel "brick-and-click" strategies are likely to dominate the market (Steinfield *et al.* 2002; Steinfield 2004). In the competition with e-retailers, incumbent distributors benefit from built-in advantages which most of them draw from their specific "embeddedness" in territory and local markets (Wrigley and Currah 2006): the purchasing power, the brand, the visibility, the backing of online operations by a physical network of outlets—people are thus able to chose replacement tires at Walmart.com and have them installed at a local Wal-Mart (Maguire 2002). Facing tough competition in a market which is becoming mature, pure players have tried to diversify their business. In September 2005, eBay bought Internet telephone-service provider Skype Technologies for $2.6 billion (Hof 2005). In 2006, Amazon started to host digital download services, in an effort to make the most of its tremendous storage and computing capacity. It also started to commercialize warehousing capacity from its enormous fulfillment centers worldwide (Hof 2006).

However, some sectors sell well on the Internet. One category is comprised of standardized, commoditized items from which few bad surprises are expected after unpacking, such as DVDs, CDs, and books, the main market of Amazon.com whose organization we survey below. The Internet has also proved a very efficient channel in services such as travel, presented in a following section.

A case study of e-retail: Amazon.com, Inc.

From its founding in 1994 by Jeff Bezos (still President and CEO, as of 2007) to its current position of an actual global brand, Amazon symbolizes both the strengths and weaknesses of BtoC e-commerce. Amazon's annual corporate reports provide a deep insight in the company, and a fine understanding of e-commerce practices.

Headquartered in Seattle, Amazon started web-based sales in July 1995. In 1996, it sold $15 million of goods. In three years, annual sales increased by 100-fold to $1.6 billion in 1999. In years 2000 and 2001, Amazon's stock market value fell 90 percent from its height of December 1999 (Zook 2005). But the growth of sales continued. The year-over-year increase in net sales was 22.7 percent in 2005, at $8.49 billion (Table 5.4). After years of continuous losses, a frequent situation in the dot.com industry, Amazon began to make a profit in 2003 (the 2005 net income was $359 million). Today, Amazon employs about 9,000 full-time and part-time employees worldwide.

A first conclusion can be drawn from these elementary data: although Amazon's achievement is laudable, it falls well short of establishing the company as a world superpower in the corporate retailing landscape. Wal-Mart's global sales for fiscal year 2005 (ending 31 January 2005) exceeded $312 billion, with an all-time record net income of over $11 billion (Wal-Mart Stores Inc. 2006a). In terms of market capitalization, Amazon remains similarly a medium-sized corporation. As of September 30, 2007, it was valued at less than a fourth that of Wal-Mart.

Table 5.4 ***Selected data, Amazon.com, Inc.***

		2005	*2004*	*2003*	*2002*	*2001*	*2000*
Net sales		8,490	6,921	5,264	3,933	3,122	2,761
Year-over-year growth (%)		22.7	31.6	33.8	25.9	13.0	
Regional shares of sales (%)	North America	55.5	55.6	61.9	70.2		
	International	44.5	44.4	38.1	29.8		
Net income (loss)		359	588	35	(149)	(567)	(1,411)
Segment operating income	North America	296	321	283	179		
	International	270	169	78	435		
Cost of sales		6,451	5,319	4,006	2,940		
Gross profit		2,039	1,602	1,257	993		
Gross margin (%)	Consolidated	24.0	23.1	23.9	25.2		
	North America	26.9	26.6	26.6	26.8		
	International	20.4	18.8	19.5	21.5		
Shipping revenue		511	420	372	364		
Outbound shipping costs		(750)	(617)	(508)	(404)		
Net shipping cost		(239)	(197)	(136)	(40)		
Segment operating expenses (percentage of net sales)	Fulfillment	8.6	8.5	9.1	10.0		
	Marketing	2.3	2.3	2.3	3.2		
	Technology and content	4.8	3.6	3.9	5.5		
	General and administrative	1.7	1.6	1.7	2.0		

Selected and adapted from: Amazon.com, Inc. 2005, 2006, data in US $ million

Note: Data in brackets: loss or cost

Amazon's financial data demonstrate that success in e-commerce is a matter of performance in inbound logistics (from suppliers) and—more critically—outbound logistics (to customers). In 2005, Amazon net shipping costs amounted to $239 million, which represents the difference between shipping fees charged to customers and actual shipping costs charged, notably, by third-party logistics providers. Such an increasing loss reveals that Amazon, in order to remain competitive in terms of price, must absorb a significant fraction of shipping operation costs, notably through special offers or even free shipping offers which are also used as marketing operations.

Inbound logistics achievement allows the shortening of inventory, which results in a negative operating cycle, also a distinctive feature of Dell's business model we surveyed in Chapter 4. Thanks to the fast replenishment of Amazon's warehouse shelves, inventory turnover in 2005 was 14 days (18 in 2003), while payments to suppliers were due by 54 days (50 in 2003).

Amazon's data pinpoint two key issues in e-commerce operations: *fulfillment* and *technology*. In 2005, the cost of payment operations, credit card fees, and guarantees for third-party seller transactions accounted for 8.6 percent of net sales, while marketing expenses represented 2.3 percent. The second expense segment is technology (websites and linked databases), which represented 4.8 percent of net sales. While the share of other segments remains stable, this notable increase over year 2004 (3.6 percent) reveals the emphasis the company puts on technology as a key element of its global growth strategy.

Another piece of evidence of Amazon's increasing involvement in IT is given by gross fixed assets data (Table 5.5). The evolution of IT-based assets is clearly greater than those of other assets. From 2004 to 2005, technology infrastructure increased by 80 percent, the software and Web apparatus by 75 percent, while the value of fulfillment and customer service infrastructure rose by only 17.5 percent. The result is a sharp increase of the share of IT-based assets, from 22.3 percent in 2003 to 36.3 percent in 2005.

> Our spending in technology and content will increase as we add computer scientists and software engineers to continue to improve our process efficiency, enhance the customer experience on our websites and those websites powered by us and invest in several areas of technology including seller platforms, search, web services, and digital initiatives.
>
> (Amazon.com, Inc. 2006: 26)

Amazon does have a geography (Dodge 2001). It is organized into two main geographic segments. North America (55.5 percent of total sales in 2005) and international. The share of international operations rose steadily from 2002 to 2004, then stabilized: 30 percent in 2002, 38 percent in 2003, 44.4 percent in 2004, and 44.5 percent in 2005.[2]

A more thorough analysis of data shows that globalization of e-commerce operations is not achieved without costs. While Amazon's gross margin in North America has been steadily established between 26.6 and 26.9 percent from 2002 to 2005, the international segment has not performed so well, and even decreased from 21.5 percent in 2002 to 18.8 percent in 2004 (20.4 in 2005). Two plausible hypotheses may explain these figures: 1) a smaller market share in foreign countries which diminishes scale economies and network effects—a fact compounded by lower ratios of PC and Internet users who are likely to buy online—and 2) the marketing costs of international expansion, and the competition with local e-retailers.

The geography of Amazon facilities (Figure 5.4) shows that the company is far from being footloose. It rents 11.9 million square feet (more than 1 million m^2) worldwide,

Table 5.5 ***Amazon's gross fixed assets, 2004-2005***

	2005		*2004*	
	$ million	*year-over-year change (%)*	*$ million*	*year-over-year change (%)*
Gross fixed assets	571	35.0	423	13.7
Fulfillment and customer service	309	17.5	263	9.13
Internal-use software, content, and website development	138	74.7	79	38.6
Technology infrastructure	69	81.6	38	46.2
Other corporate assets	55	27.9	43	–8.5

Source: selected and calculated from Amazon.com, Inc. 2005, 2006

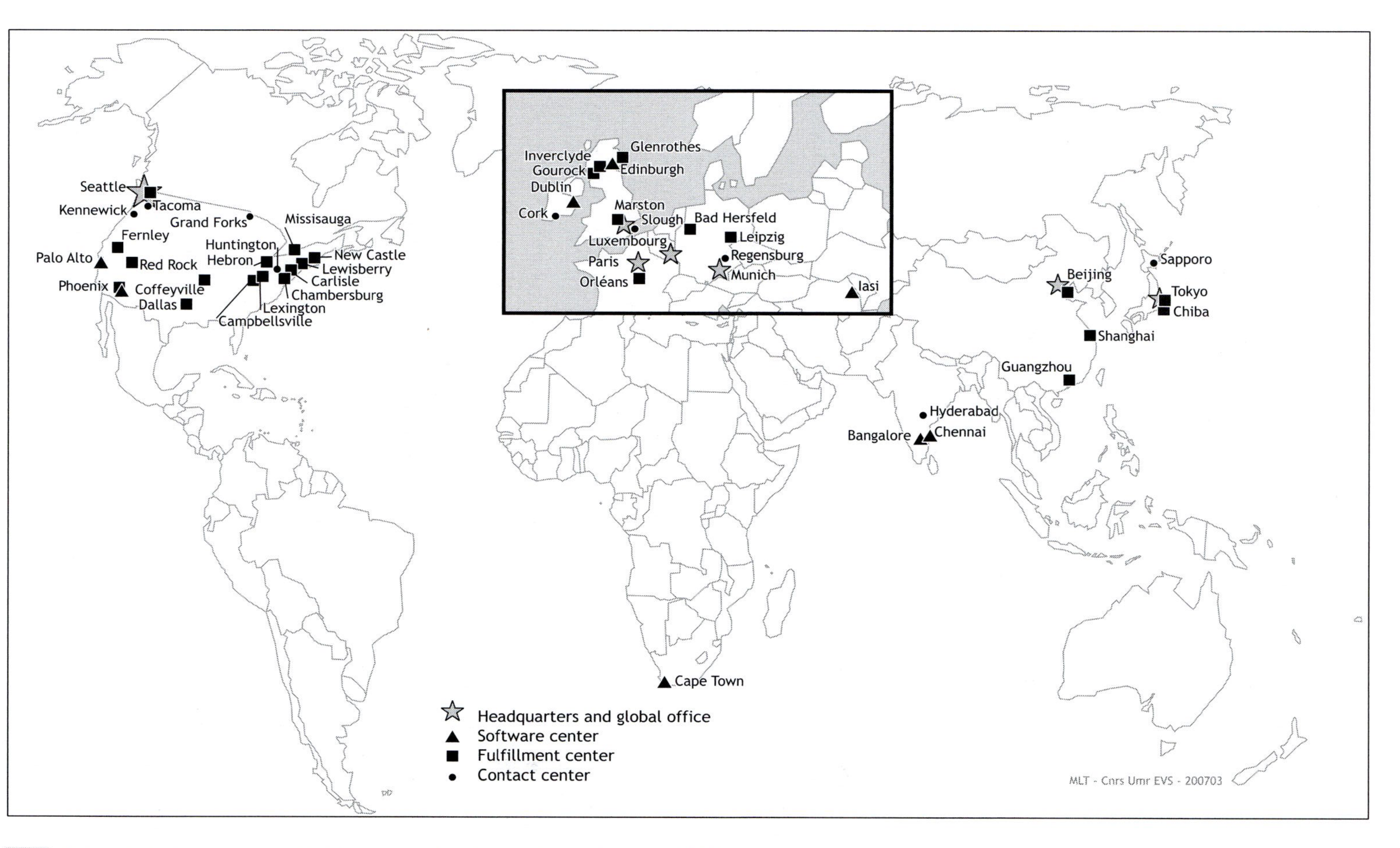

Figure 5.4 ***Amazon's worldwide geography of operations, January 2007***

Source: Amazon.com, Inc. 2005, 2006 and Web research

mostly warehouse and fulfillment operation facilities (the company does not own real estate). Amazon's geography is archetypal of location tendencies in the digital economy. While global and regional headquarters are located in first-tier business cities, fulfillment facilities (including call centers) and warehouses are often located in third-tier cities. Two main factors explain this geography: the search for lower costs (salaries and real estate) and for best air freight accessibility. Warehousing activities should not be located near busy passenger hubs, such as Chicago, Atlanta, or London, but near freight airports which serve as hubs for logistics companies. In the US, a potential location factor is the desire to avoid unionized states: none of Amazon's employees is represented by a labor union (Amazon.com 2005: 6).

Amazon's geography also illustrates the capability of e-commerce to exploit tax loopholes. For example, like other retailers such as Tesco or Woolworths, Amazon has set up a base in Jersey (Channel Islands, UK), from where customers in the UK are allowed to purchase free of VAT items such as CDs, DVDs or video games, provided they cost less than £18 (Amazon Jersey Ltd. 2007).

Travel: a sector where e-commerce is likely to dominate

With the exception of finance, no economic sector more clearly demonstrates the pathbreaking power of Internet-based electronic platforms than the travel industry. As *The Economist* (2005c: 65) writes, "the industry was ideally suited to be an internet business." Online travel penetration rates in 2006 would reach 50 percent in railways, 43 percent in airlines, and 30 percent in hotels and car rental.

The airline sector has long been digitalized at the professional level. Individual airlines or consortia of companies established sophisticated electronic reservation systems (ERSs) in the early 1970s (Wilson 2000). Today, four major ERSs serve the world market: Sabre Travel Network, Worldspan, Amadeus and Galileo (Table 5.6).[3] Each of these ERSs is linked to virtually all airline offices throughout the world. For example, the 490 airlines bookable through Amadeus represent 95 percent of the world's scheduled airline seats (www.amadeus.com). In addition to airline offices, ERSs are connected to tens of thousands of travel agencies, car rental offices and hotel properties. State-of-the-art data storage capacity, processing power and broadband telecommunication capacities are critical for such an industry. Few business sectors have established data centers larger than those of ERSs. Worldspan's mainframe systems, located in Atlanta, reportedly "occupy space equivalent to the size of two football fields" (www.worldspan.com). Amadeus boasts running in Erding (Germany) "one of Europe's largest civilian data processing centers" (www.amadeus.com).

The very revolution that the Internet has set in motion is the convergence between ERSs and BtoC e-commerce platforms. Web-enabled technology has made it possible to create an interface which is suitable not only for travel agency clerks, but also for end-users or customers. Therefore, online travel sales increase competition, because customers can easily compare offers on dedicated CtoB websites, such as Kayak.com or SideStep.com. Increased productivity, the disintermediation process, scale economies and network effects benefit customers, transportation companies, and new online intermediaries. From the company's point of view, direct digital sales reduce selling

Table 5.6 ***The four largest electronic reservation systems***

	Sabre Travel Network www.sabretravel network.com	*Worldspan www.worldspan. com*	*AMADEUS www.amadeus. com*	*GALILEO www.galileo. com*
Creation Date	1960	1968	1987	1971
Connected to travel agencies	1976	1976	1992	1976
Founder(s)	American Airlines	Delta, Northwest, T.W. Airlines	Air France, Iberia, Lufthansa, SAS	United Airlines
Current owner	Sabre Holdings Corp.	Worldspan Technologies Inc.	WAM Acquisition SA	Cendant Corp.
Headquarters	Southlake, TX	Atlanta, GA	Madrid, Spain	Parsippany, NJ
Workforce	1,600	2,066	5,362	2,000
2005 revenues	$1,600 million (2005)	$953 million (2005)	€2,400 million (2005)	–
Data center	Tulsa, OK	Atlanta, GA	Erding, Germany, Sophia Antipolis, France	Denver, CO
Facilities connected:				
Travel agencies	50,000	16,000	76,345	43,500
Airlines bookable	400	–	490	460
Hotel properties	60,000	–	53,484	58,000
Car rental companies and locations	41	–	44 29,681	23

Source: Wilson 2000 and companies' websites (accessed 25 November 2006)

costs. In the airline industry, global electronic ticketing by 2007 would save $3 billion a year (*The Economist* 2005c), by eliminating printed document processing costs and by reducing expenses in real estate and clerical workforce induced by physical presence in airports and downtown.

In this context of fierce competition, clients benefit from a share of these savings in terms of lower prices and wider choices. For example, while SNCF, the French railway company, has inexorably been reducing desk ticketing, travellers increasingly benefit from a series of bonuses and special offers, notably on TGV trains, if they book many weeks in advance on www.voyages-sncf.com. Such a reorganization allows transportation companies to make the most from "yield management" methods.[4]

A certain number of "pure players" have entered the fray with great success (processes of reintermediation). Some e-travel websites are well known: Orbitz, Travelocity, and Lastminute.com. In 2004, Expedia, the world leader, sold $13.2 billion in bookings. Since the beginning of the 2000s, the industry has undergone a major consolidation process, which accelerated in 2004–2005. Expedia, which became independent in August 2005,[5] controls TripAdvisor, Hotwire, Hotels.com, and owns a stake in the Chinese company eLong. Perhaps the most striking fact is the increasing integration between travel e-retailers, travel companies, and ERSs. In July 2005, Lastminute.com was bought by Sabre, which also owns Travelocity (*The Economist* 2005c). Chicago-based Orbitz, the world's third-largest online travel agent, was first created by a group of airlines, then sold to Cendant, the holding company which also controls Galileo. The tendency for vertical integration in the travel e-commerce industry seems logical, since integration at the legal level would facilitate integration of data at the technical level, the commercial Website becoming the upper layer of a single system.

IT implementation in the travel reservation sector generates an increase of productivity that is revealed by statistics. It must be recalled that, on a world scale, the whole travel industry is booming, in terms of both output and job figures. In the US, while sales from firms in the NAICS 5615 "travel arrangement & reservation services" grew steadily between 1994 and 2004 (despite the deep crisis of 2001), and are projected to increase even more by 2014, employment in the sector has shrunk since 1994 (Table 5.7). One travel agency out of six has shut down between 1997 and 2002, from 36,578 to 28,406 (US Census Bureau 2004). Some job losses should be attributed to the offshoring of back-office and customer support functions, because most US and British major airlines have call centers in India. However, IT implementations have played a major role in the rationalization of the travel sales sector, with heavy consequences on domestic employment figures.

Electronic finance: globality *vs* place-embeddedness

> Out of 20 billion transactions a year which are performed by HSBC's clients, only 5 percent are made through face-to-face contact. Ten years ago, 75 percent of our clients were going to our branch offices. Today, they are a mere 12 percent.
>
> (Interview of John Bond, president of HSBC, in Lemaître and Michel 2006)

Finance is the business sector which has *par excellence* supported the idea of the shrinkage of distance and space. This myth has been compounded by works with provocative titles such as *Global Financial Integration: The End of Geography* (O'Brien 1992). Therefore, we can see a paradox in the fact that during the two last decades, financial activities have become an important subject in the mainstream of economic geography and regional studies (Clark and Wójcik 2007; Corbridge *et al.* 1994; Graves 2004; Leyshon and Thrift 1997; Martin 1999; Tickell 2000; Wójcik *et al.* 2007). However, it is precisely the digitalization of money and the financial industry which has reinforced

Table 5.7 ***Employment and sales in the travel reservation sector in the US***

	1994	*2004*	*2014 projected*
5615—Travel arrangement and reservation services			
Sales ($ billion)	23.4	27.8	37.6
Employment (thousands)	271.2	225.7	233.0

Source: Berman 2005: 66

their theoretical interest as a field of study for economic geographers. At first sight, the financial sector trades an almost fully digital output, therefore almost completely ignoring "transportation" costs. Thanks to digital convergence phenomena, such as Internet access on mobile handsets, financial services have become seemingly ubiquitous (*The Economist* 2007a). Nevertheless, the financial industry shows a very complex geography, which is still very influenced by local business conditions, such as local market idiosyncrasies and national regulations, and by the imperative of face-to-face contact in complex transactions which carry both some risk and tacit content.

IT and "a borderless world of hyper-mobile capital"?

Thanks to electronic networks, enormous amounts of digital money roam or "dance" around a seemingly borderless world (Warf 1999: 230). Like air travel reservations, finance is the kingdom of globe-spanning networks managed by international consortiums such as CHIPS or SWIFT, which operate in 136 countries and link together 7,800 financial institutions. During the year to October 20, 2005, SWIFT traded 1,859 million messages (www.swift.com). In 2004, $345,793 billion ($345 trillion) travelled in CHIPS' network (www.chips.org).

A prominent feature of electronic financial markets—also found in the call center industry—is the capacity to make the most from worldwide space–time arrangements and seamless telecommunications networks in order to process round-the-clock operations. The system operates from East to West. Markets open in Tokyo, then shift operations to London, then to New York and so on.

The myth of a global, distance-free financial industry has been fueled by the virtualization of stock exchanges (Lo and Grote 2003). Fully automated market places or electronic communication networks (ECNs) provide buyers and sellers the ability to trade without regard of their locations, given they benefit from an Internet connection. The most renowned ECN is NYSE Arca, founded in December 1996 as Archipelago Exchange, and merged in March 2006 with NYSE Group, the parent company of the New York Stock Exchange. NYSE Arca is now the electronic trading arm of NYSE Euronext, the world leading exchange group.

In 2005, NYSE Arca generated $493.4 million in revenues, almost exclusively from transaction and market data fees. "Through NYSE Arca, customers can trade over 8,000 equity securities and over 175,000 option products" (NYSE Group 2006: 4). NYSE Arca's trading platform, based on Sun Microsystems servers, "is designed to accept up to 16,000 orders per second and to provide up to 1,000 simultaneous customer connections." In 2005, "it handled an average of approximately 38.1 million orders daily" (NYSE Group 2006: 24). A notable consequence of ArcaEx's growth was the closure of the San Francisco exchange trading floor. There are other ECNs. In the US, it is worth mentioning the Boston Equities Market or BeX, the online trading arm of the Boston Stock Exchange, opened in November 2006. There are specialized ECNs such as MTS Group, owned by 51 European financial institutions, which is dedicated to the market of European fixed-income securities. Thanks to disintermediation and network effect, ECNs provide several benefits: decrease of transaction fees, increased liquidity, improved transparency, and fast, more efficient price discovery.

Such "hyper mobility of capital" (Warf 1999) is likely to catch the imagination. Nevertheless, "the end of geography" in financial services is a stale theory, which has been challenged by most authors. Despite a notable fraction of transactions processed from abroad, even if fully automated, financial activities remain embedded in locality and territory. Retail banking must keep a local presence within reach of customers. Advanced financial services show strong agglomeration effects and large scales of operations, but do not fully ignore the imperatives of a local presence.

The resiliency of brick-and-mortar and "human" banking

Since the beginning of the IT media hype, prophets of doom have anticipated a collapse of the banking industry. The whole business would shift towards electronic platforms and networks, front-office activities would become obsolete, hence leading to massive consolidation of branch offices and to mass-redundancy of white-collar workers. The crisis would be deepened, if possible, by the outsourcing of most remaining back-office activities—something which partly materialized, as we suggest in Chapter 6.

But no disaster occurred. On the contrary. The evolution of the US banking industry analyzed by De Young *et al.* (2004) is highly revealing. The deregulation completed in the 1980s and 1990s caused a dramatic consolidation process (from 14,434 commercial banks in 1980 to 7,887 in 2002). However, "their new-found geographic freedom" influenced the banks to increase vastly the number of branches (by 70 percent since 1980, 20 percent between 1994 and 2003), despite the huge implementation of automated teller machines (ATMs) and Websites. And "the spatial density of deposits in the 50 largest metropolitan areas has remained remarkably stable over time" (De Young *et al.* 2004: 45).

In fact, much of retail banking has entered the digital age. ATMs have become a commodity. In France, where Internet debuts were timid, credit card terminals are found in nearly all shopping places. The merchant in a modest market stall, who accepts credit card payment through a mobile terminal, is linked to the worldwide digital financial network.

Nevertheless, pure electronic retail banking—that is, "virtual," online-only banks—has so far remained marginal and globally unsuccessful. The combined story of Egg (UK) and Zebank (France) is revealing. Egg Banking plc was launched in 1998 as a listed company. In 2004, it was the world's largest pure player with over 3.6 million customers. In February 2001, French investors launched Zebank, which turned out to be a fiasco, and was sold to Egg in May 2002. Zebank's losses culminated to €133.5 million in 2004. Finally, in July 2004, Egg announced it was closing its French subsidiary (Desautez 2005). On February 20, 2006, Egg was delisted and became a fully owned subsidiary of Prudential plc, a British financial group. According to Prudential (2007) Egg's operating losses for year 2006 amounted to £145 million (nearly $300 million); meanwhile, the banking sector in most countries announced record high profits. Finally, on January 29, 2007, Prudential announced it was selling Egg to Citigroup, a major US bank, which sought to reinforce its presence on the British market.

The explanation of this double failure, and the resilience of branches, is linked to the very nature of financial services and their evolution. The banking industry, spurred by competition, has developed a complex bundle of higher value-added services, which still require in-person contact between bank employees—now "financial consultants"—and customers: credit management, life-insurance, tax consulting, securities trading. This leads to the explanation for the overall failure of pure players in online banking, which is the lack of trust that only social contact can provide. People are more likely to conduct routine transactions on their bank website when they can keep a certain degree of personal contact with their banker.

However, the effects of IT uses on local bank presence vary from country to country. While bank branches in Germany and the UK have seen a steady numerical decline, their French and Spanish counterparts have increased. And, as we suggested above, technology did not deter US banks from increasing the number of branches.

The impact of IT use on the financial sector must be appreciated on various scales. Branch closures, and the rise of ATMs, websites, and offshore call centers may have consequences for local communities, especially for people who are less prepared to embrace the digital society, because of age, culture, or purchasing power. But on a more global scale, regional consequences of digitalization in the financial sector are difficult to capture, because employment in this sector remains vigorous, and projections in the middle term remain rosy. In the US, for example, employment in financial activities increased by 17.25 percent between 1994 and 2004, more than the overall US employment growth of 12.65 percent (Berman 2005). Projections for 2014 expect an increase of 849,000 jobs, a 10.5 percent global growth which indeed would be consistently inferior to the whole US employment growth (14.3 percent). Actually, the overall output of the financial sector increased faster than the employment between 1994 and 2004 (55.4 percent) and is expected to increase of 43 percent by 2014, nearly the US economy average (42 percent). It is trivial that IT uses and business concentration have increased and will increase productivity. But rising productivity also results in part from the offshoring phenomenon, a subject we develop in Chapter 6: while high value-added services continue to grow at home, menial tasks in back office and customer support tend to be resourced abroad, notably in India.

Finally, but we are far from facing the complete IT-driven upheaval of the banking industry which had been forecasted by pundits and media in the early times of the Internet era. Is the matter different for high level, professional services?

Advanced financial services: a local–global geography

> Travelling at the speed of light, as nothing but assemblages of zeros and ones, global money dances through the world's fiber-optic networks in astonishing volumes.
>
> (Warf and Purcell 2001: 227)

Since the Internet makes it possible to trade financial products in real time, and without regard for location, one might have imagined that advanced financial activities could ignore the common rules of the geography of professional services.

However, most of the scholars who studied the geography of financial services and financial centers have debunked the fallacy of a fully globalized, footloose industry.

On the contrary, we are seeing a process of geographic concentration or *clustering* of financial firms in a limited number of business cities. The most studied example is London, which has become since the "big bang" of October 1986 (the liberalization of trading) "a textbook example of an economic cluster, in which businesses locate close to one another because they gain from proximity" (*The Economist* 2006c: 84), "a vital point in time and space for the global financial institutions that manage on a 24-hour basis flows of capital and transactions around the world" (Clark 2002: 450).

In fact, finance illustrates the "double-edged" geography of most contemporary service sectors, which is strongly influenced by both centripetal and centrifugal forces (Poon 2003). Centrifugal forces are described as market access costs and coordination, rent-seeking and political Intervention, and localized information (Gehrig 2000). Centripetal forces include network effects, agglomeration economies, informational spillover, and market liquidity. "Financial firms cluster in London because they derive external economies of scale. By thronging together, they create large, liquid markets that drive down trading costs and reduce risks by allowing large deals to be handled" (*The Economist* 2006c: 84–5).

Network effects and agglomeration economies

There is a positive feedback effect in the relation between IT implementation and horizontal concentration of financial markets. Digitalization made it possible for markets to increase in size and efficiency. In turn, semi-automated or automated markets even more benefit from size, through network externalities or network effects (Lo and Grote 2003) As in the eBay scheme, it is a basic application of Metcalfe's law: the value of the market increases with the number of potentially interconnected people.

Evidence of the utmost geographic concentration is suggested by the credit derivatives market (Agnes 2000). Derivatives are complex financial products, usually contracts (the right or obligation to buy or to sell) which *derive* from more basic, usually traded instruments such as stocks, bonds, currencies, commodities etc. The rising complexity of these products, and the gigantic volumes traded, explain—and in turn are explained by—the fact that it is "perhaps the area of finance most reliant on high technology computing and telecommunications" (Power 2001: 244). According to a survey released in September 2004 by the British Bankers' Association (BBA), the City of London trades over 40 percent of the world market, which is expected to rise to $8.2 trillion in 2006 (British Bankers Association 2004).

Agglomeration economies lead to the concentration of financial markets in a small number of cities because of informational complexity. In finance as in other business sectors, competitive advantages often derive from informational asymmetries (Pohl 2003). While data travel nearly at light speed and at zero cost, strategic information is obtained from informal, sometimes unplanned meetings which are part of the buzz of financial districts such as the City of London (Storper and Venables 2004). Financial centers produce an "idiosyncratic noise" which is carried through formal or informal channels (Gehrig 2000).[6] Some financial activities, such as initial public offerings (IPOs) and mergers and acquisitions, are more likely to require information which derives from personal relationships (Pohl 2003).

Another factor for geographic centralization is the need for cutting-edge technologies and highly skilled people in computer systems, financial mathematics and econometrics. "Underpinning the hub is access to talent. Firms locating in London can tap into a huge specialist financial workforce drawn from both domestic and foreign sources" (*The Economist* 2006c: 85). Therefore, there is evidence that advanced finance benefits from "untraded interdependencies" which are a key constitutive element of local economies (Storper 1997). The geography of advanced financial services is "production-driven," rather "than market-driven" (Sassen 2001b: 104): firms tend to locate in close proximity to their competitors.

On the contrary, when information complexity is linked with the market context, it should be regarded as a centrifugal factor. Gehrig (2000: 431) regards "local information" as a "major centrifugal force for financial activity." While studying the swaps market, Agnes (2000) explains why regional hubs, such as Sydney, have survived. Cultural factors ("market feeling"), and the need for instantaneous decisions, seem to be decisive. Extreme sophistication of interbank telecommunication systems has not made local floor trading obsolete, because "live," human voice-based interaction significantly outperforms long-distance telephone and other means of telecommunications.

> There are windows of opportunity for trading a particular swap at a particular price. In Sydney, everyone has voice boxes and you press a button, and just shout. This takes three seconds. You would not get this information offshore, because you are sitting alone. It might take five minutes to get a hold of the bloke in Hong Kong or Tokyo . . . and by the time I get them, they have missed that window anyway.
>
> (A trader interview, in Agnes 2000: 356)

Clark and O'Connor (1997) deliver a theoretical interpretation through the distinction between "transparent" and "opaque" products. Transparent products are those whose features and behaviour are well known, such as currencies or global companies' stocks. In these markets, whose reach is global, information technology has created an actual dissociation between the dispersed location of the buyers and the concentration of the exchanges, well illustrated by the recent merger between NYSE Group and Euronext. "Opaque" products, such as private equities or medium-sized firms' stocks, are known nationally or locally, and incorporate more tacit knowledge. They are traded by a greater number of smaller exchanges, whose brokers only master local idiosyncrasies and information asymmetries needed to trade efficiently in these markets. The geography of these secondary marketplaces, therefore, has been less influenced by IT.

In the same vein, Klagge and Martin (2005: 413) explain why the delivery of financial services to small and medium-sized firms requires close proximity with institutions and local markets:

> As far as new and small enterprises are concerned, the scale and nature of the risk attached to such firms, the need for detailed and regular information on their performance and activities, and the importance of fostering and supporting entrepreneurship, all put a premium on the value of close co-location of financial institutions and markets on the one hand, and client SMEs on the other.

Therefore, there is room for the existence of second-tier and third-tier financial centers, which host subsidiaries of large banks headquartered in tier-one centers, or small local financial firms dedicated to local business needs. This tendency for a limited degree of concentration in financial services leads to a hub-and-spoke hierarchy, with the domination of world-class financial centers such as London, New York, Tokyo, and Frankfurt over other marketplaces with functions merely regional or national, such as Paris, Amsterdam, Zurich, Singapore, Hong Kong, Shangai, and Sydney (Langdale 2001; Poon 2003).

Conclusion

The conclusion of this chapter is well illustrated by the slogan of HSBC, the third largest banking group in the world in 2005: "The world's local bank" (www.hsbc.com). HSBC's marketers pinpoint the main geographical character of finance, its embeddedness in places and communities. Actually, most activities we analyse, despite being fully IT-enabled, have not been freed from geographic constraints or become "footloose."

As in Chapter 4, the key word is complexity. The old fallacy of the "end of geography" derives from the simplistic thinking that commerce trades only goods, and that finance trades only money. In such a case, warehouses, websites, ECNs, call centers, and ATMs would have proved sufficient, and we might have seen the end of the shop and the end of the bank branch. However, the actual service merchants and bankers sell is a much more complex mix of experience, product discovery, commercial advice, and information collection and analysis. Living customer experience, and face-to-face exchange of non-codified and tacit content play key roles in the transactions with the highest value-added. For example, Sidaway and Bryson (2001: 412–13) report that "UK-based fund managers spend between two and six weeks a year visiting companies in emerging markets," in order to obtain "a 'first-hand' 'feel' for the current condition of a country."

In both e-commerce and financial services, we are witnessing the domination of multichannel processes and institutions which require a mix of electronic interactions and material transportation of goods and people. "Pure" e-commerce is the exception rather than the rule, and might not be a sustainable business model: on 24 May 2007, Dell announced it will sell PCs at 3,500 Wal-Mart stores in North America. The scale of operations is rarely global. Amazon.com operates from seven national websites (US, Canada, UK, Germany, France, China, and Japan) which mirror the cultural and linguistic diversity of its customers. An empirical study by Williams (2004) of the mutual fund industry similarly challenges the "hyperglobalist" view of a "borderless world of hypermobile and homeless money." It shows that mutual funds assets in major economies are heavily concentrated in domestic markets (85 percent for Japanese funds, 84 percent for US funds).

This complex geography is the result of both centrifugal and centripetal forces. In the following chapters, we deal more closely with the spatial patterns and organizations which result from these two opposed forces. Centrifugal effects lead to dispersion tendencies which can be analyzed on two scales. The global scale is those of the offshoring of business functions, studied in Chapter 6. The local or regional scale is represented by telework, the subject of Chapter 7. Centripetal forces—agglomeration effects—result in the clustering of high-tech, innovative activities we survey in Chapter 8.

6 SPLINTERING THE ECONOMIC SPACE

The offshoring of corporate services

> Because packets of digitized information play the role that boxes used to play, many more services are now tradable and many more will surely become so. In the future, and to a great extent already, the key distinction will no longer be between things that can be put in a box and things that cannot. Rather, it will be between services that can be delivered electronically and those that cannot.
>
> (Blinder 2006: 115–6)

In Chapter 4, we have explained the role IT plays in the emergence of global *product lifecycles*, or, more generally, *global value chains* (Gereffi 2006) which include a complex mix of manufacturing and informational tasks. These value chains are subjected to *global sourcing* phenomena, that is the processing of tasks, or the purchase of parts and capabilities (in the case of outsourcing contracts) in places whose location depends on the best mix of prices, quality, and skill availability offered by local conditions. This much finer spatial division of labor is characterized by a trend towards vertical disintegration (or "value chain unbundling"), and towards horizontal reintegration and concentration.

The purchase of industrial goods in *offshore*, low-cost countries is an elementary scheme of the international division of labor which prevailed in the late twentieth century. However, the continuous decline of manufacturing employment in Western countries—also fueled by the increase of productivity—was over-compensated by the growth in service jobs. The great change we are witnessing since the beginning of the twenty-first century is that service functions and jobs have joined the offshoring process. Does this trend lead to "the next industrial revolution" (Blinder 2006)?

As Blinder underlines, information technology plays a critical role in global sourcing. Firms are no longer compelled to locate design, R&D, administrative functions, customer support, data analysis, file treatment, or "back office" tasks in the same building, in the same city, or even in the same country, because digitalization has made business information highly standardized and tradable. Thanks to powerful, cheap computing, unlimited data storage capacity, integrated software, and versatile Internet-based applications, companies can reassemble the pieces of the jigsaw (that is to say the dispersed amounts of value-added created worldwide throughout the global network of production). Add the fall of long-distance telecommunication fees mentioned in Chapter 3, and there are virtually no geographic barriers to business service offshoring.

The perspective of a "job shift" in services has generated some turmoil in developed countries, where services represent the bulk of employment and GDP. According to Farrell *et al.* (2005a: 34), "of the 1.46 billion service jobs worldwide projected for 2008, about 11 percent on average—the equivalent of 160 million employees—could theoretically be performed anywhere."

The evidence that several major firms have entered a process of business function relocation has received important media coverage, such as 'The New Global Job Shift' (*Business Week*, February 3, 2003), 'Relocating the back office' (*The Economist,* December 11, 2003), and 'The new jobs migration' (front cover, *The Economist*, February 21, 2004). The subject has entered the public policy arena. In Fall 2003, the UK Prime Minister was called on at the Parliament after HSBC, a world leading bank, revealed a program of 4,000 job cuts in back-office centers in the UK, and of new call center openings in India and China. In 2004, the US Congress "prohibited federal agencies from outsourcing some kinds of work to private companies that use workers abroad" (Baily and Farrell 2004: 1). The 2006 article by Alan Blinder in *Foreign Affairs* has stirred up a controversy among US economists, policy-makers, and would-be presidential candidates, about the scale and reality of the threat, and the possible remedies to implement, including a certain degree of protectionism (Wessel and Davis 2007).

Beyond political debates, the offshoring of services is a provocative subject for those who seek to understand the evolving geography of the global economy, because it is a perfect embodiment of the splintering of the economic space, in the context of an information-based, digital economy. The debate comprises several interlinked questions:

- Why are services going offshore?
- What constraints and limits does offshoring face?
- Which kinds of services go offshore?
- How many service jobs are actually going offshore?
- Where are they going?
- What consequences may reasonably be expected in the medium term for both developed and emerging countries?

There are few simple, one-sided answers: recent, intriguing evidence of moves from emerging countries back to developed countries has been observed, and Indian outsourcers have started to create call center and back-office jobs in Europe. Even more surprising, thousands of American citizens now work in India for Indian IT companies.

Offshoring strategy: benefits, risks, organization

The phenomenon of offshoring, globally seen, is driven and shaped by a bundle of drivers and facilitators, which interact with a series of costs, constraints and risks (Table 6.1).

The main factor: labor costs

While technology acts as the great facilitator, the most critical driver of offshoring is the search for lower labor costs. A study of 104 US companies by Lewin and Peeters

Table 6.1 ***Offshoring drivers, facilitators, and drawbacks***

Drivers	*Drawbacks (costs and risks)*
Abroad (pull factors)	**Abroad**
Availability of talent	Political and social hazards
Low-cost labor force	Unstable regulation
Flexible labor regulation	Lower productivity
Low taxes	
Incentives	
At home (push factors)	**At home**
Shortage of talent	Political backlash
Over-heated labor market	Backlash from customer
"Bandwagon effect"	Backlash from employees and unions
	Damage to the firm's image and brand
"Facilitators"	**Additional factors for in-house offshoring**
Free trade regulation	Higher training costs
Information digitalization	Higher establishment costs
Low-cost, reliable long distance telecommunications	
	Additional factors for outsourced offshoring
	Leakage of knowledge
	Contractual hazards
	Lock-in with vendors

Source: Partially adapted from Ge *et al.* 2004

(2006) shows that cost reduction is targeted by nearly all the firms surveyed (Table 6.2). Labor is the main segment in the cost breakdown of IT-enabled service operations, such as back-office tasks and transaction processing. In the US, the labor cost "equals 60 to 70 percent of the annual call center operating budgets" (Anton *et al.* 2002: 7).

Logically, the search for lower labor costs leads companies from wealthy, expensive countries to target places in the developing world. The average monthly salary of a call center employee in Morocco is €450 (benefits included), 44 worked hours a week, against €2,000 in France, 35 hours a week (Dumout 2004). An Indian teleoperator gets under $1 per hour, while the same job in the US pays $12.50. The gap decreases in relative terms when the level of qualification rises, but remains huge: an accountant receives between US$6 and $15 per hour in India, compared to $23 in the US (Bardhan and Kroll 2003).

The low standard of living of most preferred offshoring destinations is intrinsically a factor of low wages, because of the high ratio between actual purchasing power and nominal salaries, which is between two and three in Eastern Europe, over four in the Philippines, and nearly five in India and China. An Indian programmer who earns a nominal $9,000 a year actually makes $45,000 in terms of purchasing power. A call center employee in Dakar (Senegal) gets three times the national average salary.

The second main factor in offshoring strategy—interlinked with the labor cost factor—is the actual availability of ad hoc human resources in emerging countries, compounded

Table 6.2 ***The objectives of offshoring***

Offshoring objectives	*Percentage of companies that cite the objective as a factor of decision for offshoring*
Cost reduction	97
Growth strategy	73
Competitive pressure	71
Access to qualified personnel	70
Industry practice	52
Improved level of service	50
Business process redesign	48
Changing rules of the game	37
Increased speed to market	35
System redundancy	33
Access to new markets	25

Source: adapted from Lewin and Peeters 2006: 22

by shortages of the required skills in developed countries. Large labor pools only offer *scalability*, which is the possibility of increasing quickly the scale of operations. Recent or planned recruitments by leading companies in India and China are made on large numerical scale—often several thousands. Even basic tasks such as customer support require college education. This explains the success of India we emphasize below, which offers both low costs and a massive supply of young, well-educated and English-speaking workers. Batt *et al.* (2005) showed that the typical Indian call center worker has a university degree. The Philippines offer a similar profile, notably English fluency of most educated people. In some regions such as North and Western Africa, recruitment is made even easier by the extremely high unemployment rate among young people with university degrees. Employers in low-cost countries also benefit from flexible labor laws. Redundancy compensations are minimal, and night or Sunday work shifts pose little problem.

> What makes this trend so viable is the explosion of college graduates in low-wage nations. In the Philippines, a country of 75 million that churns out 380,000 college grads each year, there's an oversupply of accountants trained in US accounting standards. India already has a staggering 520,000 IT engineers, with starting salaries of around $5,000. US schools produce only 35,000 mechanical engineers a year; China graduates twice as many.
>
> (Engardio *et al.* 2003: 58)

Ge *et al.* (2004) identify a primary factor: the "bandwagon effect": that is, firms go offshore because other firms went, as a consequence of competitive pressure, notably from shareholders of publicly traded firms. In the case of outsourced offshoring, Maskell *et al.* (2005) suggest that companies face a kind of collective learning process, which results in an upgrading of practices. There also are factors acting not as causes, but as facilitators. Besides IT, this is the case for free trade regulations, which allow firms to import software and equipment free of duty, or to repatriate profits made by foreign subsidiaries.

An important facet of the digital economy we do not fully explore in the present book is related to tax havens or *offshore financial centers*, such as Cayman Islands, the British Virgin Islands, Dubai, the Canton of Zug (Switzerland) and so on. These places have secured a tremendously important position in the hosting of cross-border assets, the most visible consequence being a dramatic loss of tax revenue for governments and communities (Brittain-Catlin 2005; Palan 2003). This is a true world-scale governance issue. Undoubtedly, IT and the Internet have played a critical role in the development of financial services in tax havens (Cobb 2003; Ramos 2007). However, the present chapter focuses on value-added services based on human resources, like those found in Ireland or India, rather than those based almost exclusively on the peculiar niche of money flows and tax dumping.

Drawbacks and hazards

Attractiveness factors are mitigated by drawbacks and risks. Political and social risks in emerging countries are usually greater than in developed countries. Business regulations in Asia or Latin America may be volatile. Social or natural hazards must not be underestimated: religious conflicts, earthquakes or hurricanes, and diseases such as SARS. Prevention of technological risks (power outages, breaks in telecommunications) leads to additional costs. These problems must be considered even more seriously by firms which want to locate abroad those critical functions whose continuity is essential, such as air travel reservation, sales support or IT management. This fact explains the success of Eastern Europe as a location for mission-critical services: wages are low, but distances are small, people are skilled and the EU membership guarantees a business-friendly, safe and stable environment.

Increasingly sophisticated outsourcing contracts raise the risk of knowledge leakage or even fraud and robbery. Contractual hazards increase with the geographic, cultural, and political distance between buyer's and vendor's countries. In case of litigation, trade, or intellectual property laws are likely to be more easily enforced if the vendor's home country has a similar and reliable system of jurisdiction.

Last, firms must consider political risks in their home country. The public exposure of huge offshoring practices, concomitant with mass redundancies at home, may damage the reputation of the company, discourage local employees, and generate retaliation by customers, unions, and governments.

The drawbacks of offshoring, and the fact it is not a unique, miracle "solution" are well illustrated by the example of Everdream, an IT services provider based in Fremont, California (Silicon Valley). Seeking "scalability," with help desk positions hard to fill at home, it contracted in 2002 with a Costa Rican company. It had to train the Costa Rican team. Nevertheless, "they were never able to get to that next level required." Costa Rican employees held college degrees which proved inferior to their US equivalents. And they were unable to meet Everdream's expectations, given the variety of questions and answers which "made it impossible to script support responses." Everdream's customer satisfaction ratings fell by six points. In the end, it moved the work back home, and started to consider outsourcing in a US city with costs lower than Silicon Valley (Rosenthal 2004).

Organization vs geography

Thanks to technological improvements, and to the upgrading of skills in host countries, the scope of service offshoring is broadening in both sectors and locations. However, that does not mean that every company, in every business segment, will offshore the full list of the aforementioned tasks. In addition to considerations of costs and skills availability, which often dictate the feasibility of a project, an offshoring decision is a matter of strategy within the firm.

A useful introduction to offshoring strategy is to consider the now classical 2 × 2 matrix of outsourcing and offshoring—which we enrich with real business cases (Figure 6.1). Companies face four basic possibilities in the spatio-organizational arrangement of a given task or production process. In the real business world, rather than an abrupt, two-way alternative (offshore or abroad), companies which are considering an offshoring

		Governance scope (control)	
		Inhouse (captive)	**Outsourced**
Geographic scope	**Onshore** (at home)	(1) Processed in-house and locally American Airlines answers customer queries from its call center in Cincinnati (Ohio)	(2) Outsourced to a local service provider Schneider (France) outsources IT management to Capgemini (France)
	Offshore (abroad)	(3) Processed abroad in a fully-owned facility or controlled subsidiary Dell processes US customer support in its contact centers in Bangalore (India) Global Business Service, a subsidiary of Procter & Gamble, processes P&G's administrative functions for the Americas in a shared service center in Costa Rica	(4) Outsourced to a foreign-based service provider Bank of America outsources software development to Infosys (India). Delta Airlines outsources reservations to Spectramind (India), a Wipro subsidiary. Rhodia – a French chemical producer – outsources accountancy to Accenture in Prague (Czech Republic)

Figure 6.1 ***Outsourcing and offshoring classification: the classical 2x2 matrix***

Source: Adapted from Ge *et al.* (2004: 6), Kraemer and Dedrick (2004: 1), Farrell *et al.* (2005a: 15), Olsen (2006: 7), UNCTAD (2004: 148)

decision face a vast range of possibilities, which may be measured by a degree of remoteness—combined with a certain degree of control. Some authors have refined the analysis by promoting the term of *nearshoring*, to describe location in neighboring countries such as Canada for US firms (Aron and Singh 2005; *The Economist* 2005e). The introduction of the intermediary concept of out-tasking, presented in Chapter 4, would therefore lead to a more sophisticated 3 × 3, nine-cell matrix.

Dossani and Kenney (2003: 24–6) identify six factors which may impact the offshoring decision:

1) The knowledge component of the activity: highly knowledgeable tasks are less likely to be transferred abroad, notably if they represent the core activity of the firm, because of intellectual property concerns and the scarcity of top-level talent in low cost areas. However, a few emerging countries, notably India, are now able to deal—to a certain point—with cutting-edge activities such as R&D in software and computing.
2) The interactive component: tasks which require a high level of interactivity—and especially face-to-face interactivity—between vendors and clients, or between teams of different producers, are difficult to locate abroad.
3) The level of separability of the process (which is close to tradability). It derives mostly from digitalization within the production cycle. The possibility to perform IT-enabled, sequential tasks on digitalized material (for example, client files) allow geographically separated treatments.
4) Savings from concentration, which derive from scale economies. In a large, single facility, highly productive equipment and software are likely to be used more efficiently by a specialized staff. This is the very *raison d'être* of call centers and shared service centers, as we describe below.
5) The importance of re-engineering in the transfer process. The transfer of tasks abroad may be an opportunity for business process reengineering, if the need exists for it. The potential benefit of such a reorganization can emerge from the offshoring preparation itself. In the end, "the workers in the new location are met with a fait accompli" (Dossani and Kenney 2003: 25), while the upgrading of the existing process at home could have been a complex and lengthy task.
6) The time-sensitive nature of the work, which is the logical outcome of a customer-centric, digital economy. IT helps companies to free themselves from the barriers of "time zone economies" (Zaheer 2000). An offshoring decision may be considered in order to make profit from time difference to shorten delays in the processing of operations such as radiology diagnosis or insurance claims. The latter case proved a decisive factor for the growth of financial service in Ireland (Grimes 2003a).

Many business institutions, including India's National Association of Software and Service Companies (NASSCOM) make the distinction between *IT services* and *ITES* (IT-enabled services). *IT services*, narrowly conceived, are the primary market of software and computer services firms such as IBM, SAP, Oracle, and Capgemini. *ITES* encompasses a wider range of tasks in which computers and digital networks are not the target of the service, but the tool which helps to perform the job and to deliver it to remote clients: back office function processing, data and imaging processing, customer relationship management (CRM) and all kinds of knowledge-based processes. ITES are often performed in *call centers* and *shared service centers*.

However, the classification of offshored services is made difficult by a process of convergence and integration between IT service firms, on the one hand, and ITES or business process outsourcing (BPO) firms, on the other. This trend is epitomized by the emergence of global IT corporations, such as Accenture, IBM Global Services, and Genpact, or fast-growing Indian outsourcers such as Wipro, TCS (Tata Consultancy Services), and Outsource2india, which deal with a wide spectrum of services, presented in Table 6.3. This convergence between IT and BPO industries is illustrated by the 2004 acquisition by IBM of Daksh, an Indian outsourcer, which delivers BPO and CRM solutions.

Table 6.3 ***The scope of service offshoring: Outsource2india***

Software services	*Data and document management*	*Knowledge process outsourcing services*
Telecom software	Electronic document management	Database creation
Travel applications	Document conversion	Catalog indexing
Dating applications	Forms processing	Market research
Recruitment websites	XML conversion	News abstraction
Workflow automation	Data entry	Company profile
Healthcare software	Document imaging	Desk and Web research
Portal applications	Internet publishing	Information extraction
Smart card software	Photo enhancement	
Embedded software	Image editing	**Animation and Architectural**
Software testing services	Litigation support services	CAD conversion and drafting
Business analysis	Data cleansing	2D and 3D animation
ASP programming	Audio transcription	Architectural rendering
Coding services		Medical animation
J2EE programming	**Additional Services**	Digitization/vectorization
Website maintenance	Web analytics	Online games development
Network maintenance	Creative writing	Architectural design
	Internet marketing	Engineering services
E-Learning Solutions	Human resource	
Instructional design		**Financial Services**
Corporate training	**Healthcare Services**	Tax preparation
Product demos	Billing	Financial analysis
	Coding	Bookkeeping
Call Center Services	Transcription	
Inbound call center	Claims processing	
Technical support	Teleradiology	
Outbound telemarketing	Clinical services	
Disaster recovery		

Source: Selected and adapted from www.outsource2india.com, accessed 14 December 2006

UNCTAD (2004: 164) delivers data for an elementary analysis of export-oriented foreign direct investment (FDI) projects in services for years 2002–2003. Out of 1,849 projects, 632 were in IT services, 513 in call centers, and 138 in shared service centers. The remaining (565) is made of regional headquarters activities. Table 6.4 presents the distribution of call center and shared service center projects among industry sectors. The third category of projects, IT services, is not relevant for this analysis: almost all the projects (97.8 percent) logically come from the IT industry itself).

IT industries, broadly conceived, dominate the scene of service offshoring. Most major computer manufacturing or IT service firms from OECD countries have established a huge presence overseas, notably in India. The second sector in offshoring practices is finance, notably in shared service centers (29 percent). A 2006 survey of 156 financial service firms by PricewaterhouseCoopers suggests that "50 per cent offshore IT functions or plan to in the next three years" (Thomas 2006). The telecommunication sector is a major user of offshored call centers, because it is a business where customer support has become essential, notably with the rise of complex multifunction handsets and multiple calling plans.

Many sectors unmentioned in the statistics above have successfully entered the offshoring era, although to a lesser extent. Offshoring leads to interesting patterns of "global sourcing" across the medical value chain. Ge *et al.* (2004: 3) report about US hospitals sending digitized X-rays in Israel, where radiologists make the diagnosis. Indian firms such as Wipro Technologies convert 2D MRI images into 3D images. Decision support systems are operated in Ireland. Pharmacy claims are processed in the Philippines. Indian typists transcribe physicians' voice recorded notes. Insurance companies may ask a panel of doctors in Israel for advice before expensive surgery authorization.

Architecture also is subjected to offshoring practices. Engardio *et al.* (2003) report about a San Francisco architect who transfers two-dimensional layouts in Budapest (Hungary) where a colleague makes up blueprints and 3D models he then delivers to the contractor. The architect in Hungary charges $18 an hour, while a US specialist

Table 6.4 ***Export-oriented FDI projects in call centers and shared service centers: distribution by industry (2002–2003)***

Business segments	*Call centers*		*Shared service centers*	
	Number of projects	*Percentage*	*Number of projects*	*Percentage*
IT and software, Internet	166	32	34	24.8
Business services	116	23	24	17.0
Telecom equipment and services	50	10	3	2.0
Electronics	42	8	6	4.4
Finance	30	6	40	29.0
Transport equipment	30	6	6	4.4
Energy	14	3	5	3.6
Other	65	12	20	14.8
Total	513	100	138	100

Source: Adapted from UNCTAD 2004

would charge $65. Remote skills are also required in large engineering projects, like those coordinated by Fluor Corp, the largest US engineering and construction corporation.

> Fluor Corp. employs 1,200 engineers and draftsmen in the Philippines, Poland, and India to turn layouts of giant industrial facilities into detailed specs and blueprints. For a multibillion-dollar petrochemical plant Fluor is designing in Saudi Arabia, a job requiring 50,000 separate construction plans, 200 young Filipino engineers earning less than $3,000 a year collaborate in real time with elite US and British engineers making up to $90,000 via Web portals.
>
> (Engardio *et al.* 2003: 59)

The breadth and diversity of a new global industry

The rise of call centers (CCs) and shared service centers (SSCs) goes far beyond the field of IT industries. It illustrates vividly the concept of global sourcing in service functions.

Call center operations

Technology improvements have revolutionized the old way of doing telephone-based marketing and customer relations, and have made the CC industry an enormous business, which employs millions of people worldwide (Table 6.5).

The search for productivity improvements has led firms to abandon their old switchboard facilities. CC operations are monitored through automatic call distribution (ACD) systems which dispatch incoming calls to available employees. Interactive voice response (IVR) systems sort out calls through automated recorded responses and voice recognition programs. Thanks to computer telephony integration (CTI), employees can

Table 6.5 ***Selected call center data***

Region	*Number of call centers*	*Agent positions (if not mentioned)—employees*
US (2004)	50,600	2.86 million
Canada (2004)	4,500	212,000
The Caribbean and Latin American (2003)		336,000
Mexico (2006)		33,500
UK (2004)	5,980	435,000 employees
South Africa		47,000
Morocco (2005)	180	18,000 25,000 employees
India (2005)	425	160,000 employees
Philippines (2004)		67,000
Australia (2002)	3,900	140,000 225,000 employees

Source: International Customer Management Institute (ICMA) (2006)

immediately refer to clients' files. The management of modern CCs requires complex software, using the mathematical theory of queueing (how to adapt the staff to achieve a given service). Multichannel "Web call centers" generate new labor requirements, shifting from language skills to keyboard capabilities.

CC activities may be roughly divided into two main categories. The treatment of inbound calls deals with customer claims and after-sales service for clients, and with technical support for professionals. Outbound calls include the broad and often unpopular area of telesales and telemarketing (customer surveys and market enquiries).[1]

Call center operations have been subjected to outsourcing tendencies. This trend has given birth to large national or transnational companies, not to mention numerous SMEs which operate on a regional basis. Convergys' 65,000 employees serve customers in 30 languages from 65 centers. It is headquartered in Cincinnati, Ohio and had $2.6 billion of sales in 2005. Stream, headquartered in Dallas, has 24 centers worldwide, with 10,000 agents. Paris-based SR.Teleperformance (Figure 6.2) is the world's second largest CRM outsourcer. It operates 266 facilities and 45,600 workstations in 39 countries, with annual sales in 2005 worth €1.3 billion (source: companies' websites, accessed 1 November 2006).

The CC industry, we have stressed, is labor-intensive. Even if some facilities employ highly skilled people, such as engineers, the bulk of the industry consists of routine tasks (telesales) or, at best, semi-skilled tasks (hotline). Therefore, given the fall of long-distance telephone costs, the sector has become a highly favorable field for low-cost-based overseas operations.

Language is a decisive location factor. India offers a pool of 300 million English speaking citizens. In 2005, India's CCs employ 160,000 people, working for foreign companies or Indian outsourcers (Outsource2india 2006). The Philippines has also become a preferred location for CCs. In 2004, the country's centers had 67,000 employees, up from a mere 2,000 in 2000 (Landingin 2005). South Africa, located in the European time-zone, is regarded increasingly favorably by companies which seek an English-speaking alternative to India. It offers also competencies in Dutch and German—Lufthansa (the major German airline) operates a CC in Cape Town (Benner 2006). US firms that want to serve Spanish-speaking customers are likely to locate or to outsource in Latin America, notably in Mexico. The expansion of Teleperformance-Mexico Group (formerly Merkafon), headquartered in Monterrey (5,860 workstations in seven sites) is mainly based on the rapidly growing market of American Hispanics (source: company's website).

Eastern Europe is a fast emerging destination for companies which target the European market. A young and rather cosmopolitan culture has made some capital cities such as Prague or Budapest very attractive places. For example, Genpact's employees in Budapest come from 29 countries and speak 15 languages. Because of long-established historical links with Germany, which materialize in millions of working emigrants in this country, Turkey now has plenty of German-speakers, and is becoming a target for German companies: in 2003, Lufthansa located a 130-agent CC in Istanbul (Van Marsh 2004).

France-based firms, which endeavor to offshore, face a quite narrow geographical scope, by comparison with US or UK firms. Given its proximity to France and local linguistic skills, Morocco has secured 50 percent of the French-speaking offshoring market, with about 25,000 employees, working in 180 call centers—including a recent

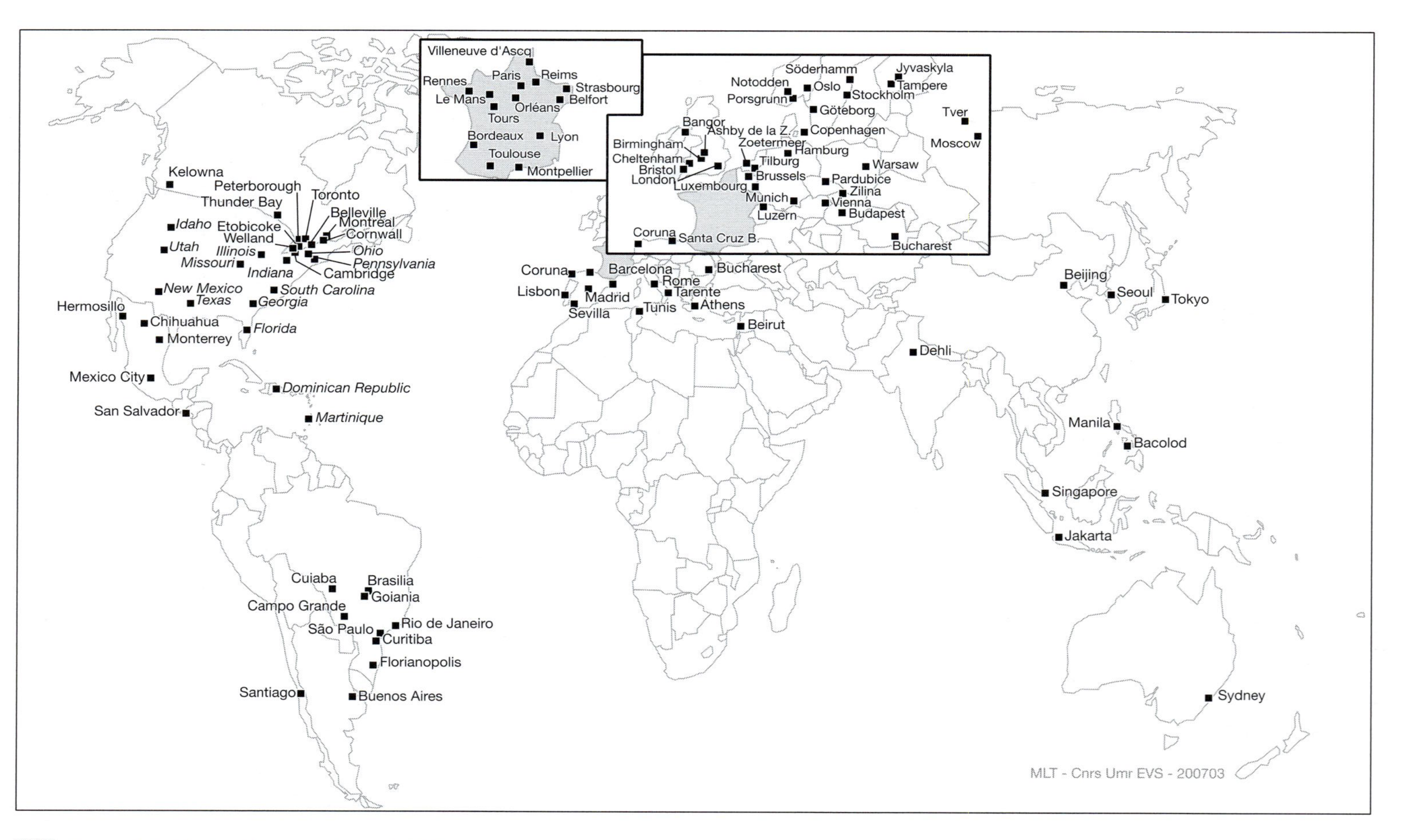

Figure 6.2 ***Worldwide operations of SR-Teleperformance, January 2007***

Source: Company's website

Dell facility (IT Maroc.com 2006). Tunisia, Senegal, and Mauritius are secondary locations.

Premium Contact Center International (PCCI), based in Dakar, Senegal (West Africa) is an archetypal examples of CC outsourcing and offshoring. Founded in 2002, the firm is supported by a syndicate of major French banks such as Société Générale and LCL (formerly Crédit Lyonnais), but the holding is located in London. The company's main output is made of prospecting/selling telephone calls for French corporations (which do not want to be named). PCCI is the leading private employer in Senegal. It recruited about 1,000 agents, most of them former students at the University of Dakar (a city where unemployment among graduates is over 30 percent). Dakar and Paris are in the same time zone, which makes it possible to work within French business hours. It also allows PCCI's staff to have their lunch break in front of the French TV broadcast: the main recruitment criterion is fluency with the least local accent. Employees must use French names when they are online with clients. In the Parisian commercial office, 5,000 km away, PCCI's clients—corporate executives—may see and listen to, in real time, African employees selling their products. Video conferences and the flow of calls travel through a trans-oceanic cable; the company rents 420 private lines for a $60,000 monthly fee (Josselin 2004).

Shared service centers (SSCs)

According to Oxford Intelligence and IBM Business Consulting Services (2004), 80 percent of Fortune 500 companies have now implemented, or have access to shared service centres. The growth of SSCs is the logical organizational and geographical outcome of value chain unbundling, vertical disintegration and horizontal reintegration processes, as presented in Chapter 4. Most companies now seek to concentrate on their core business, the series of tasks at the top of the "value hierarchy of processes" (Aron and Singh 2005). In a nutshell, a car maker would like to specialize in automobile design and manufacturing, not in secretarial work and computer maintenance. The digitalization of the value chain has generated the separability and tradability of a myriad of routine and semi-routine tasks which were previously embedded in the production process, and now may be identified as "modules" amenable to reorganization, outsourcing or relocation: payroll management, billing and invoice processing, data entry, accountancy, purchasing, and computer and telecom network maintenance. These tasks incorporate little tacit content and are easy to codify.

A survey of 248 European centers by Oxford Intelligence and IBM Business Consulting Services (2004) shows that financial functions are by far the main task performed by SSCs: 95 percent are involved in financial tasks as a primary or secondary function. Other functions are IT management (23 percent), human resource (21 percent), CRM (17 percent), logistics (7 percent), marketing (5 percent), and purchasing (4 percent).

Scale economies, and the search for lower operation costs in remote areas, are the *raison d'être* of SSCs. The geographic concentration of a company's clerical functions in a single facility gives the opportunity for a re-engineering of these functions: IT equipment and software may be consolidated in a single, more powerful integrated system, labor management and work distribution may be more efficient. Companies have implemented administrative sites since the beginning of the telephone era. The originality of the concept of SSCs is found: 1) in the geographic scope of the reorganization process, and 2) in the possible outsourcing of such peripheral activities.

1) Geographic: the advent of SSCs represents a considerable widening of the scale of operations. Of the 248 European SSCs mentioned above, eight percent are global (worldwide), more than a half are regional (pan-European), 16 percent are sub-regional (Northern Europe, Southern Europe), 21 percent are national (e.g., France, Germany). When the search for lower costs is taken into account in the consolidation process, there is not only concentration, but concomitantly a geographical shift to a nearshore or offshore location. While TNCs maintain high-tech, highly productive manufacturing lines or R&D centers in developed countries, they target low-wage countries to locate support functions which often require less skilled, less paid people.

2) Outsourcing: like call centers, SSCs fit the 2 × 2 matrix of outsourcing and offshoring presented in Figure 6.1. Companies depicted as global "business process outsourcers" have entered the market. IBM Global Service (190,000 employees worldwide) has in charge human resource functions Procter & Gamble had concentrated in three SSCs—see the case study below. Genpact—formerly GE Capital International Services, has 16 "global delivery centers" worldwide, manned by a staff of 23,000 (19,000 in India). It is expanding in China (Dalian and Changchun), the Philippines, and Eastern Europe (Budapest, Bucharest, and a project in Poland).

In 2006, Accenture (formerly Andersen Consulting) had 40 SSCs worldwide in 20 countries, with a combined revenue of $5 billion in BPO, 37 percent of the company's global sales. Accenture has a strong presence in India (14,000 employees), in China, and is rapidly strengthening in Eastern Europe with centers in Prague, Bratislava, Riga, Warsaw, and Bucharest, the newest one, which opened in Summer 2006 and is planned to host 1,000 professionals. Accenture's slogan, "the geography of high performance," illustrates its spatial strategy and the value-added it is likely to deliver to its clients. "The Accenture Global Delivery Network comprises carefully chosen strategic locations to match culture, language and time zone parameters with cost and skills requirements. With its offshore, nearshore and onshore options, the network has the redundancy to reduce geopolitical or even unforeseeable environmental risks" (Accenture 2006).

Offshoring measurement and consequences: a controversial subject

Offshoring measurement is a tricky issue where confusion reigns supreme. The word "offshoring" is often taken for "relocation" or geographical shift of existing activities to remote places, usually in developing countries. The subject has become politically sensitive. Aspray *et al.* (2006: 76) underline that public concern in the US was triggered by a Forrester Research (2002) report "indicating that 3.3 million service jobs would be lost to offshoring by 2015."

According to Farrell *et al.* (2005a), 2003 worldwide exports of service amounted to $1,697 billion, of which 3 percent ($51 billion) originated from low-cost locations. With a 2008 world market projected at $2,370 billion, a 30 percent compound annual

Box 6.1 A shared service center: Procter & Gamble's Global Business Services for the Americas in San José, Costa Rica

Global Business Services is Procter & Gamble's (P&G's) worldwide shared services organization. It provides back-office support to nearly 98,000 employees in over 80 countries and comprises three centres: Manila (Philippines), Newcastle (United Kingdom) and San José (Costa Rica). In the process of selecting these three cities as locations for its Global Business Services centres, the company reviewed more than 120 cities worldwide. The key reasons for choosing Costa Rica were the pool of highly educated and skilled labour, the country's long-standing democratic tradition, an attractive cost structure and an investment-friendly approach to foreign investors.

The San José service center started operations in late 1999. By 2004, it was providing 28 different services to 63,000 P&G employees in 22 countries in North and South America. This includes serving 58 plants and 15,000 retirees. Services delivered from Costa Rica include:

- Closing the books for 132 legal entities and managing 310 bank accounts in 35 different banks across 22 countries.
- Payroll and salary planning and compensation for 57,000 P&G employees.
- Annual processing of some 2.5 million invoices and managing accounts payables in the order of $24 billion.
- IT support to 5,000 sales representatives in the United States.

Global Business Services has created 1,300 high value-added service jobs in Costa Rica, thereby helping to mitigate the risk for brain drain from the country. The local operation has also promoted the transfer of skills through intensive training programmes. The company has "raised the bar" on recruiting and educational standards, reviewing over 12,000 résumés and requiring applicants to demonstrate proficiency in English and international accounting standards.

The company has recently become involved in global negotiations with strategic partners concerning the outsourcing of some functions previously handled by P&G Global Business Services. The first strategic partnership implemented was with Hewlett Packard with regard to IT support services starting 1 August 2003. From 1 August 2004, Hewlett Packard will also handle the accounts payable services. In November 2003, a real estate company, Jones Lang LaSalle, took charge of the facilities services and, since 1 January 2004, IBM has provided employee services to P&G. These partnerships will allow the centre in Costa Rica to attract higher value-added work more concentrated in its core business activities. P&G expects more and more sophisticated services to be handled by its Global Business Services centres in the future.

Source: UNCTAD (2004: 175)

growth rate (CAGR) from 2003 to 2008 would establish the world offshoring market of services at $237 billion, 10 percent of projected world service exports. The same authors surveyed eight major business sectors, totalling 244 million jobs (15 percent of the world's nonagricultural employment): package software, IT services, banking, insurance, pharmacy, automobile, health care, retail. They estimate that in 2003, 565,000 jobs in these sectors were performed in low-wage countries to serve demand from developed countries, 0.6 percent of developed country employment in the eight aforementioned industries; the highest share of offshored jobs, 7 percent, is found in IT services and packaged software. The number of would-be offshored jobs in these sectors is projected to be 1.2 million in 2008, or 1.3 percent of the total.

Extrapolating to the global economy, this report estimates that in 2003, 1.6 million jobs were resourced offshore by firms in developed countries, 0.5 percent of the global employment in services. This number was expected to increase to 4.1 million in 2008, a 22 percent CAGR. Offshoring assessment must also be done in terms of potential. This can result in enormous figures. In the eight sectors studied, Farrell *et al* (2005a) identified 18.3 million jobs "that could be performed anywhere in the world," equivalent to 7.5 percent of total employment in those sectors (Table 6.6). Blinder (2006) estimates that "the total number of current US service-sector jobs that will be susceptible to offshoring in the electronic future is two to three times the total number of current manufacturing jobs (which is about 14 million)."

How many jobs are lost to offshoring?

The measurement of the actual consequences of the phenomenon on employment in developed countries is such a tricky task that nobody knows by a reasonable margin of error the features, the statistical extent, and the geographic map of job losses (Aspray *et al.* 2006). The same authors regard most of the data from consultancies as "suspect."

> The studies measure different sectors of the economy use different measures of the extent of offshoring (e.g., jobs lost, percent of jobs lost, value of jobs

Table 6.6 ***Potential for employment offshoring worldwide in eight business sectors, 2008***

Industry	*Number of jobs amenable to offshoring (millions)*	*Share of employment amenable to offshoring (percent)*
Package software	0.3	49
IT services	2.8	44
Banking	3.2	25
Insurance	2.3	19
Pharmacy	0.2	13
Automobile	0.3	11
Health care	4.6	8
Retail	4.5	3
Total	18.3	

Source: Adapted from Farrell *et al.* (2005a: 35)

> or business lost), and different start and end points. Few of the studies provide details on how they developed their projections which makes it difficult to assess the reasonableness of the assumptions and the soundness of the methodology.
>
> (Aspray *et al.* 2006: 85)

However, offshoring consequences on OECD countries employment can hardly be evaluated by the evolution of official labor statistics, because the causal link between employment decline in a given sector and offshoring in the same sector is not clearly demonstrated. It is simplistic thinking to conclude that "one job performed offshore" equals "one worker fired or unemployed at home." The decrease of employment in a given industry is often the result of productivity growth, rather than of offshoring practices. In US call center activities for example, automation of inbound calls processing, consolidations, and the do-not-call registry are responsible for a large share of employment decline.

> Many of the jobs in India today are viable only in a low-wage environment and would not exist in the United States. That half a million people are now employed in India's outsourcing industry does not mean that there could be 500,000 more jobs in the United States. Without offshoring, companies would scale back or stop offering services like 24/7 customer help. Companies are also using technology to replace many of the jobs at risk in the United States. Automated voice response units are replacing call-center workers, online hotel and airline booking systems are replacing live operators and travel agents, and imaging software is replacing data-entry workers.
>
> (Baily and Farrell 2004: 7)

Indeed, occupations threatened by offshoring processes are often situated at the lower-end of the skills spectrum (Kirkegaard 2004). US employment data projected for 2014 show that a series of IT-based routine jobs is expected to decline: computer operators, data entry and information processing workers telephone operators, telemarketers, legal examiners, abstractors, and searchers (Table 6.7). The number of programmers is projected to remain stable. As a matter of fact, these occupations process tasks that constitute the typical offer of foreign—notably Indian—outsourcers.

However, even if we accept the hypothesis of employment declines—current and future—resulting from offshoring phenomena, they are of little numerical importance by comparison with the robust employment growth in highly skilled occupations which is presently observed, and which is foreseen in the medium term. The projected loss of 63,000 computer operators and data entry and information processing workers would be overcompensated with the nearly one million projected increase of computer specialists, notably software engineers, systems analysts and administrators, and database administrators (Hecker 2005). Likewise, declines in basic clerical jobs are by no means comparable to the overall projected growth of financial specialists.

All in all, the IT-based industries which are theoretically amenable to offshoring are among business sectors with the highest employment growth perspectives. The examples

Table 6.7 ***US employment 2004 and projected 2014 in selected IT-enabled occupations***

Codes	*Occupations*	*Employment*		*Change*	
		2004	*2014*	*Number*	*Percent*
15-1000	Computer specialists	3,046	4,003	957	31.4
15-1030	Including: computer software engineers	800	1,169	369	46.1
15-1021	Computer programmers	455	464	9	2.0
43-2021	Telephone operators	39	25	–14	–35.7
43-9011	Computer operators	182	151	–31	–16.7
43-9020	Data entry and information processing workers	525	493	–32	–6.1
13-2000	Financial specialists	2,497	2,941	444	17.8
43-4041	Credit authorizers, checkers, and clerks	67	39	–27	–41.2
43-4071	File clerks	255	163	–93	–36.3
43-4151	Order clerks	293	230	–63	–21.4

Source: Selected from Hecker (2005: 82), data in thousands

of offshoring in health services such as teleradiology, however spectacular they are, remain anecdotal by comparison with the massive growth which is expected in the medical sector for the next decade in wealthy countries. In the context of demographic ageing which characterizes Europe, Japan, and, to a lesser extent, the US, the real threat is not the shortage of available jobs, but the shortage of talent, which will compel health institutions and firms to resort to both the offshoring of tasks which do not require face-to-face interaction *and* the immigration of foreign professionals.

Finally, offshoring contributes to the global process of "job churning," that is to say the decline of some low-paid, routine-oriented jobs, which are replaced by automated digital systems or foreign low-paid workers, while overall employment in tasks in which more "richness" is embedded, whether they are amenable to distance processing or not, continues to grow steadily, notably in the US.

The geography of service offshoring

Where do offshored business processes go? Companies which seek to locate or to outsource IT and ITES abroad face a large choice of destinations. UNCTAD (2004) gives a preliminary estimation. In 2002 and 2003, Asia was the most important region for export-oriented foreign direct investments (EO-FDI) in call centers, SSCs, and IT services (516 projects, out of a total of 1,284). Latin America, the Caribbean, Eastern Europe, and Africa stayed far behind, but since have seen some promising development; 18 countries have attracted 79 percent of the projects, nearly equally distributed between call centers, shared service centers and IT services. Figure 6.3 shows the very global footprint of the phenomenon: UNCTAD found 78 host countries for EO-FDI projects, including 55 developing countries.

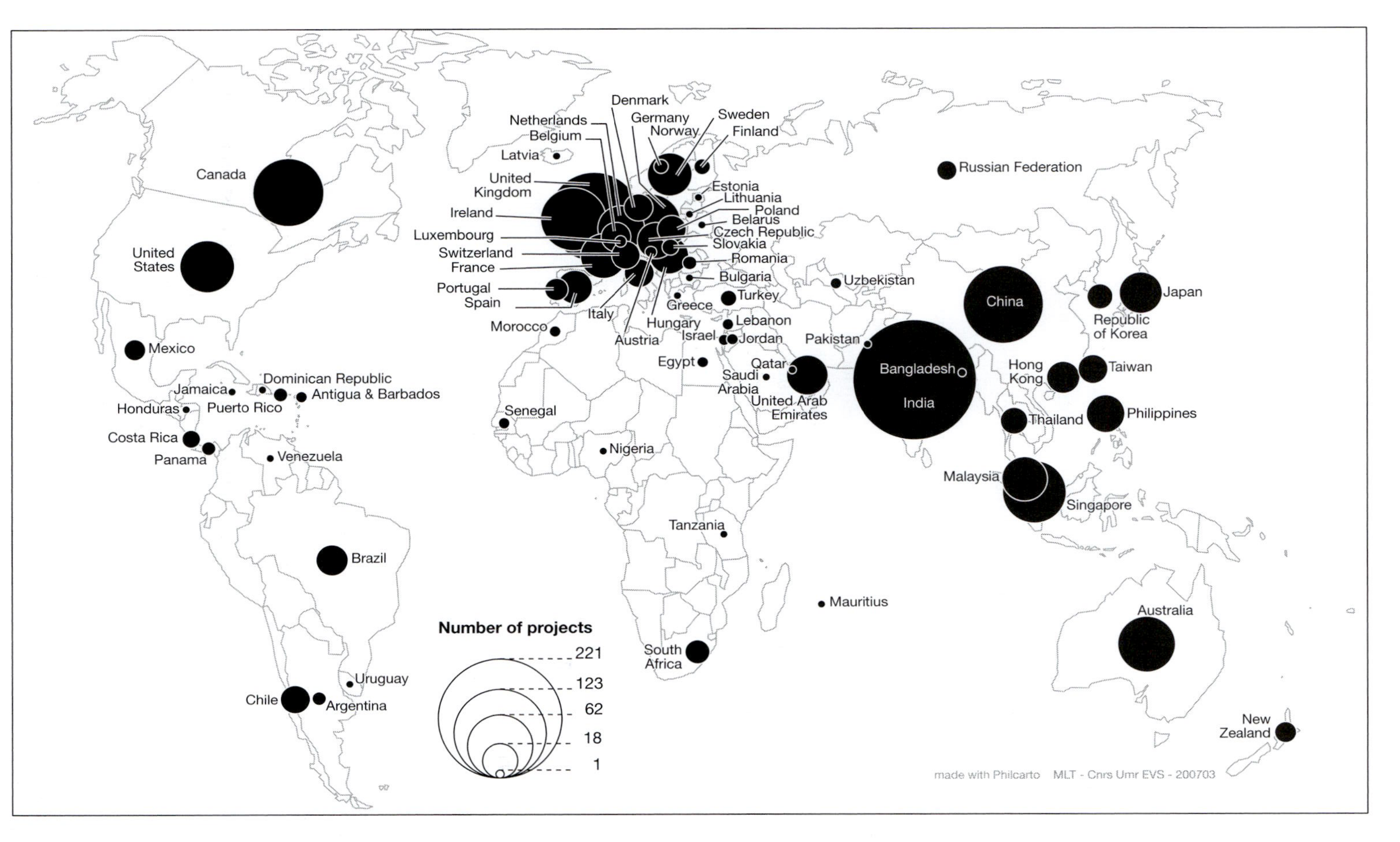

Figure 6.3 ***Host countries of export-oriented FDI projects in call centers, SSCs, and IT services, years 2002–2003***

Source: UNCTAD (2004: 162)

Aspray *et al.* (2006) suggest four categories of countries as ITES offshoring destinations:

1) those that have a large capacity of highly educated workers and have a low wage scale (e.g., India, China);
2) those that have special language skills (e.g., the Philippines can serve the English and Spanish customer service needs of the United States by being bilingual in these languages);
3) those that have geographic proximity ("nearsourcing"), familiarity with the work language and customs, and relatively low wages compared to the country sending the work (e.g. Canada accepting work from the United States, the Czech Republic accepting work from Germany); and
4) special high-end skills (e.g., Israeli strength in security and anti-virus software).

The graph by A.T. Kearney (2004: 4) presents the 25 "most attractive offshore destinations for services" (Figure 6.4). The first interest of the figure is to show the tradeoff between costs on the one hand, and skills and business environment on the other hand. Cost is a relative concept: Canada is a cheap location by comparison to its American neighbor. Ireland has a highly skilled workforce, but has passed its prime as an offshoring destination, since labor costs have surpassed those of all the other members of the European Union, except Luxembourg. At the other end of the spectrum, we are witnessing the emergence of new places whose development, up to now, has been plagued by poor business conditions—weak infrastructure, hazardous political environment—but present attractive features, in terms of cost, and/or human resources: Russia, Turkey, Vietnam.

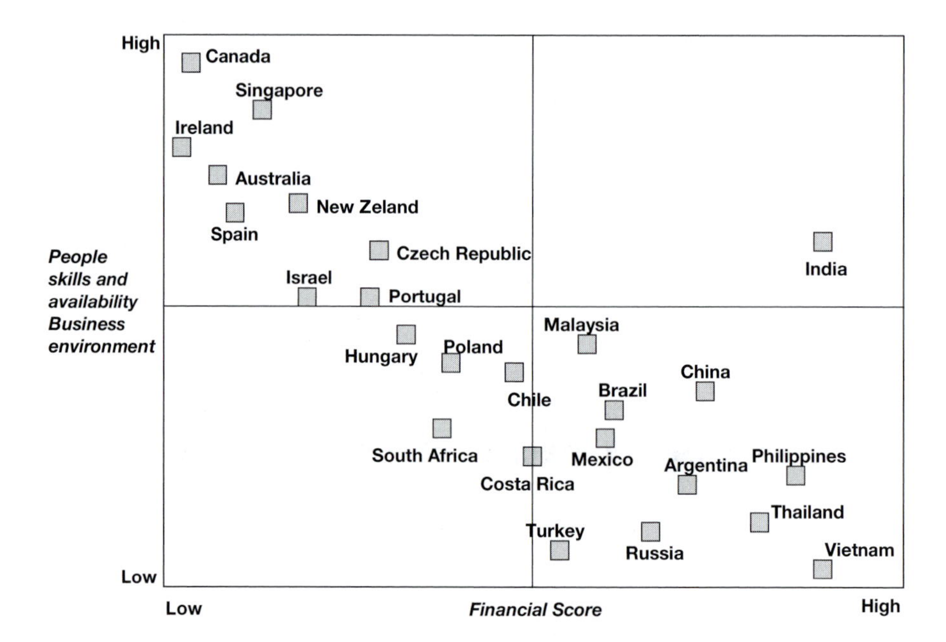

Figure 6.4 ***The 25 most attractive offshore destinations for services***

Source: Adapted from A.T. Kearney (2004: 4)

Several countries that are absent in the graph may be added: Belarus, Romania, and Cambodia.

Another advantage of Figure 6.4 is to show how unique the situation of India is, and to explain, finally, why India is by far the world leader of ITES offshoring, thanks to the concomitant presence of low labor costs *and* of an enormous, well-educated workforce, including top level engineers and scientists. This exception deserves a full section, which we present on p. 142.

China, which has a massive and cheap workforce, too, is in the same category, but lags behind. This fact contrasts with its status of "world manufacturer" (an area in which India, conversely, remains far behind China). Two features, rooted in recent history, explain part of China's rather timid entry into IT-enabled service exports:

1) the recent, enormous educational effort at university level, mainly state-led and technocratic, has so far produced a mass of professionals with a fairly good theoretical knowledge, but little actual experience of the business environment, notably because of a lack of vocational training (*The Economist*, 2006a); and
2) Chinese professionals often lack effective fluency in English, which is a key feature of most of their Indian counterparts.

Nevertheless, pundits foresee China catching up with India in the middle term, at least in high-end, technological specialities. Figure 6.5 suggests that China has already twice as many young engineers as India.

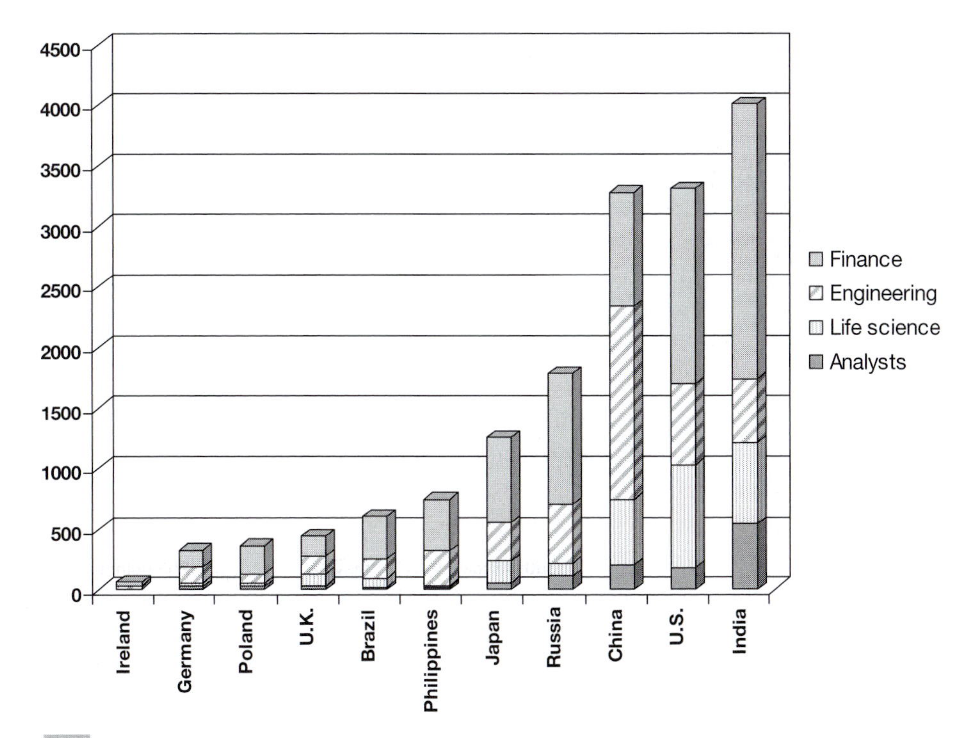

Figure 6.5 ***Labor supply in some occupational groups: number of young professionals with less than seven years experience (2003, thousands)***

Source: Adapted from Farrell *et al.* (2005b: 37)

The role of distance: offshoring or nearshoring?

A.T. Kearney's global index cannot render the full reality of business offshoring, because enterprises in diverse activities, and from diverse countries, have very different needs. A locational decision must take into account the full complexity of the business–geography interlinkage.

Technology has made the marginal cost of distance for the transportation of digitalized information fall to nearly zero. Nevertheless, offshoring practices have not become distance-neutral. Distance increases contracting, control, training and management costs, notably through additional travel time and expense for executives, or expatriation costs for managers. Companies which create captive offshore sites often have to send local employees to facilities in the home country for several weeks of training. Distance also may raise investments and maintenance costs, because most of the hardware is imported from developed countries.

Cultural distances have even heavier consequences than geographical distance. Dell has notoriously suffered from a backlash by dissatisfied customers who could not tolerate the accent of Indian call center agents. The imperatives of cultural distance are especially important in voice-based activities (call centers) where linguistic skills are critical. But the cultural issue is not confined to the linguistic problem. Some basic knowledge or "feeling" of the client's culture is required. Naming Indian or African employees Nancy or Patrick often proves insufficient, and is not enough to lure customers, who easily uncover the trick. Call center agents who ignore client culture and customs rely entirely on a script—and must watch television shows to learn the accents and idioms of the country of customers. The door is fully open, therefore, for dissatisfaction and complaints.

For this series of reasons, companies prefer locations geographically and culturally close to their home country, a practice pundits describe as *nearshoring*. Near Shore(r) was registered as a trademark in 1996 by Softtek, a Mexico-based outsourcer which made location close to the US market and the benefits of the North America Free Trade Agreement (NAFTA) its main marketing argument (www.softtek.com, www.nearshore.com).

Companies that go nearshore not only minimize distance-based costs, and time difference problems, but often find a business culture closer to those in their home country, and abundant linguistic skills. Canada, Mexico and the Caribbean are *par excellence* the nearshore destinations of US firms. In the Czech Republic or the Baltic States, German companies not only find large numbers of German speakers and skilled IT people, but also the German way of doing business. Indian outsourcers themselves are now locating call centers and SSCs in Eastern Europe, in order to be more closely attuned to the needs of European clients.

> "The quality of work matches and sometimes even surpasses the best we do in India," says Amitabh Chaudhry, chief operating officer of Progeon, a division of India's Infosys. It has opened a centre in the Czech city of Brno, with roughly 100 people working in 13 languages. Mr Chaudhry also highlights the

> friendly time-zone, good political and regulatory environment, multilingual workers and "cultural affinity". "If I want a small controllable team of telesales people, or back-office workers with a cultural fit to the continental EU or Britain, then it can be much more appropriate to choose an east European country."
>
> (*The Economist* 2005d: 66)

The Irish Republic, now a wealthy and costly place, is no longer considered as a prime offshoring destination. On the contrary, Northern Ireland, a part of the UK which lags behind in terms of overall development and associated costs, has become a much desirable spot for both UK and foreign-owned investments in ITES (including those by Indian service providers—see below). Another striking example of the role played by history and geography in current nearshoring processes is Dalian, a north-eastern Chinese city strategically located on the Yellow Sea, where many locals speak Japanese and Korean, and which has become, therefore, a prime destination for Japanese and Korean firms (*The Economist* 2006a).

More and more peripheral

Given the skyrocketing growth of service offshoring in the last decade (30 percent annually on average), even the largest labor pools show signs of labor shortage, which results in rising salaries and growing attrition rates. The labor markets of IT and ITES-suitable people in first-tier business cities of India, China, and Mexico have become over-heated. Employers have had to diversify their locations, targeting more peripheral countries, such as Mauritius, Vietnam, and Cambodia, or less central cities.

Softtek has a delivery center in Aguascalientes, a mid-sized city halfway between Mexico City and Monterrey, where it boasts an attrition rate of 4 percent (while 30 to 45 percent are common in the business). In India, activities are growing in cities usually unknown to Westerners such as Chandigarh, Indore, Ahmedabad, Cochin, Jaipur, Rajkot, and Nagpur (Bhagowati 2005). A similar phenomenon is seen in China, where the costs in inland cities such as Xian, Chengdu, and Wuhan are 40–50 percent lower than in coastal metropolises such as Beijing, Shanghai, and Guandzou (*The Economist* 2006a).

An important factor induces companies to discover new, unspoilt labor pools: the "first-mover advantage." All other things equal, as long as a company remain alone or with very few followers, it is likely to receive "the red carpet treatment" from local authorities, and the benefit of low absenteeism, low attrition rates, and low salary pressures. This is typically the case of PCCI in Dakar (Senegal) mentioned above.

However, secondary locations face two major problems: poor infrastructure, and lack of senior and middle management employees, who often do not want to relocate away from major cities (this problem also arises in developed countries). Therefore, exotic locations have so far drawn mainly routine occupations, such as data entry or low-end call center activities. And these places do usually not offer the potential for scalability that only a handful of metropolises in large countries provide to global corporations and major service vendors.

A focus on India

The emergence of India as a major power of the digital economy has received wide media coverage. Public attention has been caught by the contrast between, on the one hand, mass poverty in a mostly rural country with a billion-plus population and, on the other hand, fast growing, impressive performance in IT-based industries. Kash *et al.* (2004) suggest the existence of an "exceptional development pattern." What is exceptional, indeed, is that India challenges advanced countries in IT and ITES industries while remaining a very poor country by nearly all standards. The average annual income per capita in 2004 stood at $630. The *Human Development Report 2005* ranks India 127 out of 177 countries, behind Gabon, Morocco, and Namibia.

Nevertheless, India's IT economy presents some features that usually characterize advanced economies. In 2005, India was the world's eighth-largest exporter of "other commercial services" (which exclude travel and transportation, and include mostly IT and ITES exports). Given its exceptional current growth rate, it should become fifth in 2007 or 2008. Actually, only the US, the UK, and Germany are unsurpassable in the short term (Table 6.8). India is the undisputed leader for export-oriented FDI in IT services and ITES and dominates the market of offshored BPO. From 1999–2000 to 2005–2006, the number of IT and ITES workers increased fivefold, from 284,000 to 1,287,000. The *NASSCOM-McKinsey Study 2005* foresees that by 2010, Indian exports of IT and IT-enabled services will increase to about $60 billion.

As a matter of fact, India has become a "hot spot" for the world's major IT companies, which are on the way to implementing a major geographic redeployment of their operations (Table 6.9). India is increasingly central in IBM's operations and strategy (*The Economist* 2007b). IBM's local staff of 53,000 (in 2007) is surpassed by no country except the US. India is also the main place of operations outside the US for Accenture and Genpact. Among other industries, which see a huge foreign presence for back office processing or customer support, we must mention finance, computers (examples of firms in Table 6.9). In 2006, Dell released plans to increase its presence in India to nearly 20,000 people, notably for customer support and transaction processing (Kaka 2006).

Even more remarkable is the emergence of Indian companies as major players in the business of IT services and BPO. This is very different from Ireland, where the IT

Table 6.8 ***Leading exporters of "other commercial services" (2005, billion dollars)***

United States	188.8
United Kingdom	127.1
Germany	81.4
Japan	59.7
France	47.7
Ireland	46.1
Netherlands	45.1
India	*43.8*

Source: WTO (2006)

Table 6.9 ***A sample of leading foreign firms with offshored activities in India***

Company	*Market segment*	*Indian workforce 2006—planned*
IBM	IT services and software	53,000 (2007)
Genpact	BPO	19,000
Oracle	IT services and software	17,000
Accenture	BPO	16,500—30,000
HP	IT services, computers	15,000 (2005)
Dell	Computers	10,000—20,000 (2009)
HSBC (UK)	Finance	10,000
EDS	IT services and BPO	8,700* (2004)
Siemens (Germany)	IT services and software	4,000
Cap Gemini (France)	IT services	6,000—10,000
Microsoft	software	4,000—7,000 (2010)
Covansys	IT services, BPO	5,700
American Express	Finance	4,000 (2005)
Standard Chartered (UK)	Finance	4,500 (2004)
JP Morgan Chase	Finance	4,500 (2005)—9,000 (2007)
SAP (Germany)	IT services and software	2,800—5,600 (2011)
Cisco	Network equipment	2,300—4000
Apple	Computers and electr.	1,500—3000
Delta Airlines	Transportation	1,000 (2004)
Bank of America	Finance	1,000 (2004)

Source: Web research. If nationality is not mentioned: US firms

* before acquisition of Mphasis, an Indian outsourcer

Table 6.10 ***Major Indian IT companies***

	Headquarters	*Revenue 2005–2006 ($ millions)*	*Employees*
HCL (Hindustan Computers Ltd)	Noida	3,300	34,000
TCS (Tata Consultancy Services)	Mumbai	2,970	70,000
Wipro Technologies	Bangalore	2,390	50,100
Infosys Technologies	Bangalore	2,150	58,000
Satyam Computer Services Ltd	Hyderabad	1,430	29,800
Cognizant Technology Solutions	Teaneck (NJ), US	885	32,000
Xansa	Reading, UK	674	8,000
Patni Computer Systems	Mumbai	450	12,000
i-Flex Solutions Ltd	Mumbai	320	7,000
24/7 Customers	Bangalore	—	7,000
ICICI Onesource	Mumbai	123	8,000

Source: Companies' websites

industry is dominated by foreign firms (Grimes 2006). Five India-headquartered IT companies have had revenue for fiscal year 2005–2006 exceeding one billion dollars: HCL, TCS, Wipro, Infosys, and Satyam (Table 6.10). In software, of the top 20 exporters in 2001–2002, only five were subsidiaries of foreign firms (Athreye 2005). Several Indian companies have achieved international recognition for quality of service: 17 Indian companies are ranked among the "50 best managed global outsourcing vendors" by *The Black Book of Outsourcing* (Brown and Wilson 2005), with six firms in the top ten: Satyam, Cognizant, Infosys, Patni, TCS, and HCL.

Up to now, the Indian IT industry has located in a small number of cities. Five city-regions concentrate 83 percent of the headquarters of NASSCOM's 980 members: the National Capital Region (23.1 percent, in Delhi, Gurgaon, and Noida), Bangalore (22.6 percent), Mumbai (17.5 percent), Chennai (10.4 percent), and Hyderabad (9.7 percent) (NASSCOM 2006).[2] This spatial concentration is understandable. Given the country's average level of development, very few cities offer the business environment which is suitable to IT industries, notably in terms of infrastructures: telecommunication networks, roads, and above all, electricity, still today a major weakness of India. Another major point is the availability of talent: the best and largest universities and engineering schools are concentrated in the aforementioned major cities.

Why has India become a major player in IT services and BPO?

India benefits from low costs. But Bangladesh and Vietnam are even cheaper, and present no similar achievement. The roots of India's success must be found in a series of features that are unique among developing countries: it is a democratic country (despite a history of strong central planning), it never fell into the communist sphere, and it kept alive a local elite under the British colonial domination, whose linguistic inheritance plays a critical role today. India developed since its independence a strong local culture of capitalism and entrepreneurship, which grew under the umbrella of state regulation and protectionism, and is epitomized by the existence of large family-owned groups such as Tata Group, the parent company of TCS.

The existence of a local elite proved a fundamental factor. The development of an Indian higher education system was encouraged by the British occupants. Also a legacy of colonial domination was the custom to send young local elites to follow courses in the best European institutions, mainly in the UK. Nowadays, Indian IT companies host plenty of managers and professionals who have been educated and trained for several years in the US or the UK. Many of them have returned to India with skills and knowledge of international business as well as capital to invest (Saxenian 2006).

> India has a strong and highly competitive K-12 educational system emphasizing science and mathematics. Today the higher education system includes more than 300 universities, 15,000 colleges, and 5,000 training institutions. Some of the schools, such as the Indian Institutes of Technology and the Indian Institutes of Management are world-class; but the quality falls off rapidly after the top 15 schools.[3]

> Total bachelor and master degree production in the computing and electronics fields is approximately 75,000 per year. There are also some 350,000 students in other science and engineering fields at universities and polytechnics receiving degrees each year, and many of them enter the IT industry upon graduation.
>
> (Aspray *et al.* 2006: 35)

However, current results could not have been achieved without a deep reform of the heavy, bureaucratic regulation which had been inherited from "Indian socialism." Since 1991, a series of reforms have been successfully lobbied by the powerful NASSCOM, including:

- income tax exemption from profits of software exports;
- reduction of import duty on computer software (from 114 percent to zero);
- amendment of copyright laws;
- deregulation of the telecom market.

(NASSCOM 2006)

Limits and handicaps: is the Indian IT industry oversold?

It is not so spectacular, for a country of 1 billion inhabitants, to compete with a country of 4.2 million (Ireland) in terms of services exports. Major Indian IT players such as HCL, TCS, or Wipro are a tenth the size of world leaders such as IBM, HP, Dell, Oracle, and Microsoft. It is the contrast with the mass poverty that gives the success of Indian IT corporations a kind of "overbrilliance." Brand new IT parks in Bangalore or Hyderabad are intriguing to foreign eyes because we know that millions of city dwellers live in slums nearby. Enthusiasm must be tempered by quantitative and qualitative facts.

The first consideration is that the growth of India's IT industry could be not sustainable in the long term, because wage inflation is eroding its competitive advantage over other countries, and because labor shortage could occur in the short term. According to Farrell *et al.* (2005b), wages in India have inflated dramatically since the beginning of the 2000s. The annual wage for an entry-level software developer has increased from $4,080 in 2000 to $6,600 in 2004, and project manager salaries have rocketed from $13,600 to $31,100. The risk is that India may be caught in a pincer between: 1) emerging countries with extremely low costs for low skilled tasks, and 2) nearshore places or second-tier and third-tier low-cost cities in developed countries, which take advantage of geographical and cultural proximity.

Another threat is the shortage of talent. The Indian labor pool is not inexhaustible, especially in high value-added occupations (Sengupta 2006). Some projections estimate that the need of new qualified people to work in the IT and BPO industry will be 1.5 million by 2010, but that only 1 million suitable people will graduate during this period (*The Economist* 2005e). The McKinsey Global Institute conducted thorough analyses which suggest that in some places, the demand of highly skilled professionals already exceeds availability. In Hyderabad, the projected demand for new engineering graduates would exceed the supply by 38 percent in 2008 (Farrell *et al.* 2005c).

Another obstacle for India on the way to become a "global hub" of the new economy (Rajan 2005) is the lack of modern infrastructures. Power outages remain frequent, and firms must have their own generators to guarantee business continuity. Poor road urban networks and obsolete public transportation systems plague commuting (Hamm 2007). In recent years, for every dollar India has invested in roads, power generation, electric grid, seaports, and airports, China has spent seven (Bradsher 2006). In 2005, the federal government recognized the need for $150 billion to be spent by 2010 to "develop world-class infrastructure" (*India Daily* 2005a).

However, the big issue is how the IT industry as a whole could upgrade its business. Indian IT companies have so far proved low in innovativeness, remaining mainly service providers. The example of the software industry is revealing. Up to now, Indian IT companies have licensed very few software packages. They do "D rather than R" (Cukier 2005). This fact is recognized by Indian academics themselves. Indian service exports consist mainly "of low-level design, coding and testing" (Arora *et al.* 2001: 1273); "software development is still labour intensive and requires relatively little capital" (Arora and Athreye 2002: 255); "because there is little or no Indian intellectual content in the work, firms are unable to provide value beyond writing code" (Parthasarathy 2004: 677); "the Indian base for moving into the area of the innovation of complex technologies, appears strikingly narrow" (Kash *et al.* 2004: 795).

In such a perspective, the main obstacle faced by Indian companies is the lack of a strong domestic market, which compels local firms to rely on distant customers only. Because of remoteness from customers, Indian companies are unable to gain experience in design and marketing (Arora *et al.* 2001). The Indian IT industry also lacks Silicon Valley-like clustering effects, which are recognized as a key path to innovation (and which include market presence effects). The status of most Indian IT companies as service providers prevents them from creating and nurturing fruitful interfirm interaction. "In Bangalore, there were few incentives for firms to interact with one another or to create the regional institutions in which software production could be embedded" (Parthasarathy 2004: 682). And the great distance from markets makes difficult the close cooperation and brainstorming practices between clients and vendors which often characterize modern, innovative business. However, there is evidence of improvement. Grote and Taübe (2006: 1302) suggest that "upgrading to financial research might be a viable option for Indian IT clusters." Parthasarathy and Aoyama (2006: 1281) have found evidence in Bangalore of a significant upgrading "facilitated by active local entrepreneurship and supported by the gradual thickening of institutions." In Figure 6.6, we suggest that offshoring countries follows a normal upgrading process, which starts in first-tier business cities, and then spreads throughout the country—this is notoriously the case for manufacturing offshoring in China. The evolution of India IT clusters and companies epitomizes this trend.

Kash *et al.* (2004: 795) conclude that, to date, India's role in the world market rests on a "one-legged stool," or "a narrow base," especially because of its "strikingly poor performance as an exporter of manufactured goods." Indeed, if we consider the share of services in world trade, which has remained at about 20 percent for a decade (WTO), the country's path of development remains narrow. In Chapters 2 and 4, we suggested that the digital economy goes far beyond IT industries and ITES, but encompasses

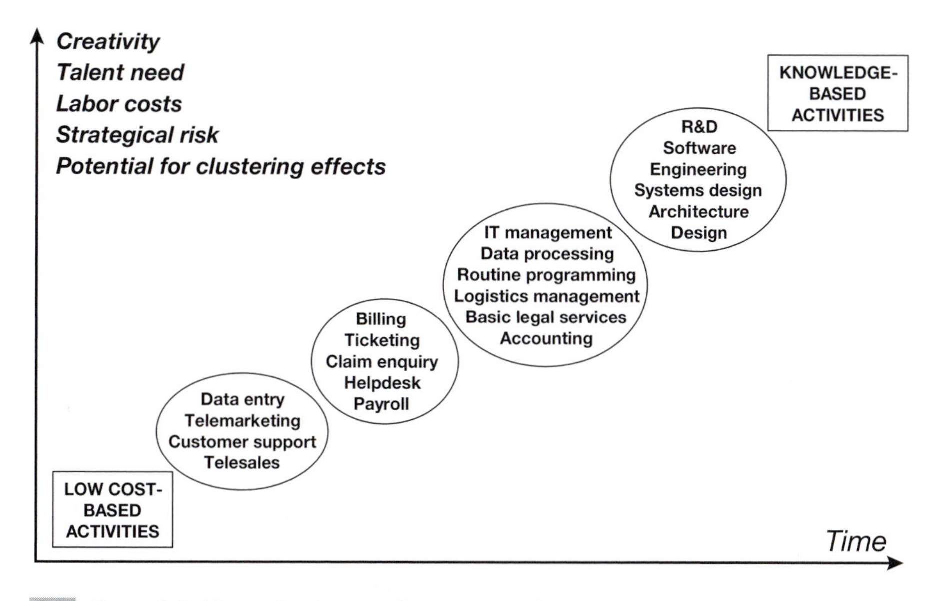

Figure 6.6 ***The regional upgrading process of IT-enabled industries***

Source: Authors' own research

manufacturing sectors such as capital equipment goods, aerospace, electronics, automobiles, and textiles.

The contrast between the slums of Kolkata or Mumbai and the so-called Indian Silicon Valley of Bangalore has fed typical "digital utopian" thinking, such as the leapfrogging phenomenon that would see such an immense, mostly rural country jumping straight into the informational, computational era. This is wishful thinking. What India lacks is literacy, good roads, modern utility networks, better water management, and mass commodity production. Kash *et al.* (2004) demonstrate that India remains virtually non existent—while China markedly emerges—in high-tech manufacturing sectors, which make the bulk of international trade and remain the economic base of industrial giants such as the US, Germany and Japan. There are good reasons for thinking that the IT sector will not be the decisive tool for India's global development.

However, it is not a question of "throwing the baby out with the bath water." India shows impressive, real IT developments, which make it a very intriguing case study of economic geography, notably given the local context of global poverty and remoteness from the core of the world economy.

Conclusion: will IT services and ITES go back to home countries?

The idea that an offshoring decision risks being a flop has been rampant in business circles since the end of the 1990s. It has been fuelled by actual examples. In Summer 2003, GE Capital announced it was repatriating its Australian Customer Service operations from New Delhi back to Melbourne (Kjellerup 2003). *The Economist* (2003) reported a debate about Dell's plan to move technical-support jobs from India back to Texas, after a surge of customer complaints. In 2005 and 2006, Dell has been assailed by bloggers reporting complaints from discontented customers.

As a matter of fact, 2006 has seen several important cases of *backshoring*, which challenge commonplace ideas about the collapsing of territory in business.

In June 2006, the whole Indian industry of ITES fell into perplexity when Apple suddenly announced it was closing its brand new Bangalore call center, where recruitment had begun (the facility was expected to employ 1,500 people at the end of 2006, with plans to expand to 3,000). Without any warning, fired employees were told their jobs were returning to the US (Kripalani and Burrows 2006).

On June 15, 2006, the British electricity company Powergen—a part of the German giant E.ON—announced the withdrawal of its Indian-based customer service functions. All the inbound calls must be transferred to UK-based call centers in Bedford, Bolton, Leicester, Nottingham, and Rayleigh. By the end of 2006, nearly 1,000 additional employees will have been recruited in these centers, 450 of these new roles being a direct consequence of the removal of operations from India. This change is a part of E.ON's policy of upgrading its CRM functions, by targeting service quality rather than focusing on low cost only. Powergen will also be returning outbound calling and back office work (E.ON 2006).

> Offshore call centres may have their place for certain industries. However, we believe that we can best achieve industry-leading customer service by operating solely in the UK. When customers contact us they need to be confident that their query will be fully resolved quickly. Although the cost of overseas outsourcing can be low, we're simply not prepared to achieve savings at the risk or expense of customer satisfaction.
>
> (Nick Horler, Managing Director of Powergen, in E.ON 2006)

The story of Abbey's call center in Bangalore presents similar features. Abbey was one in a wave of British financial companies that offshored a large number of back-office jobs in India in the first half of the 2000s. Abbey relocated operations in India in 2003, closing positions in UK call centers. But in 2004, Abbey was taken over in a £9.5 billion deal by Grupo Santander, a Spanish company, whose direction decided that the move was "a costly failure," because "customers were unhappy with the service from India, some complained of language difficulties." In the end, the workload will be transferred to Abbey call centres in Milton Keynes, Teesside, and Glasgow. "The group is said to believe that switching jobs to India does not necessarily cut costs—and may be more expensive in the long run because of reduced efficiency and customer complaints" (*India Daily* 2005b).

A common, intriguing feature of these two business stories is that the backshoring move was decided after a critical change of management, following a takeover by a foreign company. Perhaps this is an indication that the effective implementation of the offshoring model (like other branches of management in general) is not dictated by pure economic pragmatism only, but also by business culture—which varies from one country to another—and subjective considerations, including the "bandwagon effect." Likewise, many pundits and managers regard offshoring as a speculative, hazardous change, whose long-term returns are uncertain. Customer satisfaction and the risk of damage to the firm's image are notoriously difficult to calculate. In the end, the benefits of offshoring—especially outsourced offshoring—in terms of productivity and overall returns remain uncertain, and "much seems to depend on both sector and firm-specific characteristics" (Olsen 2006: 28).

At the same time, we are witnessing the interesting phenomenon of "reverse offshoring," notably by Indian firms that seek to increase their market share and to compete with Western companies by locating closer to their clients, in Europe or in America. In June 2006, ICICI OneSource announced it was planning to create 1,000 jobs in Northern Ireland over the next two years (ICICI OneSource 2006). We mentioned above the establishment of Infosys in Brno, Czech Republic. Even more intriguing are some cases of "reverse brain drain." Indian firms have started to hire US expatriates, and to train young American graduates at home. More than 10,000 American citizens reportedly work in IT companies in India (Lewis 2006). This phenomenon partly derives from immigration issues, notably recent restrictions on H1-B visas for foreign professionals (TCS only has a US staff of 9,500, two-thirds of whom are Indian people with foreign visas). Therefore, foreign firms have had to change their strategy, and seek to recruit and train at home.

> Infosys Technologies Ltd. will spend $100 million over the next year to hire and train 25,000 workers and college graduates culled from around the world, including from Massachusetts Institute of Technology and Harvard University. Tata Consultancy Services Ltd. of Bangalore will add 30,500 employees over the next year, including 1,000 from the United States.
>
> (Lewis 2006)

The title of this section is provocative. The examples above do not mean that a complete reversal of current flows, or, at least, a situation of equilibrium, will occur in the short term. If companies such as Dell seek to improve customer support, they will have to recruit more agents. The solution is the upgrading of Indian or Filipino employees' skills, rather than backshoring in costly areas (Kaka 2006). Nevertheless, we are invited to abandon simplistic, one-way thinking in offshoring analysis. Reverse offshoring and reverse brain drain are elements of a wider phenomenon, that is the rise of some emerging countries and their firms as huge economic powers and active players in global business (Woodall 2006).

In this context, there is little probability that the most fearful scenarios will materialize, because place and territory remain powerful components of business, even in the domain of intangible goods and IT-enabled processes. The growth of local markets in developing countries will give business opportunities, not only to local outsourcers, but also to Western companies which are able to design high-quality, innovative goods and services. Companies from developed countries which want to keep back office and customer support at home at reasonable costs may target small and medium-sized communities. Some of them have even implemented "virtual call centers" operated by home-based teleworkers. The growth of telework, as an actual complement to offshoring, is another intriguing feature of the current splintering of the economic space.

TELEWORK/ TELECOMMUTING

Time and space flexibilities in work and business organization

7

The words *telework* and *telecommuting*—almost synonymous—were coined in 1973 by Jack Nilles, a physicist and missile expert, when he was director of interdisciplinary research at the University of Southern California. They were also popularized by Francis Kinsman in *The Telecommuters* (1987). The importance of the problematic of telework for geographers and planners is captured by the domain name of ITAC (International Telework Association and Council): *www.workingfromanywhere.org*. Indeed, the very nature of telework, from which results its social and economic impacts, is to provide firms and workers with time and space flexibilities in labor organization (Moriset 2004).

Telework is a key element of the cyber-utopia promoted in the 1990s by "cybergurus" such as Nicholas Negroponte, founder of the Massachusetts Institute of Technology (MIT) "Media Lab" and *Wired* magazine, and author of *Being Digital* (1995). Telework would critically diminish the constraints of social synchronism that requires people to crowd in the same place at the same time. Traffic jams and overcrowded buses and metros would disappear. Transportation infrastructures, office buildings, and holiday resorts would no longer have to be oversized in order to meet the demand during peak hours or holidays. Because face-to-face meetings would no longer provide any advantage, the need to build or to live in cities would vanish. Salomon (1998: 34) suggests that telework "offers a 'romantic' image of returning to the pre-industrial family or community-centered society." Hence the "myth of the electronic cottage" (Forester 1988).

Since the beginning of the 1970s, attention to the phenomenon in political, academic, and business circles has been fueled by the rampant environment and energy crisis. In the 1990s, the IT hype reinforced interest in telework. Salomon (1998) depicts the growth of "cookbook monographs" and the existence of "snowball citation methods." However, since the crisis of the digital economy in 2000–2001, utopian discourses about telework have largely faded. Media coverage has decreased, and some analysts have sometimes regarded it as an old-fashioned, declining practice. Most telecottage-like experiences—notably in the UK and Scandinavian countries—have failed (Blanc 1998). On 25 June 2001, a *USA Today* headline read: "Telecommuting Gets Stuck in the Slow Lane: Working from home loses appeal for harried employees and skeptical bosses."

The apparent crisis of telework derives from a general misunderstanding. In the late 1990s, popular and scientific literature have focused on the newsworthy version of telework, that is rural small office-home office or SOHO (self-employed, home teleworkers), and various kinds of micro-enterprises (Clark 2000; Gillespie and Richardson 1996; Molini 1997; Moriset 2003a).

At the turn of the century, this version of telework was suffering from such heavy drawbacks that it was regarded as doomed to failure or marginality. Cheap broadband lines in peripheral areas were not available. Internet access was not yet a commodity. Both firms and unions were not keen to encourage a practice which could weaken well-established networks of power inside their organizations. In fact, the restrictive, idealistic image of "lone eagles and high fliers" in rural places (Beyers and Lindahl 1996) must not obscure the reality: a majority of teleworkers has always been—and remains—made of urban people, who telecommute only part-time.

Nevertheless, Gordon (2004) is right to say that "telecommuting is alive and well and growing." As a matter of fact, the rise of telework is a fundamental economic trend, driven forward by several intertwined factors:

- the organizational patterns of production and labor in the digital economy
- the coming to maturity (in terms of affordability and technical efficiency) of home office and mobile technologies we have studied in Chapter 3 (DSL lines, personal computers, laptops, wireless connections etc.)
- emerging features of urban society and geography
- the global context of energy-related, environmental and political threats.

A "multifaceted" concept which confounds analysis

A problem of definition

> The lack of a concise and universally-accepted definition of telecommuting has confounded research and policy-making since the beginning . . . ambiguity is inevitable precisely because of the complex, multifaceted nature of telecommuting as a social phenomenon, and because of the volatility of the technological, institutional, and social environment in which it is occurring.
>
> (Mokhtarian *et al*. 2005: 425)

Telework and telecommuting remain fuzzy concepts. However, there is a lowest common denominator in all definitions: they refer to IT-based work performed "at a distance" from the normal workplace (Mokhtarian *et al.* 2005: 426). But the issue is not so simple. A large fraction of people who actually telework do not fit telework measurement criteria. This is the case of managers who connect to the company's intranet after office hours. Likewise, millions of teachers who connect from home to the Internet to prepare a course—etymologically speaking, they telework—will not be counted as teleworkers. But disabled teachers who work at home for telelearning institutions will.

A precise definition of telework/telecommuting must be operational, in the way that it captures its actual social and economic significance, with regard, notably, to the transportation issue. From this point of view, we recommend the definition suggested by Benhamou *et al.* (1998: 5): "the partial or total substitution of telecommunications technology for the trip to and from the primary workplace along with the associated changes in policy, organisation, management and work structure." The key word in this definition is *substitution*.

According to Breton[1] (1995):

- telework must be a customary (full- or part-time) labor practice;
- it must be done during open hours; this is consistent with Mokhtarian *et al.* (2005: 427) who "do not consider after-hours work to be telecommuting, if the employee still spends a full day at the regular workplace";
- it is executed outside the employer or customer premises and out of physical control from clients, managers or supervisors; and
- it requires the use of telecommunications and a computer.

Therefore, telework does not encompass all kinds of teleservices and e-business. Call center employees, bank traders, or managers who hang on all day to the telephone, because they commute daily, are not regarded as teleworkers. Neither are consultants or salesmen who work inside client offices. Therefore, from a locational point of view, we may identify four categories of telework practices (which often enter into combination):

- *home-based telework*—the most widely known—is practiced by salaried or self-employed people; this is typically a case for using the word "telecommuting" (Allenby and Roitz 2003);
- *nomadic work* considers professionals working with computers and telecommunications in locations other than normal office, clients' premises, and home (vehicles, hotels, airports), or during vacation stays;
- *mobile work* is a subset of nomadic telework, which considers teleworking during a journey, using wireless connections: typically, a manager aboard a train who connects to his company's intranet.
- *work in telecenters* (or "telecottages") and *satellite offices*, which are working facilities specifically designed and equipped for clerical tasks by public authorities or corporations, in order to free employees from exhausting telecommuting, or to provide self-employed people with technical support and shared, high-grade equipment they cannot afford on their own (broadband line, laser color copier and printer).

A detailed taxonomy of telework should also take into account organizational patterns. The main distinction is between "organizationally-affiliated" (employees) and self-employed, or "freelance" (Garrett and Danziger 2007). If we add the variable frequency of the practice, the vast range of task performed, and the way IT is used, we obtain a complex area.

Mokhtarian *et al.* (2005: 426) have suggested that "the definition of telecommuting should be narrower than that of teleworking in the broadest sense." However, as Mokhtarian *et al.* (2005: 426) themselves acknowledge, the common language, as well as the US administration and legislators, have made telework and telecommuting almost synonymous.

> Following are terms and definitions used in and for the purposes of this bulletin: *Telework and telecommuting are used interchangeably* and are defined as the act of performing all or a portion of work functions at an alternative worksite, such as working from home or a telework center, under circumstances that reduce or eliminate the employee's commute.
>
> (General Services Administration 2006: 13845) (emphasis added)

"Telecommuting" is more widely in use in the US, while people and institutions in Europe often use "telework" for the same purpose. Finally, uses and definitions are a matter of context, "one reason for the numerous definitions of telecommuting—and a reason why consensus is unlikely—is that the 'best' definition varies with the focus of interest" (Mokhtarian *et al.* 2005: 426). People interested in transportation issues often use telecommuting. Those interested in labor organization—notably corporate management—preferably use "flexible" or "alternative work arrangement," which sound less romantic and utopian.

A hardly measureable, albeit massive phenomenon?

How to measure (and measure the impact of) something people do not do, without entering in hazardous speculations? Travel flows, office work, are easy to survey efficiently, because these are positive phenomena. But telework is intrinsically a non-phenomenon, which is resistant to measurement.

This problem has been thoroughly addressed by Mokhtarian *et al.* (2005). Telework measurement is a tricky task complicated by the lack of a unanimously recognized definition. Because the measurement is usually performed through surveys, an unknown number of teleworkers remains hidden. The methodology of popular surveys, like those conducted by ITAC varies from year to year. The most critical issue is the frequency threshold: how many times a week (a month, a year) must people telework to be considered by surveys? The most popular threshold seems to be one day per month (Pratt 1999). Ory and Mokhtarian (2005) require "at least two days a month for a period of at least three consecutive months." Therefore, figures vary enormously and the margin of error remains large (Table 7.1).

Table 7.1 ***Example of conflicting figures in home-based work measurement***

ITAC 2003	*Bureau of Labor Statistics 2005*
Today ITAC . . . revealed significant increases in home-based telework in the United States . . . The number of employed Americans who work from home during business hours at least one day per month *has increased by nearly 40 percent since 2001; for the self-employed, the equivalent increase is almost 18 percent (emphasis in original).*	These workers, who reported working at home at least once per week, accounted for about 15 percent of total nonagricultural employment in May 2004, *essentially the same percentage as in May 2001 (emphasis in original).*

Source: ITAC (2003); Bureau of Labor Statistics (2005)

In the US, perhaps the most reliable source is an extensive survey of work at home by the Bureau of Labor Statistics (BLS) (2005). As of May 2004, it identified 20.7 million people who did some work at home at least once a week, 15.1 percent of non-agricultural, civilian employment (Table 7.2). But the application of widely recognized definitions presented above significantly reduces this number. Teleworkers must use computers and a mean of telecommunication, and relevant telework measurement must be restricted to situations when there is *substitution* to normal workplace practices; therefore there is a bias in the use we made from the survey's data: it is not certain that people targeted by the survey do not commute during the day they do homework.

However, an elementary calculation based on these statistics provides a fair measurement of telework. Reasonably, among home-based salaried workers, only those who have a formal arrangement with their employer to be paid for the time worked at home can be regarded as teleworkers (those who do not have such an arrangement are likely to be regular, full-time or part-time salary workers mentioned above, who bring some work at home after normal office hours, and do not fit the definition.

Therefore, 84.8 percent (using a computer) of 3.35 million salary people who do paid work at home (once a week) represent 2.84 million salaried teleworkers. To this figure must be added 76 percent (using a computer) of 6.96 million home-based self-employed people, which represent 5.29 million. This would result in a total of 8.13 million home-based teleworkers in the US, 5.95 percent of the total civilian, non-agricultural employed labor force. But it must be recalled that salaried teleworkers usually work at home only part-time, 19 hours a week on average, according to the BLS. Even if we admit that home-based self-employed teleworkers work 100 percent at home (while the reality is probably below, because the self-employed may have to visit clients), the share of total work which is actually done at home as a substitution of normal office work falls to about *4.9 percent.*

This estimation does not include work in satellite offices, or nomadic work, which feature impressive, although fuzzy statistics (see below), and are often the normal way of doing telework in the corporate milieu. But the conservative figure we suggest above is important with regard to the energy and environment-related issues, and to the wider concept of sustainability in economic development. The political, social and economic relevance of telecommuting must be found in the reduction of travel to work. This is

Table 7.2 ***Work at home and usage of IT equipment (May 2004)***

		Percentage of use for work		
	Total (thousands)	*Computer*	*Internet or e-mail*	*Telephone*
Total	20,673	80.6	69.6	78.0
Wage and salary workers	13,678	83.0	71.5	73.4
Paid work at home	3,349	84.8	78.3	84.6
Unpaid work at home	10,189	82.8	69.6	69.7
Self-employed	6,960	76.0	65.8	87.0
With a home-based business	4,627	74.0	63.2	86.9

Source: adapted from Bureau of Labor Statistics 2005, table 6

the very reason for which people who perform some work at home after their work day at the office are not fully relevant for telework analysis. Nomadic work, although important from the business point of view, is of little help to fix the sustainability issue (while work in a satellite office, which is likely to diminish travel hours, must be considered). The very problematic of telework has always been addressed in terms of hours of travel and gallons of petrol saved. In the years to come, it will also be addressed in terms of tons of carbon dioxide emission saved. These questions are discussed later in this chapter.

Logically, broader definitions lead to larger figures. The *American Interactive Consumer Survey* conducted in August 2005 by the Dieringer Research Group for the ITAC "asked respondents to check up to 13 different locations where they may have conducted work in the past month." ITAC (2005) found that out of 135.4 million American workers:

- 45.1 million worked from home;
- 24.3 million people worked at client's or customer's place of business;
- 20.6 million in their car;
- 16.3 million while on vacation;
- 15.1 million at a park or outdoor location; and
- 7.8 million while on a train or airplane.

In the European Union, a survey conducted in 2002 by Empirica (2003) suggests that 13 percent of the workforce was made of teleworkers, a large increase over the 3 percent found in a 1998 report by the European Commission (Johnston and Botterman 1998: 28). These data are calculated from the largest definition (including nomadic work): the US is given 24.6 percent of teleworkers in the workforce (Table 7.3).

Northern Europe, which is well ahead in various indicators of the digital economy, shows the highest figures, while Southern Europe is lagging behind. It should be noticed that the small, very urbanized Netherlands dominates large, sparsely populated Scandinavian countries, as well as the US. In fact, telework diffusion is by no means a matter of remote wilderness. Southern Europe's backwardness fits the global landscape of the digital economy in these countries, whose IT indicators often lag behind.

Table 7.3 ***Share of employed population who are teleworking in EU-15 countries and in the US, 2002***

The Netherlands	26.4	Greece	11.1
Finland	21.8	Belgium	10.6
Denmark	21.5	Italy	9.5
Sweden	18.7	France	6.3
UK	17.3	Spain	4.9
Germany	16.6	Portugal	3.4
Austria	13.8	EU-15	13.0
Ireland	10.9	US	24.6

Source: Empirica (2003)

Note: Includes home-based, mobile, or SOHO-based self-employed teleworkers

The pros and cons of telework

The very reason for the telework media hype is that it "promises something for almost everyone" (Handy and Moktarian 1996). Green, cost-efficient, family-friendly (especially for workers with children), reducing stress, raising productivity and creativity, ad hoc telework practices seem to be the ideal response to many threats our society increasingly faces.

> The $150 million in annual business advantage AT&T receives from telework is small, however, compared to the benefits that can accrue to society as a whole from these new models of work, education, recreation and community. For example, our teleworking employees avoided commuting over 150 million miles to work in 2002, saving millions of gallons of gasoline and thousands of tons of pollutants from being exhausted into the air. Families and local communities benefit from increases in employee job satisfaction and work/family balance.
>
> (Roitz *et al.* 2003: 3)

Most reports and surveys are quite unanimous about the benefits of telework for both employers and employees (for a detailed analysis in Europe, cf. James 2004). The Telework Coalition, a non-profit organization, conducted in May 2003 the *Greater Washington DC Telecommuting On-line Survey*. Employers were asked the importance of the benefits of telework to their business. Table 7.4 presents those rated as "very important" or "important."

Mokhtarian and Salomon (1997) present a detailed analysis of telework drivers, which are strongly linked to workers' behaviors and attitudes. The potential positive effects of telework are strongly linked to both the employees' and the employers' personalities, natures and attitudes, and cannot be generalized. The consequence is a frequent occurrence of symmetry in the cost–benefit analysis synthesized in Table 7.5: a telework feature which benefits one person or organization might prove ineffective or even harmful to another person or organization, and vice versa. Another source of complexity in telework cost–benefit analysis is that most of its elements are intertwined,

Table 7.4 *Telework benefits to business*

Benefits rated as "very important" or "important"	*Percentage*
Increases workforce morale	91
Improves productivity	78
Provides essential component of operational continuity plan	76
Makes it easier to attract and retain quality employees	68
Improves employee concentration	67
Reduces unexcused absences	61
Reduces office space requirements	55
Reduces employee commuting costs	47

Source: *The Greater Washington DC Telecommuting On-line Survey*, The Telework Coalition 2003

Table 7.5 ***The pros and cons of home-based telework***

Target	*Benefits: Drivers*	*Costs and hazards: Constraints*
	Flexible working time	
Worker	Improves family life and facilitates: – child care – disability/parental leave	Lifetime/worktime confusion – workaholic syndrome – spoils personal and family life – endangers health
Firm	Decreases employees' absenteeism Improves employees' loyalty Decreases attrition rate	
	Private workspace—Work freedom	
Worker	Out of sight from colleagues and managers Relaxed behavior Quiet environment Improves efficiency Greater feeling of responsibility	Lack of visibility from management Fears regarding career advancement Cost of additional home space for separate work room Legal uncertainty regarding labor contract
Firm	Raises productivity Reduces office space Decreases business property expenses Favors business continuity	Lack of business interaction Decreases productivity Lack of control Legal uncertainty regarding labor contract Network and computer costs
	Reduced travel to work	
Worker	Saves time for personal and family life Saves money Reduces stress, fatigue Lessen occurrence of car accidents Improves health Makes home location choice easier	
Firm	Reduces insurance premiums Decreases absenteeism Enhances productivity Favors business continuity	
Community	Decreases energy consumption Reduces pollutants and greenhouse gas emissions	Possibly increases urban sprawl

and are difficult to consider separately. Times and places of work and private life become overlapping, giving telework a systemic character.

The largest series of telework advantages derives from the decrease of travel and the flexibilities associated with commuting (in this section we emphasize home-based telework). Reduced daily travel time would logically decrease fatigue and stress, therefore improving morale and productivity, which also benefit the company. A major benefit for workers with young children is the flexibility it gives for managing day care and school schedules, or dealing with health problems, all tasks which are the plague of young working parents' lives. Employees' health is also concerned: telework may reduce notably sick or accident leave. The immediate benefit for the employer is the reduction of absenteeism, and the increase of employees' loyalty.

This issue was particularly relevant at the end of the 1990s, when the economies of metropolitan areas in Western countries were nearly over-heated, especially in IT-based sectors, and when the balance of power between firms and highly skilled workers was, therefore, tipped in favor of professionals. Faced with a shortage of talented people, companies sought to retain valuable employees while giving them more freedom in their work arrangements (in addition to other facilities such as child-care service and fitness rooms in the workplace).

More generally, telework may give more flexibility and pay benefits for all the occupations of private life. Teleworkers are free to organize their work schedule, as long as the job is done. Because workers may perform part of their professional duty in the evening, or during the weekend, they can play tennis or swim during normal working hours when courts and pools are deserted, they can go the cinema in the morning at reduced prices, and they can avoid the horrible experience of the shopping mall on Saturday afternoon. Finally, they can escape from many alienating, sheeplike behaviors that characterize standardized urban life.

Augmented freedom is also found in locational behaviors. When it is time to rent or purchase a home, teleworkers who are less dependent on commuting constraints can consider a wider geographic spectrum, and are more free to select a neighborhood according to property prices, environment, conveniences for children's education, and the location—if need exists—of the partner's job.

Teleworkers often claim higher concentration for creative tasks, thanks to the quiet environment of their home office. This may also benefit the employer, in terms of productivity. However, we enter here into a highly subjective field: some people would find their personal equilibrium by spending 60 hours per week in their office, and would hate working at home. It must be pointed out that all people are not suited for telework arrangements, which require a high degree of autonomy, self-organization skills, and the capacity to focus on business objectives rather on the daily timetable.

Some arguments favoring telework may have a negative face. Telework is very suitable for people with children. But the working environment may be not so quiet, nor the productivity so high, when children are at home. A separate office space is usually required. The lack of dedicated workspace, or the disturbing features of family life are arguments which typically favor the use of telecenters or "satellite offices" (see p. 164).

But the big threats, and the main deterrent factor against the diffusion of telework, are isolation and the consequent lack of social and professional interaction. Leamer and Storper (2001) emphasize the role of "shaking hands" in business relationships. In the same vein, Storper and Venables (2004) analyze the "buzz" that characterizes dense urban business milieus. Indeed, full-time home-based telework would deprive the employee from up-to-date, possibly vital business information which is produced by the "buzz" of brain concentrations typically found at company premises. The risks include a fast obsolescence of technical skills, and the lack of strategic information—a problem that is even worse for many self-employed persons. On this point, the interests of the employee and those of the employer are clearly convergent. Most companies that do not favor telework invoke the need of teamwork and collective brainstorming, and sometimes fear a lack of control. In addition, would-be teleworkers often fear some lack of visibility from managers. In Japan and Korea, despite a world-class IT infrastructure, the tradition of long hours of physical presence in the office, and the premium given to collective behavior by the corporate culture, still hamper the development of telework (Ihlwan and Hall 2007). "Out of sight, out of mind": bad management practices favoring cronyism and fawning remain common in every country.

But the consequences of remoteness from the office go beyond technical or strictly business-related issues. The job is among the most powerful structuring elements of being and personality, because it is at work where people are faced with *otherness* (the workplace plays the role that kindergarten, school and university played when the employee was young). Work is a key determinant of social inclusion, and the job is an important part of social and personal status (notoriously the fundamental part for Japanese "salarymen"). Home-based teleworkers, and more generally all people who benefit from flexible work arrangements risk being suspected—at best—of enjoying a long weekend.

Telework is likely to diminish the stress generated by strenuous commuting, overload of work, busy work environment, and even harassment. But teleworkers could lack the good stress of office work—maybe of competition at work—which is likely to create a feeling of responsibility and to generate higher productivity and self-being. In such a case, a full work day at home may be a boring, less productive experience.

The emergence of telework has also been hampered by the legal issue. Home work (and, to a lesser extent, nomadic work) generates ambiguities in the legal status of the worker, which result from several intertwined problems (Vatier 2005):

- the enforcement of labor health and labor accident laws (if an employee is burned by the battery of his laptop, how to prove if he was using it for professional or private means?);
- the status, financing and use of professional IT equipment, with related responsibility and security issues;
- the enforcement of working time limitations;
- the tradeoff between respect for privacy and work control and monitoring;
- the application of collective agreements regarding training and workers representation.

In the EU, an attempt to fix collectively this problem has been the Framework Agreement of July 16, 2002 (European Trade Union Confederation 2002). From the company's point of view, an important driver of telework implementation is the decrease of business property expenses, also aided by nomadic work practices. Employees who telecommute and work at clients' premises or during travel are no longer given the exclusive use of an office, but find on demand a desk and an outlet to plug their notebook computer when they visit the company's building. However, this solution has its own disadvantage. Employees like to be assigned a workplace for their exclusive use, which they often personalize with family souvenirs. This is an element of professional and personal status. By default, the link between employees and the company risks being weakened, which is contradictory to the increase of loyalty expected from telework implementation. The rise of productivity per employee results also from the reduction of commuting time, which may be partly reinvested in working hours.

An additional benefit companies may find in telework—at home or nomadic—is the improvement of business continuity in case of a catastrophic event. In the survey quoted above, this criterion is ranked third among telework benefits. The survey was conducted in Washington, DC two years after 9/11, a fact which could explain the high rating of the continuity plan criterion. *The Economist* (2006d: 78) reports that "99 percent of the business interruption plans in the financial industry are based on telecommuting." Several firms have already experimented with some telework devices in traders' homes. In November 2006, 70 financial companies in the UK simulated a flu pandemic, with nearly half the workforce absent from the office. Indeed, they observed the common drawbacks of home-based telework, including little trust from clients and supervisors, and the lack of back-up power generators!

Actually, teleworkers could operate even when a region is crippled by transportation blockade. It must be recalled that recent terrorist attacks have often targeted commuters who were using public transportation systems, with heavy human losses: a toxic gas attack in the Tokyo subway (20 March 1995), a train bombing in Madrid (11 March 2004), and a subway bombing in London (7 July 2005). Natural disasters such as earthquakes, hurricanes, or the fear of pandemic outbreaks (SARS, bird flu) are also likely to raise attention about advantages of telework for business continuity. These events could have consequences on employees' willingness to commute and, therefore, could be regarded as telework drivers from the employee's point of view.

Telework may benefit the whole community. The limitation of synchronism lessens the impact of peak hours on transport infrastructures. It helps firms in retail trade, the tourism industry, and other personal services sectors to spread out the demand. And telework is likely to decrease petrol consumption. These issues have played a great role in the emergence of the telework utopia. Given that they are linked to the impact it would have on urban life and urban forms, we develop it below in a separated section.

The last benefit of telework implementation we may envision is as a counterweight to the economic and social effect of the aging of the population, notably in Europe and Japan. In January 2006, the "IT Strategic Headquarters" (the national committee for IT established in 2001 within the Cabinet of the Prime Minister) released the *New IT Reform Strategy*. Taking into account that "Japan's society is aging and its population of children is dropping at a pace unseen in other countries around the world," the report targets to "increase teleworkers to 20% of the working population by the year 2010" (IT Strategic Headquarters 2006: 40). Through this program, the government endeavors to ease the working conditions of senior workers, and the care of both children and the elderly.

A sign that the conservative corporate culture we have mentioned above is changing: in April 2007, Matsushita Electric (76,000 employees) announced "it would begin a telework program for nearly half of its employees in Japan as part of the private sector's effort to cope with an aging society" (Forbes.com 2007). Thanks to high speed Internet links, computers, and video cameras loaned out by the firm, applicants might work one or two days a week at home.

Telework in practice

Rather than a simple, unique phenomenon, what we are witnessing is the emergence of a bundle of intertwined, context-dependent practices which expands the spectrum of labor in terms of time, space, and capabilities. However, the main conclusion of most analyses is that telework is often a part-time practice, especially for salaried employees.

In Chapter 4, we have seen that digitalization leads to greater tradability, portability, and scalability of business information. The use of IT favors a very fine division of tasks at all stages of the extended production process which, therefore, can be split in both organizational and geographic terms. Telework is an outcome of these organizational trends. The list of computer-aided tasks—creative or menial—which could be done remotely by individuals or small businesses is virtually endless, given there is no need for face-to-face meetings with colleagues, clients or providers: reporting, data analysis, typing, publishing, translation, secretarial work, accountancy, legal consulting, software writing, computer graphics and design, and architecture. It is worth noting that telework potentially operates nearly in the same skill spectrum as offshoring, and may even be seen, in some cases, as a substitute for offshoring. Even the call center industry requires home-based teleworkers. Only the scale of operations differs notably. But "virtual office" technology allows the distribution and interaction of work among spatially dispersed teams of professionals (this is typically the case in "virtual call centers").

Technology improvements, in the context of the digital convergence, have greatly expanded the possibilities and scope of telework. Home-based teleworkers now benefit from the fall of IT equipment and telecom prices. DSL lines at home with business-effective bandwidth (over one Mbps) are becoming a commodity. Thanks to voice over IP (VoIP) technology, bundled Internet subscriptions often include free, unlimited national—and sometimes even international—telephony, at least between fixed lines. Nomadic and mobile telework have benefited from the improvement of advanced, wireless telecommunication systems available on notebooks and sophisticated handsets (Chapter 3), which allow virtually ubiquitous connection to business databases, telephone talks, and voice conferencing.

Corporate telework: mobile work and "flexible work arrangements"

> Large numbers of AT&T employees are moving out of traditional offices and into virtual offices as a way of increasing productivity, work/life balance and their quality of life. They rely on a structure that is more and more 'net-centric'—organized around networks instead of buildings.
>
> (Roitz et al. 2003: 1)

There is recent evidence that telework, as an innovative method of management, is considered increasingly seriously by majors companies worldwide. Several leading corporations, such as IBM, AT&T, BT, and Alcatel, have implemented extensive telework schemes. In early 2007, Renault (Lauer 2007) and Matsushita Electric announced new telework plans. According to the 2002–2003 AT&T Telework survey (Roitz *et al.* 2003), 57 percent of managers in the company telework, and 17 percent of them divide their working time equally between home and client sites. IBM claims that 42 percent of its workforce "is covered by flexible workplace practices" (Ehrle 2006). However, these figures remain fuzzy statistics, given the lack of common definitions and survey methodologies.

Telework could be logically considered as the ultimate outcome of the advent of the networked—or virtual—enterprise. With the exception of the decrease of real estate costs, and possible improvements in business continuity, the benefits that telework implementation provides to the firm are mostly indirect, and therefore are strongly linked to labor market features. Companies in some business sectors and located in certain places are more likely to favor telework in order to increase employees' loyalty during periods when talent is scarce, wages rising, and recruitment and retention are difficult. This was the case in the 1990s in the IT industry (computers, software, telecommunications) which was facing a nearly permanent shortage of highly skilled professionals. C. Beveridge (2003), chief human resource officer of Sun Microsystems, estimates that, thanks to the work-from-home aspect of its iWork program, the company "has kept 680 employees who otherwise would have left."

In addition to human resource-related benefits mentioned above, two other drivers explain why companies in IT industries are at the vanguard of telework implementation:

- most employees in these firms are knowledge workers and IT users, able to get advantage from telework tools;
- these firms regard telework as a field of experimentation and a full-size showroom for their own products. Like British Telecom, they want "to demonstrate that teleworking can work so that other organizations will be encouraged to adopt it" (Hopkinson *et al.* 2002: 6). Logically, rising adoption of telework would increase the global market of IT products and services.

Telework-friendly companies do often not use the words "telework" or "telecommuting," which are too much connoted with "home work" and negative images. Business circles prefer "flexible work arrangements," "mobility programs" or "e-working." Actually, these terms best mirror the reality of corporate telework which is usually a mix of desk-sharing practices, home-based work, and nomadic work, during travel or in client's premises.

In March 2002, Hopkinson *et al.* (2002) surveyed 1,874 participants in the "Workabout Program" at British Telecom, the largest employer of teleworkers in the UK. Working arrangements among participants in the program may be distributed as follows:

- 31 percent work mainly at home, with sporadic presence in BT's offices
- 25 percent split working hours equally between home and BT's offices
- 19 percent work equally at home, at the office, and in client premises
- 12 percent visit clients during opening hours, while early morning and late afternoon are worked mostly at home, and presence at the office is occasional.

Therefore, there is some evidence that most salaried teleworkers try to keep a strong physical link with the company's office. According to Miller (2004), "roughly 20% of these people telecommute on a full-time basis." Miller reports a speech in October 2003 by Intel's CEO Craig Barrett explaining that "while Intel promotes telecommuting, they limit working at home to one or two days a week," because people "need that physical contact to be effective." And telework is often a temporary arrangement (for example, parents who would like to take care of young children). A full-time physical absence from the firm would hamper the return to a "normal" career.

A major benefit of flexible workplace practices, such as Sun Microsystems' iWork program, is to reduce real estate costs, by reducing vacancy rates which are usually "around 10 to 15 percent." Sun Microsystems determined a ratio of 1.8 users per office for salesmen. In 2003, the company saved $50 million in rental fees and operating costs, and it expects $140 million a year when the program reaches full implementation. The reduction of travel to work is a win–win formula. Sun has located in residential neighborhoods a network of "drop-in centers" (satellite offices) where "some seats are available on a first-come, first-served basis and some can be reserved." These centers help users to save "6.6 hours of driving time a month and give 4 hours of additional work time to Sun," while "work-from-home participants save 12.4 hours a month and give 8 of them to Sun" (Beveridge 2003).

In 1999, IBM France launched the "Mobility Program," which involves desk sharing for 4,500 "IBMers" in Paris. Employees may "book" an office space at the headquarters in the CBD of La Défense, one month in advance, up to a maximum of eight half days. The remaining working day is spent in clients' premises, at home, or in one of eight satellite offices which are strategically located in Paris Metropolitan Area. In the end, this program allowed IBM to vacate two secondary buildings, with subsequent savings in real estate expenses (Biseul 2004).

"Homeshoring" or "virtual," home-based call centers

Mary Driffill has three children, ages 9, 7, and 4. Like 700 other people, many of them women with young children, she is in the call center of JetBlue Airways in Salt Lake City (Utah). However, she is able to stay near her children when she works late in the evening because, like all the 1,400 reservation agents of JetBlue Airways, she works at home in a *virtual call center* (Salter 2004). This case is not isolated. Orders to 1-800-Flowers.com are received mainly by home teleworkers. In 2005, Office Depot closed 10 of its 12 US call centers, replacing 900 full-time agents with home-based agents. Mountain America Credit Union, a US local bank, has "40 telecommuters across the state of Utah, serving members from their own homes" (Cisco Systems 2003b). The practice of "homeshoring" has entered the outsourcing market. Alpine Access Inc., headquartered in Golden, Colorado, is a virtual call center firm with about 4,000 home-based agents (Geller 2005). In 2006, about 112,000 American call center agents worked from home, and IDC Research, a consultancy, forecasts this number to increase to 300,000 by 2010 (Conlin 2006).

The use of home-based teleworkers by the call center industry is driven by technology improvements and the business context of the sector, which is labor-intensive, and increasingly subjected to relocation in low-cost countries. Therefore, home-based call center agents are regarded as an alternative to offshoring, hence the term *homeshoring*. This organization provides numerous benefits. The overall productivity is increased, thanks to a less stressing environment, improved punctuality, business continuity (workers at home are not subjected to traffic jams), and decreased attrition rates. Wages are often lower than in "brick-and-mortar" call centers, because home-based operators are more likely to live in peripheral, less costly regions. Through the development of home-based, *virtual call centers*, employers can have access to remote, dispersed workers which otherwise would have difficulties to find a job. Last, because employees are at home, companies may manage queuing in a more flexible, less costly way: home-based agents are called on demand, to accommodate peak periods. Homeshoring also favors the professional inclusion of disabled people and parents with young children.

Telework centers and telecottages

Satellite offices and "drop-in centers," popular in the sphere of corporate telework described above, are a reminiscence of telecottages which emerged in the UK and Scandinavian countries in the late 1980s (Qvortrup 1989). They are based on the same concept: the search for external scale economies in the investment and maintenance of ad hoc real estate, IT equipment and services. Telecottages and telecenters are often established by communities or public institutions, and are more or less open to a wide range of users. Telecenters try to help nomadic workers and teleworkers to avoid the difficulties of both home and office work, by providing a well-equipped standard office environment within a short journey from home. In the US, the "state of California alone has seen more than 40 such facilities established since 1991, and the US federal government is participating in the operation of more than two dozen telecenters under its Flexible Workplace Initiatives program" (Mokhtarian and Bagley 2000: 224).

In the Washington DC metropolitan area, which is known to be affected by some of the worst traffic jams in the US, the General Services Administration (GSA) has implemented the Washington Metropolitan Telework Centers (WMTC), a network of 16 telecenters in Maryland, Virginia and West Virginia, with a total of 1,200 work positions. In an October 2005 survey conducted by the GSA, WMTC users were asked why they prefer using a telework center over home-based telework? Answers and testimonials show a mix of technical and social factors (Table 7.6). One evokes "a retired husband and a teenager in the house." Another believes that "dressing professional and going out to the center" does more for self esteem and work productivity than "sitting at home in flannel pajamas." Many address the issue of technical support (Washington Metropolitan Telework Center 2006).

Table 7.6 ***Why people choose a telework center over home-based telework (%)***

Home distractions, spouse, children, housework	44
High-end office equipment not available at home	43
Prefer a professional office environment	29
Broadband not available at home	21
Supervisor prefers a telework center	21
Onsite technical support	16
Want to keep home life and work life separate	11
Isolation of home telework	11

Source: Washington Metropolitan Telework Center 2006

However, telecenters have more or less met the fate of telecottages. Many of them were closed after experimentation, being unable to achieve economic self-sustainability after the end of the public funding period (Mokhtarian and Bagley 2000). Why, in the end, have telecenters had so little success? The first explanation is that white-collar work often requires real-time interaction among companies' staffs and taskforces. Like the city as a whole, the office building generates a "buzz" which facilitates swift decision-making, and is a source of creativity and social interaction. The concepts of telecenter or satellite office do not fit teamwork requirements, which are critical in innovative activities. Although "IBMers" or Sun's employees can achieve a certain social interaction in satellite offices—although limited by irregular presence—there is little chance that any valuable professional interaction would be created, precisely because of the random character of attendance in telecenters. Another explanation is the commoditization of the technologies of telework (PC, DSL line), which makes efficient home-based telework possible with little investment. In the end, telework centers may appear as a "one-leg" solution, which provides employees neither the social and professional interaction of office work, nor the comfort of "staying at home in pajamas."

Self-employed teleworkers and micro-enterprises: an issue of the network

Freelance teleworkers, or micro-enterprises, acting as subcontractors for larger firms, should be understood as the tail end of the network economy. Once again, the key word is flexibility. Individual or small businesses usually serve clients locally, but through *ad hoc* business networks, they can make contact with remote service firms that need to subcontract a part of their work load, or which look for highly specialized skills they do not have in their permanent staff.

This business sphere remains partly unknown. Richardson and Gillespie (2000), Clark (2000), and Moriset (2003a) offer some insights, particularly in rural areas. But the literature about urban, individual teleworkers is virtually nonexistent, and there is no reliable measurement on large scales. "Lone Eagles and High Fliers" (Beyers and Lindahl 1996) are more or less specialized in four main categories of tasks: secretarial work (typing, desktop publishing), multimedia creation (web design, business communication), translation, and software design. Accountants, lawyers, and architects may also be found.

Technological and commercial isolation is the main drawback faced by self-employed teleworkers. Unlike salaried teleworkers, self-employed do not benefit from technical support and regular opportunities for professional contacts. Therefore, in the context of a network economy, the logical response of self-employed teleworkers has been to network themselves together. There is an abundance of websites, forums, and online magazines dedicated to social, technical, legal, and commercial support for teleworkers. The most renowned institution is the International Telework Association Council (ITAC), founded in 1993 (it operates mainly in the US, cf. www.telecommute.org). The portal of Gil Gordon Associates, active since 1995, also must be mentioned (www.gilgordon.com). In the UK, the Telework Association (www.tca.org.uk) has 2000 members. These structures act as both virtual communities and lobbies.

But commercial networks matter more. Some well-advised entrepreneurs have designed large electronic platforms, which meet the most crucial need of freelance teleworkers, that is commercial visibility to potential clients. Cyberworkers links 15,000 professionals from 70 countries, serving 200 specialties (www2.cyberworkers.com). In July 2003, Pittsburgh-based Emoonlighter acquired assets from GURU, a Silicon Valley-based network, creating the world's leading telework platform, with 360,000 affiliated freelancers and a portfolio of 27,000 customer firms. "Guru.com's Web-based marketplace directly connects businesses with freelancers locally, nationally, or globally who specialize in over 160 professional categories including: website design, programming, graphic design, business consulting, and administrative support" (Guru.com 2004).

Box 7.1 "Living" examples of teleworkers

Laure Bernazzani, development and innovation manager, France Télécom, Paris.
In 2002, Laure Bernazzani (37, married, one child) left Paris and moved to Troyes, a city of 121,000 inhabitants located 170 km to the east. Three days a week, she works in Paris at FT's headquarters. Two days a week, she teleworks from FT's local office in Troyes. "The overall balance is very positive, notably for my husband and my daughter who really enjoy the provincial *douceur de vivre*. The equilibrium between private life and professional life is fine. And I was promoted last year. However, it is the result of a fierce fighting, and nothing is granted forever: my employer can put an end to the experience at will. This way of working allows a fine, efficient organization, with a clear distinction between physical presence in meetings and *offline* [in English in the original version] work such as analysis and report writing. But the confusion between telework and part-time work is common. Telework is perceived as a preferential treatment. Too many people think that absence from the office does mean holidays. With the result that I must work twice harder to convince them it does not."

(*Le Journal du Net* 2004)

Shirley Bailey, owner, Quick-Type Secretarial Service, Hamilton (Ontario), Canada
"I've been operating Quick-Type for over 13 years. My computer is opposite a wall-to-wall, ceiling-to-floor window looking out on my garden and a family of cardinals. I love what I do (transcription for medical specialists) and I'm a highly motivated and disciplined person and I don't transcribe in my nightgown although I could. Working at home is of benefit to my clients. I can work 5:00 am to 5:00 pm if I need to . . . and often do. Many times I'll put in a couple hours of work on the computer following a 3–4 hour hike. I don't feel lonely working here because I'm a well-rounded social individual when the office door closes. I only wish more of the medical profession would understand the benefits I can offer as opposed to working at their location. My productivity is beyond belief. I don't find they're as receptive to teleworking as one would think. I've had the capability of receiving their files digitally which would reduce the turnaround time of the transcription considerably. I started teleworking because I could see the efficiency of it. I wish more employers would see the benefits."

(InnoVisions Canada 2007)

The energy and urban issues of telework

Telework has always been linked to the rise of environmental and sustainability issues in public debates. The current concern about "global warming" and the impact of greenhouse gas emissions, and the 2006 surge of oil prices, raise the relevance of the energy issue. The future of urban forms is also at the heart of telework rhetoric. The two questions are strongly linked: the reduction of travel to work could have consequences for metropolitan features and for dynamics such as surburbanization and urban sprawl.

More than ever, the issue is full of complexity and ambiguity. The actual or potential effects of telework in this field are context-dependent, in the same way as its impact on people's lives. The consequence is that a fully reliable measurement of actual (potential) energy savings, or of actual (potential) geographic consequences, is virtually impossible. The nexus of the problem is in the substitution concept: measurement methodologies derive travel reduction from the bet that people who telework would have traveled if they did not telework. In many cases, it is pure conjecture: if a teleworker who spends two work days per week at home is deprived of the possibility of teleworking, he or she may become a part-time worker or an unemployed person). However, the subject is intriguing and deserves an attempt at clarification.

Telework and transportation

Substitution for transportation is at the core of telework analysis (hence the name telecommuting). This issue has been the subject of a great deal of research, the most prolific author on the subject being Patricia Mokhtarian, Institute of Transportation Studies, University of California, Davis. The very question for policy-makers, planners, and environmentalists is: do telework practices actually reduce travel, and could they have a significant impact on global energy consumption and pollutant emission? The topicality of the matter is enhanced by the global increase of the journey to work, a symptom of urban sprawl and spatial "mismatch" between business and residential neighborhoods.

The present analysis is mostly based on the US example. Among major, industrialized countries, the US has the most "automobile-addicted" population, with the highest figures for driven mileage, automobile reliance for commuting and, more generally, energy consumption per inhabitant. In more dense countries, with more efficient public transportation systems, telework implementation is likely to have lesser effects on petrol consumption. However, that is not to say that telework is not suitable for citizens living in London or Tokyo: the substitution for long, exhausting train or subway journeys, while less significant in terms of energy savings, may attract people and companies in terms of fatigue, stress, and time savings.

The *2001 National Household Travel Survey* (NHTS) implemented by the US Department of Transportation (2003) provides important statistics about travel and commuting practices. Surprisingly, work and work-related trips represent only 17.7 percent of all American people's daily trips. The remaining is made of private, non-work purposes (recreational, school, church, shopping, family . . .). Without going forward in the analysis, this is a strong indicator that a dramatic impact on petrol consumption must not be expected from any realistic degree of telework implementation.

Automobiles are by far the first mode of transportation used for commuting in the US: 91.2 percent. The actual journey to work driven daily is not easy to calculate, because data often encompass all kinds of trips. Employed people drive 35.5 miles (56.8 km) daily, but unemployed people drive 16 miles (25.6 km). Therefore, a significant distance for non-work purposes must be deduced. And the frequency must be taken into account. Gordon *et al.* (2004) calculate from 2001 NHTS data a one-way commute length of 12.1 miles (19.4 km) and 23.6 minutes, a significant increase from 1983 (8.5 miles and 18.2 minutes). For a comparison, the census by the French National Institute of Statistics (INSEE) found that in 1999, one-way travel to work averaged 15 km, compared with 14 in 1990, and 13 in 1982. But these data rely on persons who do not work in the commune (municipality) they live in, who represent only 60 percent of the working population (Talbot 2001). Therefore, the nationwide average is significantly less.

All in all, on the basis that most workers commute once a day only, it could be acknowledged that in the US, all other things being equal, one day of home-based telework saves about 24 miles (38.4 km) in vehicle miles traveled (VMT),[2] with correlated savings in travel time and petrol expense. From these figures (which are nationwide), we could derive approximately the actual and potential savings in terms of petrol consumption (and carbon dioxide emission, which is proportionate).

We may consider the weekly suppression of two round trips to work for each of the 2.84 million paid (home-based) teleworkers who telework once a week, 19 hours per week on average (from BLS' conservative evaluation presented above). For each of the 5.29 million self-employed teleworkers, we may consider the weekly substitution to five commuting round-trips. Taking into account that 90 percent of workers commute by car, with a vehicle occupancy rate of 1.2, the global saving in VMT due to home-based telework could be evaluated to 44.41 billion km (on a 48 work weeks basis—calculations are developed in Table 7.7).

However, this somewhat impressive figure represents less than 1.62 percent of light vehicles' total VMT—2.74 billion km in 2004 according to the Federal Highway Administration (2005)—and therefore a similar percentage of gasoline use (taking into account a reasonable consumption of 0.1 liter per km, the saving of petrol would be of nearly 4.4 million tons[3]) to compare with the national consumption of crude oil, 1.2 billion tons in 2004, according to the US Energy Information Administration (it must be recalled that cars and light trucks represent 42.5 percent of US total consumption of crude oil).

The simulation-based analysis of Jancovici (2001) suggests equally modest results in France: thanks to travel reduction, "if half of employed people in the tertiary sector would telework from home half-time, all other things being equal, they would save 1.34 million tons of crude oil, 0.5 percent of the global energy consumption." But the actual share of telework is near 5 percent, rather than 50 percent. Admittedly, telework could result in additional, indirect energy savings, by minimizing traffic congestion. In the US, 2.3 billion gallons (8.7 million tons) of fuel is wasted by engines idling in traffic jams, according to Best Workplaces for Commuters (2006), a business-government program sponsored by the US Environmental Protection Agency and the US Department of Transportation. But it is very difficult to measure the effect of the absence of teleworkers on traffic.

Table 7.7 ***Telework impact on energy and environment***

Nationwide average one-way commute length (1)	***12 miles***
	19.2 km
Round trip	38.4 km
Percentage of light vehicle use for commuting (1)	90%
Mean car occupancy rate (1)	1.2
Average daily commuting VMT (one round-trip a day)	***28.8 km***
Estimated number of home-based, paid teleworkers (2)	***2.84 million***
Average weekly VMT reduction due to telework (two days a week)	57.6 km
Individual VMT reduction, per year (48 work weeks)	2,765 km
Total VMT reduction (paid teleworkers)	7.85 billion km
Estimated number of home-based, self-employed teleworkers (2)	***5.29 million***
Average weekly VMT reduction (five days a week)	144 km
Individual VMT reduction, per year (48 work weeks)	6,912 km
Total VMT reduction (self-employed teleworkers)	36.56 billion km
Total VMT reduction due to home-based telework	***44.41 billion km***
Nationwide global distance travelled by passenger cars in kilometers (3)	2,744 billion km
Reduction due to telework, in percentage of total VMT	***1.62%***
Share of light vehicle in US consumption of crude oil (4)	42.5%
Reduction of crude oil consumption due to telework	***0.69%***
Share of gasoline consumption in light vehicles in the total of greenhouse gas emissions (4)	20%
Reduction of greenhouse gas emissions due to telework	***0.32%***

Source: (1) US Department of Transportation, Bureau of Transportation Statistics 2003; (2) Bureau of Labor Statistics 2005; (3) Federal Highway Administration 2005; (4) Best Workplaces for Commuters 2006

An additional source of energy savings would be the decrease in business needs of office heating, ventilating, and air-conditioning (HVAC). Following the scenario mentioned above, Jancovici (2001) finds a potential saving of 3.4 million tons of equivalent oil, resulting from the reduction of HVAC. The result suggests that telework-related energy savings in the property sector could be greater than those expected in the transportation sector. But, one more time, the result is based on a highly improbable scenario: a telework adoption rate of 50 percent in the services sector, and a perfect adaptation of real estate to the decrease of office presence, which seems idealistic. Therefore, whatever the margin of error which is retained, the impact of telework on oil-related problems is, and will probably remain, very low. And the impact on greenhouse gas emission is even more tenuous, because gasoline consumption in light vehicles generates only about 20 percent of total greenhouse gas emissions. Therefore, the lessening effect of telework on these emissions with the calculation suggested above would be a mere 0.32 percent, with an actual impact on climate change virtually equal to zero.

The suggested estimation of telework numerical importance is conservative. The calculation does not take into account people who do not declare home-based work, and people who telework less frequently. Nevertheless, the result we found could even be considered as optimistic, because it is based on the assumption that self-employed people who telework at home generate a 100 percent substitution for each day's commuting. This is wishful thinking, because case studies show that an unknown, but substantial number of teleworkers would not be in the labor market—especially on a full-time basis—if they did not have the technical or legal means of IT-enabled, distance working. This could be the case of disabled people, parents with young or disabled children, single parents, persons compelled to leave a job because of the relocation of their partner's job, or compelled, for similar reasons, to live in remote, small labor pools with few recruitment prospects in their specialties.

The complexity of mobility behaviors biases calculations

Actual global benefits of telework could be even smaller than the figures presented thus far. In the analysis above, we take for granted that teleworkers, when they come to travel to work, have the same behavior as non-teleworkers, and commute on the same geographic basis. Unfortunately for telework supporters, they do not. On average, teleworkers are likely to live at a greater distance from their job location. The days they go to the office, they drive more. Mokhtarian *et al.* (2004) present an in-depth study of the behavior of 214 employees of six California state agencies that have kept active telecommuting programs. They found that, in 1998, the mean one-way commute length for non-telecommuters in the sample was around 16 miles—25.6 km, while the one-way distance commuted by teleworkers was over 23 miles—36.8 km. The result is that greater remoteness from workplace (and possibly from other facilities) tends to nullify, albeit not entirely, the benefit of home-based work on global fuel consumption.

> One-way commute distances are higher for telecommuters than for nontelecommuters . . . the average quarterly per capita total commute distances are generally lower for telecommuters than for nontelecommuters, indicating that they telecommute often enough to more than compensate for their longer one-way commutes.
>
> (Mokhtarian *et al.* 2004: 1892)

Teleworkers are able to live farther from the office, because they commute less frequently. These results are logical, but the causal link remains unclear. As knowledge workers, teleworkers are more mobile than average, and can afford travel expense more easily. But there is no evidence that the choice of the residential location is made after the engagement in telework practice, and therefore, that the latter is a driver of remoteness in housing locational choices.

> In this paper we do not argue for any particular direction of causality. That is, we are not able . . . to discern whether longer commute distances encourage telecommuting or, conversely, whether the adoption of telecommuting facilitates residential relocations farther away from the workplace.
>
> (Mokhtarian *et al.* 2004: 1893)

The systemic arrangement of telework drivers and constraints

Perhaps the best explanation of telework's mediocre impact on kilometers traveled and resultant energy consumption, and of the lack of evidence provided by research on the link between telework and residential location must be found in the concept of "travel time budget" (TTB), the subject of an important literature surveyed by Mokhtarian and Chen (2004). Several authors found that the time average individuals are willing to spend on travel is virtually constant across time and space, in the bracket of one to one hour and a half daily. Therefore, people whose commuting frequency is decreased by telework may accept longer journeys. The concept of TTB is of key importance in the equilibrium system which dictates telework implementation and practices.

Long journeys to work, with related costs, stress and time losses may be recognized as a logical driver of telework development. But a massive increase of telework practices would free some road space, traffic congestion would decrease and, therefore, the incentive to telework would shrink. Meanwhile, people (whether or not they telework) would be able to live farther from their workplace, and to commute greater distances, with the same resulting TTB. The backlash effect would be the return of traffic jams. Therefore, we are facing a typical negative feedback loop (Figure 7.1). Telework development reduces traffic, but decreasing traffic minimizes telework rewards and creates an incentive to drive and to pollute more. Salomon suggests that the lack of such global thinking, and technological determinism, are the main causes of over-optimistic forecasts which were published in the 1980s.

> Such a massive migration [to telework] may be noticed in reduced congestion, which in turn will reduce the incentive to telecommute as a means for coping with congestion. Thus, the dynamics of intervening factors may render the technological substitution model irrelevant.
>
> (Salomon 1998: 28)

This system is moderated by additional factors, such as oil and real estate prices. Oil prices, notably, have proven very elastic to demand (a small shift in demand often results in a larger price drop or increase). Therefore, it is conceivable that a more significant adoption of telework may have a moderating effect on oil prices. But

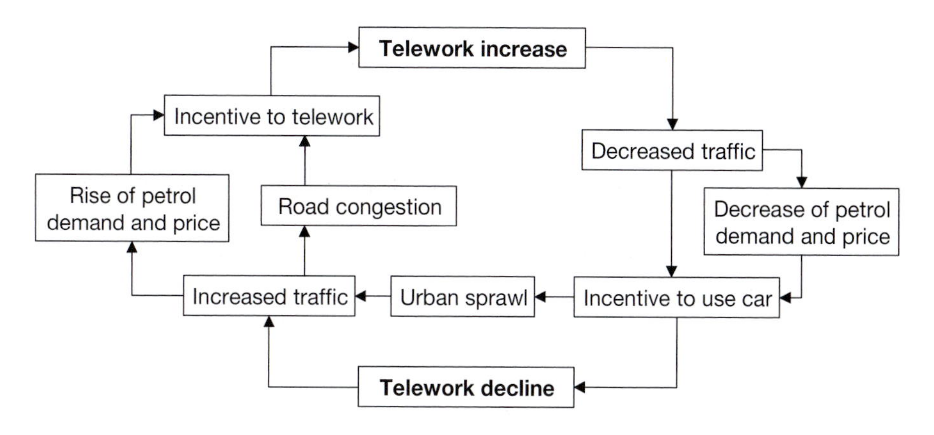

Figure 7.1 ***The equilibrium of telework drivers***

Source: Authors' own research

skyrocketing petrol prices are a driver of telework. One more time, we face a negative feedback effect: a massive increase of petrol price could spur telework implementation, which, in turn, helps to moderate petrol demand and prices.

The conclusion is in the hands of public authorities, which have some tools to break the equilibrium controlled by free market and behavioral forces. Without ad hoc regulations and planning policies, there is little reason to expect a massive increase of telework. Any telework-friendly policy should be accompanied by a bundle of correlated measures, including the reduction of road and car park surfaces, or the implementation of city toll systems such as the Congestion Charging Program of London (www.cclondon.com), which has paved the way for many large cities throughout the world (Shoup 2005).

Telework must not be considered alone, but should be linked to other means of individual travel reduction, many of them IT-related: the improvement of public transportation systems, the development of car-pooling, the rise of electronic commerce, the possible improvement of tele-learning. However, the demand for more mobility, the willingness to travel, and notably to drive, are enormous in our society. Therefore, only critical events could change the rules of the game and make telework-friendly measures politics-worthy: environmental disasters or dramatic rise of energy prices (which have not significantly changed for 40 years in real purchasing power terms). Technological changes could also impact the situation greatly. While telework, at the present time, saves less than 2 percent of petrol consumption, it is realistic to foresee a 30 or even 50 percent decrease of global automobile consumption, thanks to vehicle improvements and decrease of size (assuming changes in attitudes and behaviors, an example is the rise and then the fall of the popularity of heavy sport utility vehicles (SUVs)). Obviously, telework implementation should not be the priority in energy-saving policies.

Conclusion

In this chapter, we have addressed the fundamental "telecommunications-transportation tradeoff" (Nilles 1976) in labor organization faced today by companies, governments, unions, and individuals. We have stressed the fact that telework is an element of the "tool box" for social and economic inclusion. We have assessed the growing role of telework as a substitution to offshoring practices. In Chapter 9, we address the importance of telework implementation for individuals and communities—rural or industrial—peripheral to the core of the global economy. If no miracle can be expected from telework implementation in the field of energy savings and environment protection, telework may provide important benefits to selected individuals, companies and communities. Despite modest figures on a macroeconomic basis, telework may prove critical on a local, microeconomic scale.

Telework remains a fuzzy, somewhat hidden phenomenon, the society is embracing "walking backwards," because of the tremendous strength of cultural schemes which prevail in traditional labor and business organizations, themselves rooted in the fundamental values of our societies. A large-scale implementation of telework would mean the end of social synchronism, and would weaken the rhythm of collective experiences which is a fundamental structure of the society, in Western countries (an example: Sunday would become a working day) as well as in Asian countries (weakening of the collective values). In a more globalized world, and in a digitalized economy, synchronism is less necessary. Therefore, new ways of organizing individual, family and professional lives would probably have to be invented.

Because it often involves individual choices and practices, because it connects to the global economy millions of homes, located in thousands of localities—without mention of hotels, trains, and cars—telework may be regarded as the ultimate stage of the IT-enabled dispersal process we are witnessing today, which also comprises "global sourcing" and offshoring tendencies surveyed in previous chapters. However—and this is the very geographic paradox of the digital economy—there is equally strong evidence of geographic concentration of economic activities. In Chapter 3, we have presented the "hub and spoke" structure of the global telecommunication network, which shows the prominence of a world city network (Taylor 2004). In Chapter 4, we have reported that advanced financial services, which are almost fully digitalized, remain localized in a small number of business districts. And we have concluded that the geography of highly sophisticated business functions mirrors those of talented and creative people. In the following chapter, we analyze more thoroughly the emergence and the functioning of these creative and innovative ecosystems, whose archetype, the Silicon Valley of California, has de facto set the standards for business clusters throughout the world.

8 THE PARADOX OF A "DOUBLE-EDGED GEOGRAPHY"

Local ecosystems of the digital economy

The previous chapters have shown how digital technologies enable complex multiscalar, global business practices by placing at the disposal of organizations pools of human resources unequally distributed over space. To analyze the effects of IT on places and regions requires that we confront directly the clash between two complementary tendencies: *agglomeration* and *dispersion*. After presenting the simple dynamics that govern the processes of productive poles dedicated to the digital economy, we will see that the standard model of agglomeration—that of Silicon Valley—tends towards geographical diffusion, undergoing a process of deterioration and increasing complexity as a result of the dispersion of economic activity seen in Chapters 6 and 7.

Among the contrasts and contradictions found in the real world of places is the paradox of a "double-edged geography"—the growing significance of geography and of places at a time when distance is "dead." Space has become more "slippery" as capital investment responds to lower labor costs, new infrastructure, and pools of skilled workers that provide opportunities for profit in some locations more than in others. In response, places have had to learn to become "sticky" to attract both capital and labor—both capital investment and workers who are increasingly mobile (Markusen 1996). The geographical effect of digital technologies is not to kill distance or to annihilate space, but to make space more fragmented and complex.

> The two main economic theories are in basic contradiction. If telecommunications advances are the latest and strongest annihilators of the tyranny of distance in the saga of transportation revolutions, then economic activity should be increasingly footloose, leading to dispersal to lower cost locations or economically attractive areas, often smaller places. But against this is the classic theory of agglomeration and urbanization, that the larger the city, the greater the pool of talent, the volume and variety of supportive resources, including universities, the exchange benefits of the clustering of competing firms, the quality of conventional access (e.g., airports)—in short, the conditions that engender innovation and productivity, despite higher costs.
>
> (Morrill and Sommers 2005: 350–1)

We address several questions in this chapter. First, is economic space an archipelago, rather than becoming more uniform? Second, is the model of cluster-based agglomerations still pertinent in the IT-enabled economy? Cluster theory and its popularity among politicians everywhere is based on reported successes of some specialized business agglomerations, and is seen particularly in the worldwide evolution of the Silicon Valley model as the best example of a regional, knowledge-based cluster of IT-based industries. The imitation of Silicon Valley is overdone, since every place with a single call center proclaims itself as a technopolis, and since many creative activities (from fashion to media) have been digitalized. We then address the new IT-based types of clusters of economic activity, such as "cyberparks," which embody the attempt by public authorities to create, sometimes *ex nihilo*, "information-rich," networked places for work, trade and innovation.

Space and scales in a digital economy: "shaking hands" vs electronic transactions

> The importance of geography is not dwindling to nothing in the digital era; quite the opposite. In reality, place—geography—matters now more than ever before. If people, companies, or industries can truly live anywhere, or at least choose from a multiplicity of places, the question of where to locate becomes increasingly contingent on the peculiar attributes of any given location.
>
> (Kotkin 2000: 6)

The full story is more nuanced. The economic space (economic geography) is shaped by two opposite forces. The first is *centrifugal force*, facilitated by the constant transformation of complex and unfamiliar coordination tasks into routine activities that can be successfully accomplished at remote but cheaper locations, and thus a relentless tendency towards de-agglomeration or dispersion of production—as we suggest in Chapters 6 and 7. The second is *centripetal force*, which results from bursts of innovations that create new activities requiring high levels of complex coordination, which generate clusters of agglomeration to facilitate this coordination (Leamer and Storper 2001).

Thus, IT creates both centrifugal forces for de-agglomeration and centripetal forces for agglomeration. The result, at the metropolitan scale, is "rather than central city renaissance or absolute urban dissolution," a polycentric spatially distributed network pattern, "a regional constellation of IT agglomerations interconnected via high-speed transportation and digital networks" (Audirac 2005: 119).

The Internet is unlikely to reduce the forces for agglomeration any more than did previous rounds of innovation. Coordination of new and innovative activities depends on the transfer of complex, tacit messages, "requiring a kind of closeness between the sender and receiver that the Internet does not allow. . . . The Internet is a medium that may help to maintain relationships, but does not establish deep and complex contacts" (Leamer and Storper 2001: 643). For these contacts, face-to-face interaction is indispensable.

The most clustered economic activities are pure intellectual and creative activities, such as the arts and research and development (R&D). For example, while it has hundreds of offices and manufacturing facilities, and thousands of suppliers dispersed throughout the world, and despite intensive use of PLM software, EDI, videoconference, and other "distance-killing" applications, the car maker Renault has concentrated "all the brains" of the company in the "Technocentre," a €1 billion, purpose-built campus in the suburb of Paris. In the Technocentre, which opened in 1998, are 12,000 technicians and engineers in charge of R&D and design (2,000 CAD workstations), logistics, purchasing and IT management for the whole group (Renault 2006).

The relevance of proximity

The globalization of R&D has added costs as firms try to communicate across national, cultural, and linguistic boundaries. Reliance on IT networks must continue to be complemented by face-to-face interaction. Operating a global R&D network, with an intent to speed up innovation, places great demands on the ability to communicate scientific and engineering knowledge across physical, cultural and, at times, disciplinary distance. Face-to-face meetings, the best means of knowledge transfer, impose heavy costs in money, time, and productivity (McDonough *et al.* 1999: 384).

As companies attempt to economize on international meetings, most have taken advantage of IT to substitute for face-to-face interaction. E-mail and fax communication have advantages, particularly between team members who speak different languages. Managing knowledge, then, particularly the tacit knowledge central to a nonroutine activity such as R&D, seems to require face-to-face communication. Overall, "it is vital to maintain a critical ratio of face-to-face to electronic interaction" (Nohria and Eccles 1992: 304). Some regularity is required. "In the absence of regular face-to-face contacts which help to maintain a relationship, the usefulness of electronic communication media will be limited" (McDonough *et al.* 1999: 368). De Meyer (1993) found that within six months, personal contacts could not be maintained effectively electronically; face-to-face meetings are needed to re-establish trust and understanding.

When the knowledge being shared or transferred is complex, such as design and planning tasks early in a project, face-to-face meetings are required (Santos *et al.* 2004). A manager quoted by Sapsted *et al.* (2005: 842) quantifies the benefit: if he could have brought together all the team members for a two-week period at the beginning of the project, it would have saved six to seven weeks over the whole term. This illustrates the costs that result from restricting travel and face-to-face opportunities.

Electronic communication has not meant the "death of distance"; it merely complements and reinforces face-to-face contacts (Gaspar and Glaeser 1998; Gottmann 1970; Song *et al.* 2007). Air travel remains essential to managing global knowledge found in scattered pockets of technology and local markets, despite the increased use of the Internet and other IT. In other words, e-mail and videoconferencing help knowledge creation and communication, but they fall far short of co-location as a means of learning (Doz *et al.* 2001; Gertler 2003). Even co-location does not guarantee relational proximity, but it allows trust to be built from which network relations can develop (Blanc and Sierra 1999; Boutellier *et al.* 2000). First-mover advantages generally are found in places where face-to-face contact is enhanced: advantages in productivity, creativity, and inventiveness (Amin and Cohendet 2004; Storper and Venables 2004).

Proximity is greater in urban agglomerations, so large cities continue to be seedbeds of innovation. The basic conditions of creativity, all found most readily in large cities, include: 1) competency and traditions of knowledge; 2) capital accumulation and the physical milieu of cities; 3) communication, both face-to-face and electronic; and 4) multiplicity, variation and structural instability. In such settings, creative work is largely project-based, and highly skilled creative professionals move seamlessly from project to project (Grabher 2002; Vinodrai 2006).

It is unlikely, therefore, that cities will disappear. Larger markets have more locally targeted online content and users are more likely to connect in markets with more local online content. This suggests that the Internet is a complement to—and reinforces the existence of—cities. Overall, Sinai and Waldfogel (2005) conclude that, because of more abundant content in larger markets, the Internet is more attractive in larger places than in smaller ones. Thus, "the Internet has not rendered geography irrelevant. To paraphrase Mark Twain, reports of the death of cities at the hands of the Internet may be greatly exaggerated" (Sinai and Waldfogel 2005: 23).

Because of the concentration of expertise and creativity—a typical feature of major financial centers mentioned in Chapter 5 "the core of economic dynamism remains concentrated into metropolitan regions, especially those which have thrived as international centres of specialist expertise" (Wood 2002b: 113). Large cities continue to attract corporate investment in business services (Sassen 2006; Taylor 2004). They also attract the migrants who are most educated (Stephan *et al.* 2004), reinforcing the talent pool of urban regions.

Is economic space an archipelago?

If the digital economy were a single innovation that had diffused across the world, all places would be equally well-endowed with talent and networks. But the digital economy is far from uniform. It is centered on cities—especially large ones, which are the hubs of transportation and communication networks and which house significant pools of labor, both skilled and unskilled. Seen globally, economic space is an *archipelago* of *global city-regions* that function as the motors of the global economy (Scott *et al.* 2001; Veltz 1996). These world cities (also called global cities) are: 1) command points in the world economy, where transnational firms place their management functions; 2) key locations and markets for leading industries, especially finance and advanced producer services; and 3) major sites of production for these sectors, including production of innovations (Friedmann and Wolff 1982; Sassen 2006; Taylor 2004). They are also the principal points of production and of consumption of Internet content (Kellerman 2002; Zook 2005, 2006). As seen in Chapter 3, these "islands of the digital economy" have received a major part of the capital investment by new global providers of fiber-optic networks (Graham 1999). They are also the best-connected places in the global airline network (Derudder and Witlox 2005).

In the world city network, London and New York stand alone as the pre-eminent world cities where the bulk of capital and control resides, followed closely by Hong Kong, Paris, Tokyo, Singapore, Chicago, Milan, and Los Angeles (Taylor 2004). The relative decline of the Japanese economy has taken away Tokyo's former pre-eminence, replaced in part by Hong Kong and challenged by Singapore, resulting in a more complex system than the Triad hierarchy of two decades ago (Sassen 2001b).

Creative regions need more than infrastructure, however. They need a mix of innovative actors, both in the private sector and in universities, who conduct R&D (Feldman and Florida 1994). They also need "buzz," the interaction that generates creativity. *Buzz*, as Storper and Venables (2004) define it, incorporates the conditions of knowing what is happening and intentional face-to-face contacts, as well as the unintentional or more diffuse face-to-face "rubbing elbows" or the force of "being there." The concept of buzz was depicted decades ago by Gottmann (1970: 329), who described how "information flows criss-cross at a variety of meeting points, outside formal offices: around luncheon or dinner tables, at cocktail parties, in clubs, in the lobbies of conferences, on selected golf courses . . ." Gottmann captured the essential nature of buzz, which is the production of unforeseen and unplanned mixes of information.

The result of these forces is agglomeration or metropolitan concentration, which is particularly noteworthy in the context of innovation. Large cities are "milieux of innovation," but the innovative potential of cities is not restricted to IT industries. It extends to a variety of activities dealing with information and communication, and is essential in advanced business services, such as finance, insurance, consulting, legal services, accounting, advertising, and marketing. Indeed, a set of value-generating activities concentrated in metropolitan areas as nowhere else are the cultural industries: media, in all their forms; entertainment; art; fashion; publishing; museums; cultural creation industries (Castells 2001; Schoales 2006; Scott 2000). Kellerman (2002) ranks New York and Los Angeles as the global leaders of information production—New York the global leader in finance and information, and Los Angeles the global center for the media industry. Highly educated workers and entrepreneurs—"knowledge creators"—are attracted to a relatively small number of vibrant urban areas, such as London, New York, Paris, San Francisco, and Barcelona, where they build networks and milieux that attract additional talent (Castells 2001: 228; Florida 2002a, 2005).

The concept of the digital economy, then, is intimately wrapped up with clusters of innovation and creativity. Indeed, the top research centers are all very large urban agglomerations. Major universities and institutes, no matter how prominent, if located in small cities, rank somewhat lower. For example, Matthiessen *et al.* (2006) rank Oxford twenth-sixth and Cambridge twenty-ninth among world research centers. In effect, these clusters in large urban areas possess the ideal of agglomeration beneficial for exchange of tacit knowledge and connectivity on global networks which facilitates exchange of codified knowledge anywhere.

These urban dynamics themselves are at the core of the field of regional development, which attempts to understand why places grow. Several "territorial innovation models" have emerged over the past two decades to explain the growth of places based on innovation. The "building blocks" of these territorial innovation models include economies of agglomeration; endogenous development theory; systems of innovation, evolution and learning; network theory; and governance (Asheim *et al.* 2006; Moulaert and Sekia 2003). The regional innovation system (RIS) is a particularly powerful concept for understanding the geography of innovation (Asheim and Gertler 2005).

The pressure towards agglomeration occurs simultaneously with the ability of global production networks to extend deeper into peripheral places where low-skilled work can be done at the lowest cost. The dispersal—the *global shift*—of production of goods and services is a result of several processes and tendencies transforming the world economy. These tendencies are both organizationally and geographically uneven with no clear endpoint or equilibrium in sight (Dicken 2007).

Understanding the workings of clusters will allow us to see how the creative activities of the digital economy, particularly R&D and innovation, can be dispersed. If they can be done at many locations—in dispersed digital places—are these places the potential cores of the new economy of the coming decades?

Does the cluster theory apply to the digital economy?

The ideas of Porter have been dominant in policy circles. His concept of "clusters" has superseded the more fundamental concepts of agglomeration and localized learning. "Clusters are geographic concentrations of interconnected companies, specialized suppliers and service providers, firms in related industries, and associated institutions (universities, standards agencies, trade associations) in particular fields that compete but also cooperate" (Porter 2000: 253). Clusters increase productivity of firms in an area, drive the pace and direction of innovation, and stimulate formation of new businesses. Presence in a cluster "allows each member to benefit *as if* it had greater scale or *as if* it had joined with the others formally—without requiring it to sacrifice its flexibility" (Porter 1998: 80).

In practice, however, what has resulted is the simplistic identification of clusters based on proximity of firms that are linked in general or nationally, but without much evidence of actual local linkages and interactions among firms and between firms and institutions. Porter and others frequently merely identify the firms located in the same locale as a cluster. The institutional base simply does not exist in many cases to support the companies in the "cluster" (Scott 1996). Despite the lack of clear evidence, Porter's work has been very influential for three sets of reasons.

First, Porter has promoted his cluster concept within an overarching concern for the determinants of "competitiveness" (of firms, industries, nations, and now locations), which resonates with politicians and policymakers and their preference for micro-economic supply-side intervention and the imperatives of raising productivity and innovation (Martin and Sunley 2003). Second, Porter's clusters are framed explicitly in terms of practical business and policy strategy. Third, the generic and vague nature of clusters has provided consultants and policymakers "unlimited scope in their definition and application of the concept" (Martin and Sunley 2003: 13).

The advantages of proximity revolve around the intense communication that occurs among firms. For companies, one of the advantages of geographic clustering is the enhanced opportunity for contact with both partners and rivals. But agglomeration, even with productive specialization, does not guarantee interaction or productive linkages. But it allows trust to be built from which network relations can develop (Blanc and Sierra 1999). The impact of geographical proximity is not always direct; relational proximity may be more important (Breschi and Malerba 2001; Howells 2002).

> Geographical proximity is not so much an economic cause of agglomeration as a social effect of the embeddedness of economic relations in inter-individual relations. Face-to-face interaction between two actors cannot alone generate synergies; synergies can only develop between two individuals who belong to the same network or share common representations.
>
> (Torre and Rallet 2005: 52)

Porter treats all clusters as largely alike, and rarely explains details of the formation of a cluster and of the evolving networks, spin-offs, or long-run evolutionary paths (Asheim *et al.* 2006). But recent research has differentiated among clusters, based on the generalization that all clusters are not alike. Firms and industries vary greatly in their technological regimes and knowledge characteristics (Iammarino and McCann 2006; St John and Pouder 2006). Moreover, clustered companies face costs or "negative externalities": technological isomorphism rather than diversity, labor cost inflation, over-specialization, institutional and industrial lock-in, and local congestion.

The static, nearly all-inclusive view of clusters conflicts with the long-term process by which localized agglomerations are created and maintained. Among the most important processes is localized learning and the competitiveness that results from it. Localized capabilities translate into sustainable competencies, which enable firms to survive and thrive in high-cost locations. Capabilities can deteriorate as substitutes are found, as assets erode, as lock-in occurs, as ubiquitification takes place, and as imitators copy successful capabilities. Because of these processes—and the ever-changing dynamic of technological change—lead time matters greatly because of first-mover advantages. The essential characteristic of clusters is that they are the locations of first-movers in their industries (Maskell *et al.* 1998).

The worldwide evolution of the Silicon Valley model

Silicon Valley, located around the southern part of San Francisco Bay, is perhaps the best-known cluster in the world. Studied more thoroughly than any other economic region, it has created the expectation that clusters must be based on high-technology firms and industries, and has de facto set up the standards for all the other high-tech agglomerations worldwide, often thought of as "Siliconium" replicas (Dawson 2001).

A survey of the archetype: Silicon Valley, California

The hype about Silicon Valley is explained by its actual research and business features. No place on earth can boast similar achievements in the field of IT industries, both quantitatively and qualitatively. Small communities such as Palo Alto, Cupertino, Fremont, and Santa Clara have become famous as the birthplaces—and the headquarters locations—of emblematic IT giants (Intel, Apple, HP, Cisco Systems, Oracle, and Sun Microsystems), as well as the recent Internet generation of companies such as Yahoo!, Google, and eBay.

The digital economy as a whole is actually rooted in Silicon Valley's laboratories and research centers, such as the famous Palo Alto Research Center (PARC), formerly known (until 2002) as Xerox PARC, where laser printing, Ethernet, graphical user interface and object-oriented programming were invented. In 2005, Silicon Valley had 343,000 jobs in the high-tech and creative sectors, mainly in software, semiconductors, computers, telecommunications, biotechnology, and related services (Table 8.1). In 2005, the Valley attracted 25 percent of all US venture capital investments. The creativity of Silicon Valley's people and companies is indicated by its share of US patents granted, which grew from 5 to 11 percent between 1994 and 2004 (9,060 patents granted in 2004, a ratio of 370 patents for 100,000 people; the national average is 25).

Economic achievement has a price. The Silicon Valley cluster does not avoid negative externalities which usually result from a high concentration of people and activities. Although the average per capita income is 1.55 times greater than the US average, the standard of living is lowered by real estate inflation. In 2005, the affordability rate, which is the share of households who can afford the median priced home, was below 25 percent in Santa Clara County—from 40 percent in 1994—and 16 percent in San Mateo County. Meanwhile, the US average only slightly declined from 60 percent to 55 percent between 1994 and 2005 (Joint Venture: Silicon Valley Network 2006). Negative externalities also affect employers. The Valley is subjected to a rampant shortage of talent, with consequence on salaries and attrition rates. The phenomenon culminated at the end of the 1990s, at the peak of the dot.com bubble. The description below depicts the situation at that time:

> The cost of doing business in Silicon Valley and San Francisco is insane. For one thing, real estate is outrageous. But what I found most unbelievable was the cost of recruiting and retaining employees. You have to recruit the recruiters, and convince them to recruit on your behalf. It's exactly the opposite in Atlanta. (Interview with Jay Hall, Founder of ChemConnect, which moved from Silicon Valley to Atlanta.)
>
> (A.T. Kearney 2000: 17)

Silicon Valley and, to a lesser extent, the Boston/Cambridge region sometimes known as Route 128, have been models for other regions that seek to enjoy innovation-based economic development. Saxenian (1994) regards Boston's profile as lower, largely because its culture is more conservative and less network-rich than that of Silicon Valley, resulting in a less entrepreneurial milieu. Boston's economy, however, has

Table 8.1 ***Employment in Silicon Valley's IT industries, 2005***

Sectors	*Employment 2005*
Software	97,500
Creative and innovation services	74,000
Semiconductors and semiconductor equipment	56,400
Computers and communications hardware	55,300
Electronic component	24,100
Biomedical	21,400
Corporate offices	15,000

Source: Joint Venture: Silicon Valley Network (2006)

remained adaptive and innovative, thanks to strong universities which create and maintain both human capital and global links (Bathelt 2001; Glaeser 2005). Boston is one of the world's five major bioscience clusters identified by Cooke (2004a), as is the San Francisco Bay area.

The Silicon Valley model includes a set of institutions that promote new firm innovations: early venture capital financing, equity markets for IPOs of recently founded firms, fluid regional labor markets for global talent, fiscal policies that lower the costs of starting and operating a business, and proximity of university and research institutes within a region (Kogut 2003: 6).

The role of first-rate universities such as Stanford University, whose campus is located in the core of the valley, and nearby University of California, Berkeley, must not be underestimated. Silicon Valley's industrial saga originated in the late 1930s, when Frederick Terman, a professor at Stanford University encouraged his students to create their own businesses. Among the first were William R. Hewlett and David Packard, who developed their first product in a garage in Palo Alto. Terman also promoted the creation of the Stanford Industrial Park (Saxenian 1994).

The milieu of innovation found in Silicon Valley originated in the coincidence in space and time of microelectronics, venture capital combined with government (i.e. military) market support, and the availability of engineers from the leading electrical engineering programs at Stanford and Berkeley. Social networks of engineers, managers, entrepreneurs, and investors reinforced the research-oriented and entrepreneurial culture (Castells and Hall 1994). Its cultural attributes have given Silicon Valley "lots of chances to make it really big because the penalty for failure is relatively light" (Micklethwait and Wooldridge 2000: 214). Other important features are continuous inter-firm mobility and networks of interpersonal relationships.

This ecosystem favors not only new, small firms; it is a constellation of specialized enterprises with which large firms can link (Bahrami and Evans 1995). Miller and Coté (1987) noted two decades ago the need to maintain "technical effervescence"—a complex combination of applied research, venture capital, and institutional support for entrepreneurs. More recently, A.T. Kearney (2000) has identified for the Silicon Valley region the key characteristics of cluster regions: investors, pillar companies, government, talent, universities, and support services (Figure 8.1). Because this combination of research, venture capital, and institutional support for entrepreneurs is rare, policymakers must resist the temptation to designate too many places as innovative. For example, Tamasy (2002) concludes that there are too many innovation centers in Germany, and most of them are not competitive.

To what extent can the model of Silicon Valley and its processes of entrepreneurship and innovation be transferred to other places? The economic factors that give rise to the start of a cluster can be very different from those that maintain it. Agglomeration economies, external effects and increasing returns arise "almost naturally" after a cluster has taken off. The most difficult and risky part is to get the new clusters started. At that stage, "old economy" factors such as firm-building capabilities, managerial skills, a substantial supply of skilled labor, and connection to markets are critical for the takeoff of "new economy" clusters, including Silicon Valley 40 years ago. These initial factors are different from those that keep it going (Bresnahan *et al.* 2001).

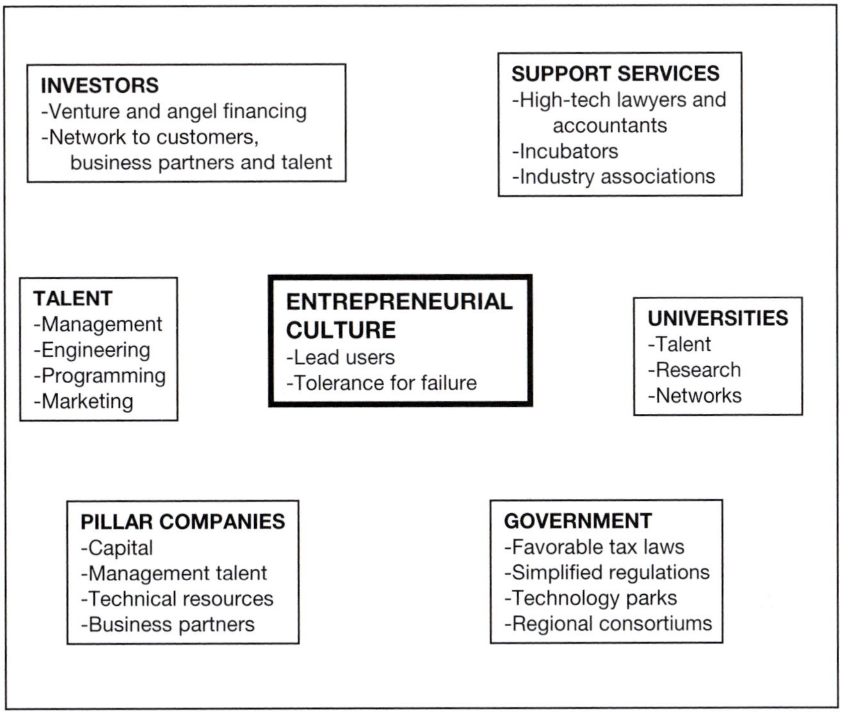

INVESTORS
-Venture and angel financing
-Network to customers, business partners and talent

SUPPORT SERVICES
-High-tech lawyers and accountants
-Incubators
-Industry associations

TALENT
-Management
-Engineering
-Programming
-Marketing

ENTREPRENEURIAL CULTURE
-Lead users
-Tolerance for failure

UNIVERSITIES
-Talent
-Research
-Networks

PILLAR COMPANIES
-Capital
-Management talent
-Technical resources
-Business partners

GOVERNMENT
-Favorable tax laws
-Simplified regulations
-Technology parks
-Regional consortiums

Figure 8.1 ***The ecosystem of Silicon Valley***

Source: Based on A.T. Kearney (2000), Bahrami and Evans (1995), and Mickelthwait and Wooldridge (2000)

Box 8.1 The 20-minute rule

The "20-minute rule" guides decisions in Silicon Valley. Craig Johnson, managing director of Concept2Company Ventures, a venture capital firm in Palo Alto, California, who has 30 years of experience in early-stage financings, said he knew many venture capitalists who adhered to the 20-minute rule: if a start-up company seeking venture capital is not within a 20-minute drive of the venture firm's offices, it will not be funded. Mr. Johnson explains that close proximity permits the investor to provide in-person guidance. At first, that may entail many meetings each week before the investor and entrepreneur know each other well enough to rely mostly on the telephone for updates. Those initial interactions are fateful. "Starting a company is like launching a rocket," Mr. Johnson said. "If you're a tenth of a degree off at launch, you may be 1,000 miles off downrange."

Sequoia Capital, another Silicon Valley venture capital firm, makes its preference for the 20-minute rule almost explicit, telling applicants whose companies are at the "seed stage" (receiving less than $1 million) or "early stage" ($1 million to $10 million) that "it is helpful if the company is close to our offices" because they "require very frequent contact."

Stross (2006)

Entrepreneurial universities, properly configured, play a central role as centers of excellence in one or more key technologies of the new economy (Cooke 2002). Along with government and industry, they form a *triple helix* of interaction, which generates a "bottom-up" outcome of innovation and "collective entrepreneurship," and creates a support structure for firm formation and regional growth. The result of these initiatives is a self-sustaining dynamic in which the role of academia and government appears to recede as industrial actors become prominent in the formation of new firms (Etzkowitz and Klofsten 2005).

Knowledge economies such as these are "localized and regionalized, clustered, collective learning systems" (Cooke 2002: 187). The landscape of such clusters will continue to be uneven, because private venture capital is so central to innovation and entrepreneurship. Public funding is able to level the playing field somewhat, by supporting the infrastructure for R&D. Therefore, only a small number of places have the network architecture, culture of innovation and entrepreneurship, and a knowledge base to attract global companies on a scale to rival Silicon Valley.

A final aspect that allows clusters to remain competitive is that the kinds of knowledge on which they are based—and which they continue to generate—is not simple *component knowledge*, but complex *architectural knowledge*. Component knowledge is largely technical, and can include subsystems but, most importantly, can be codified and become ubiquitous (Maskell 1999). *Architectural knowledge* relates to the organization of an entire system as well as the structures and routines for organizing knowledge (Pinch *et al.* 2003). It is tacit, systemic, embedded in organizations, and closely related to the synthesis of capabilities known as *systems integration* (Pavitt 2003).

Over time, mainly through mobility of workers, firms in a cluster tend to develop a cluster-specific, inter-firm stock of knowledge that is distinct from that anywhere else in the industry. Knowledge flows across cluster boundaries are minimal, leading to the creation of "spatial knowledge monopolies" (Cooke 2005). These reflect a combination of both local networks of local industrial knowledge and networks (or pipelines) to global knowledge (Bathelt *et al.* 2004). However, there is evidence that the constraining effects of physical distance have weakened, at least between innovators and collaborators (Johnson *et al.* 2006).

A certain degree of dispersion in R&D activities

The example of Silicon Valley and other knowledge-based regions has led to a "global competition of high-tech centres" (Anttiroiko 2004a). Much of the competition is based on hype, with "branding" and names similar to that of the archetype (Dawson 2001).

In part, such plans are based realistically on the widespread dispersion of R&D away from large, central facilities near corporate headquarters to new locations (Boutellier *et al.* 2000; Karlsson 2006). Two principal motivations push companies to establish R&D sites abroad: first, proximity to other corporate activities (such as manufacturing plants) and proximity to local customers and, second, the quest for technical know-how and expertise available in only a few centers of excellence around the world (von Zedtwitz and Gassmann 2002). Internationalization in research has been characterized by an increasingly selective focus on a small number of locations and the concentration of innovative activities in worldwide centers of excellence. Global firms locate their R&D laboratories "where the best conditions are to be found worldwide, both for research and also for the transfer of its results" (Meyer-Krahmer and Reger 1999: 763).

The combination of complex knowledge, shorter product lifecycles, and competition from a larger number of competitors results in a new rationale for research and development in large firms. Companies now must:

- identify, understand, select from, and connect to the wealth of available external knowledge;
- fill in the missing pieces of knowledge not being externally developed;
- integrate internal and external knowledge to form more complex combinations of knowledge, to create new systems and architectures;
- generate additional revenue and profit by selling research outputs to other firms (Chesbrough 2003: 53).

Consequently, the challenge for companies is to manage and integrate these internal and external flows into a coordinated innovation process (Chesbrough 2003; Karlsson 2006; Kuemmerle 1997). In effect, most TNCs have developed a double network organization, with an internal network and one or more external networks (Zanfei 2000).

Firms believe they must tap into many sources of expertise because knowledge is increasingly diverse and dispersed (Doz *et al.* 2001: 42). As many as two-thirds of the R&D facilities of multinational firms are now located outside the home country (Doz *et al.* 2006). No single source—even the best in the world—is sufficient. Increasingly, technologies and lead consumers are found outside the "capitals" of an industry. In order to plug into new pockets of knowledge, a firm needs a base—a location—in a source of knowledge. That might mean a science park in one or more of the dozens of major research cities. From these sites, firms can sense (i.e. identify and access) innovative technologies and local market knowledge (Doz *et al.* 2001).

The outcome of numerous individual decisions is that multinational corporations (MNCs) tend to locate their R&D activities in regions that are characterized by a richness of knowledge (Zeller 2004). The structure of inter- and intra-firm networks is shaped by the geography of talent. For example, the Swiss pharmaceutical giants have anchored themselves in regional arenas of innovation outside Switzerland, such as the San Francisco Bay Area, Boston, and San Diego. These efforts contribute to the processes of global scanning, transferring, and generating new pharmaceutical compounds and technologies.

Clusters in new economy sectors are less common than in other fields: they are found mainly in large cities and university towns, and rarely in rural areas or in industrial cities. The new economy innovation system relies more on venture capital and entrepreneurship (including serial start-ups and incubators), and much less on public sector bodies than in the "normal" regional innovation system. Examples of new economy clusters cited by Cooke (2002) include Silicon Valley, the Telecom Corridor in Richardson, Texas, biotechnology clusters in Cambridge, Massachusetts and Cambridge, England, the IT industry cluster also in Cambridge, England, and new media in London, New York, and Cardiff. The San Francisco Bay area is among a small number of the leading centers of biotechnology (Cortright and Meyer 2002), and in the newer high-tech field of nanotechnology or nanoscale science (Zucker and Darby 2005).

Cooke (2002) summarizes the characteristics of knowledge economies as five elements: research funding, a systemic method of supporting clusters of new businesses, incubators for new firms, venture capital, and policy intervention to build clusters outside large metropolitan areas. The absence of any of these, particularly venture capital and the management advice that typically comes with it, automatically suppresses entrepreneurship. And without spin-off entrepreneurship, nothing like a Silicon Valley can emerge. "Knowledge and innovation capabilities cannot be created by decree, and the processes of knowledge creation, utilization, and transformation are not top-down processes" (Schwaag Serger 2006: 258).

The presence of a major research university and a large pool of skilled workers are more important to the development of cutting-edge innovations in new industries. With few exceptions, technology locations are more important than market locations for R&D facilities (Florida 1997). Cities with a reputation as "creative" can keep and attract skilled workers more easily than those perceived as "working class" (Florida 2002a). Cutting-edge science is strongly tied to the universities and research centers where it originates. In nanotechnology, Darby and Zucker (2003: 18) conclude that:

> Where commercial opportunity is built on fast-advancing academic science it is generally more economical to establish commercial laboratories and even manufacturing facilities near the universities than to try to move the scientists and their network to an existing firm location.

Many places have some, if not all, of the characteristics of an innovative cluster. Table 8.2 proposes several categories of innovative places. The first type includes the paradigmatic innovative region, Silicon Valley, to which we add Grenoble, France, as we will detail. The other types of genuinely innovative regions have different characteristics. Innovative concentrations include some impressive and familiar clusters, such as Austin, Hsinchu, and North Carolina's Research Triangle. University-based clusters in small or suburban regions include Cambridge and Oxford in the UK as well as Turku, Finland. Large urban regions have many of the characteristics of Silicon Valley, such as Boston, Helsinki, and Stockholm. Finally, a number of places have been the products of policy attempts to build high-tech growth, based usually on science parks and related developments.

Based on decades of research on industrial districts in Italy and elsewhere, Paniccia (2006) identifies the *science-based or technology agglomeration*, which is based on venture capital and the very active role of knowledge institutions, as one of two types most able to absorb and adapt to fluctuations in the economic cycle. The other is the *canonical industrial district*, found mainly among the family-based firms in Italy. When created in remote locations, science complexes are unable to benefit from—or to create—buzz. Likewise, their pipelines to the outside world are smaller than would be the case if located more centrally in their national territories.

Castells and Hall (1994) emphasize the importance of time for a technopolis to develop—at least 20 to 30 years. Indeed, Longhi (1999) and Lazaric and Thomas (2006)

Table 8.2 ***Types of innovative agglomerations***

Types	*Alternative names*	*Examples*
Paradigmatic high-tech clusters		Silicon Valley (US) Grenoble (France)
High-tech microenvironment	Incubator, Accelerator	International Business Incubator in San Jose (US)
Research center	Center of excellence, Innovation center	Canadian Innovation Centre at Waterloo
High-tech industrial park		Zhongguancun (China), Multimedia Super Corridor [or MSC] (Malaysia)
Science park	Science park, Research park, Technology park, Technopark, Software park, Technology precinct, Knowledge park	Hsinchu, Sophia Antipolis, Cambridge, Mjärdevi (Sweden), Innopoli (Espoo, Finland)
Technopolis		Kumamoto and Hamamatsu (Japan)
Science city		Tsukuba (Japan), Daedok (Korea), Akademgorodok (Russia), Kista (Sweden), York (UK)
Intelligent city	Smart community, Learning city, Learning village, Knowledge city, Digital city	Intelligent Island of Singapore, Cyberjaya (part of MSC in Malaysia)
High-tech city	High-tech metropolitan area, Technocity	Helsinki, Munich, Stockholm, San Jose, Tel Aviv
Innovative concentrations		Austin (Texas, US), Bangalore (India), Research Triangle (US)
Suburban and university-based clusters		Cambridge and Oxford (UK), Turku (Finland)
Large high-tech complex	High-tech region, Learning region, Innovative region	Silicon Valley (US), Boston Route 128 (US)
Global or macroregional network	Associations and networks of high-tech centers, Innovation networks	International Association of Science Parks (IASP), World Technopolis Association (WTA), Telecities

Source: Adapted from Anttiroiko (2004a), Table 1

both believe that Sophia Antipolis, near Nice, France, has developed into an innovative milieu only after 30 years. Its focus to some degree on IT has led to the name Telecom Valley.

Technology-based clusters such as Silicon Valley feature two key dynamics. First, learning through networking and by interacting is the crucial force pulling firms into clusters and the essential ingredient for the ongoing success of an innovative cluster. The ways firms learn in innovative clusters include user–producer relationships, formal and informal collaborations, inter-firm mobility of skilled workers, and the spin-off of new firms from existing firms, universities and public research centers (Breschi and Malerba 2001). A second key feature of successful high-technology clusters is the high level of embeddedness of local firms in a very thick network of knowledge sharing, which is supported by close social interactions and by institutions that build trust and encourage informal relations among actors.

Cooke and Morgan (1998) provide a parallel set of findings. Focusing on Europe, they characterize Germany's Baden-Württemberg and Italy's Emilia-Romagna as "model regions" where associational activity, learning capability, and networking practices among firms and governance organizations are dense and effective. We present more thoroughly below Grenoble, in the Rhône-Alpes region (France), as a successful example of ecosystem outside the US that has parallels—in qualitative if not quantitative terms—with Silicon Valley.

Breschi and Malerba (2001: 832) suggest several lessons:

- Clusters have specific stages of development, and these are related to the knowledge and learning processes and accumulation of capability of key actors.
- The essence is the coexistence of localization and globalization.
- Clusters are embedded in specific national systems of innovation and production, and a healthy cluster is marked by international specialization and demand linkages.
- The emergence of local institutions serves to make each cluster unique.
- There is no 'complete recipe' for the replicability of successful clusters.

Entrepreneurs are a critical element in the formation of clusters. Entrepreneurs who adapt to both constructive crises and new opportunities create the factors and conditions that facilitate their business interests and these, in turn, contribute to the development of external resources. Evolution of the entrepreneurial clusters in IT and biotechnology in the Washington, DC region took, as elsewhere, nearly 30 years to mature into a well-functioning and rich innovative and entrepreneurial system (Feldman *et al.* 2005). Other examples include Denver (Colorado), a hub of IT firms—wired and wireless, cable and satellite (Le Blanc 2001), and Telecom Corridor of Richardson (Texas), near Dallas (Cooke 2002). Dense local networks are not guaranteed in such new economy clusters; large firms have many global, but few local, links (Lyons 2000).

The importance of venture capital

Zook (2005) makes a strong case for the importance of venture capital in producing the geographical pattern of the Internet industry, as well as venture capitalists' central role in producing the key firms of the emerging industry. Most importantly in the context of our earlier discussion of tacit knowledge, Zook believes that venture capital is rooted in local networks and knowledge.

Others have concluded that Silicon Valley and its success are fundamentally about both venture capital and the free-spirited entrepreneurial culture or ecosystem, which cannot be replicated elsewhere. In a study of 22,479 venture capital deals completed throughout the US between 1992 and 2001, "on average, start-ups in Silicon Valley receive VC at a younger age, complete more rounds of financing, and raise more money in each deal" (Zhang 2006a: 17). The regional advantage created by Silicon Valley's local venture capital industry is not the result solely of more money. It is "the consequence of the unique interaction of monetary, human, and social capital" (Zhang 2006a: 29).

As the phenomenon of venture capital has spread worldwide from Silicon Valley and the US, it has faced a number of institutions—relating to local norms, regulations, and perceptions—which suggest that the practice of venture capitalism will differ from country to country and from region to region (Bruton *et al.* 2005). Some of the factors hindering growth of venture capital in Europe include traditional attitudes to firm ownership and a lack of active stock markets The problem in Europe is threefold: 1) an overall low level of high-technology activity; 2) a lack of large and well-developed clusters of such activity; and 3) a corresponding lack of localized concentrations of venture capitalism (Martin *et al.* 2002).

In Finland, where high-tech activity has been above the European norm, Maunula (2006) surveyed biotechnology companies, who reported that the non-financial support from venture capitalists was important for their success. The venture capitalists provide value-added indirectly through screening and signaling, and directly through monitoring and providing non-financial support in a variety of business areas, for example, in strategic planning and in obtaining additional financing. Informal venture capitalists were perceived to provide value-added the most and were the most active in other areas.

Cooke *et al.* (2002) conclude that public research resources and private commercialization funding are central, supported by a wide array of private but few public innovation support services. Based on a number of cases from cities in less favored regions and countries where innovation has occurred, they find that the research-venture capital model is present and explains innovation in each case. In some cases, public venture capital has substituted for private capital. What is most important is the level of cooperation among regional actors—higher education and research laboratories, private industry, and governments (federal, provincial/state, local). "A high level of cooperation among the three actors is more important than which one operates as the prime mover behind development" (Corona *et al.* 2006: 214).

The emergence of new ecosystems of digital industries

The attempt to assemble and create the conditions for a successful, self-sustaining cluster is an example of "constructed advantage" (Cooke and Leydesdorff 2006). On a large scale, we see the emergence of "technopolises" or "high-tech cities," where the complex interlinkage of IT industries and supporting institutions actually works as an ecosystem. We survey two emblematic examples. True technopolises are usually not found in developing countries, with perhaps the exception of Bangalore, the site of not only call centers but a booming BPO business in software outsourcing (Chapter 6), as well as recent growth in R&D services. In developed countries, Grenoble has emerged as "the French Silicon Valley," but deserves more than a metaphoric comparison.

In some cases, the advantage is constructed in a very limited space. This is the case of science parks and high-tech campuses which have mushroomed in the main cities of emerging countries such as China and India. These enclaves are "islands where everything works," with infrastructure customized to the needs of foreign investors

(Graham and Marvin 2001: 337). In the end, we are witnessing the emergence of a "digital urbanism," which endeavors to generate systemic development effects by leveraging IT applications—especially those provided by local enterprises—in the whole range of urban functions, making the city a living laboratory of the digital economy.

R&D clusters and the case of Bangalore

The development of R&D clusters in India dates back only to 1986, when Texas Instruments began operation of a software design facility in Bangalore, a city which has attracted investment by additional global firms. This in turn led to clustering, and to the setting up of a number of entrepreneurial firms in the same cluster locations. Pune was another early cluster, and was among the first to promote industry-academic collaboration (Singhal and Rogers 1989). The result has been the stimulation of significant entrepreneurial clusters in Bangalore, Hyderabad, Delhi, Mumbai-Pune, and Chennai (Bowonder 2001). India is building many technology parks to lure back native talent and produce world-class tech companies.

Of the clusters in India, Bangalore has received the most attention and, indeed, it remains the exemplar of Indian acceptance of foreign multinational corporations (MNCs) and of long-term benefits from the embeddedness of MNCs. Part of Bangalore's base as a high-tech center is a result of Bharat Electronics Limited, a government defense firm, which has served as the "mother hen" for the electronics sector in the area (Singhal and Rogers 1989). However, while most MNCs (75 percent) collaborate, or plan to collaborate, with nearby universities, they rarely do so with government R&D institutes directly, preferring joint ventures (Fromhold-Eisebith 2002).

Bangalore is "choking on its own success," having grown from a population of 800,000 in 1951 to 5.6 million in 2001 and an estimated 7 million in 2005 (Hamm 2007; *The Economist* 2005a). "Bangalore suffers the infrastructure shortcomings common to many Indian cities: a water shortage, inadequate sewers, an erratic power supply, and pot-holed roads too narrow for the traffic they need to bear" (*The Economist* 2005a: 68)—problems noted for several years and not addressed (Audirac 2003; Heitzman 1999).

Bangalore and other regions in India remain constrained by severe shortages of infrastructure, such as transportation and electric power, and of its educational institutions (Long 2006). Local and foreign BPO firms are attracted to Infosys City, a 50-acre corporate campus, and ITPL (International Technology Park Ltd), a high-tech community with office space, production space, and 400 condos. Both provide their own dedicated utility infrastructure, including water, power, sewer, and telephone exchange, as well as satellite communications, shuttle service, commercial space, and training, recreation and entertainment facilities (Audirac 2003).

MNCs address these infrastructure shortages by improving garbage collection, police services, street lighting, and health care as well as local schools and universities. MNCs' investment in higher education and "embedding" in the locality has contributed to inculcating a "commercial culture" after the "license raj" and links to global knowledge networks, sparking a long-term regional cycle of learning and mutual support systems in Bangalore (Fromhold-Eisebith 2002).

In contrast to such positive impacts, interfirm interaction and intellectual content of the software are low—so low that Parthasarathy (2004) says that it is wishful thinking to call Bangalore India's Silicon Valley, so different it is from the California region. And Indian policy has taken a 180-degree turn from self-reliance in the supporting of private corporate campuses and technology parks where dedicated utility infrastructure (including water, power, sewer, telephone, and satellite communication), shuttle service, commercial space, and training, recreation and entertainment facilities lure MNCs to "work-live-play" fortresses in Bangalore's edge-city developments, relocating from elsewhere in Bangalore. The Karnataka state government plans to spend nearly $50 million on digital and intermodal connectivity as a string of "information technology colonies" in the IT corridor east of Bangalore near the airport (Audirac 2003).

Saxenian (2006) describes the ecosystem for technology start-ups in Bangalore as constrained both by infrastructure problems and lack of collaboration. "Local start-ups remain distant from customers and lack the interaction and learning that characterize a local technical community," compounded by "the traditional secrecy and hierarchy" typical of Indian businesses (Saxenian 2006: 320–1). Arora *et al.* (2004) believe that India can be different from the Silicon Valley model. "The typical business model for Indian software service firms does not call for venture capital" (Arora *et al.* 2004: 108) Indeed, they question whether spillovers of technical knowledge among co-located firms, which "may be critical to Silicon Valley and for clustering," might not be critical elsewhere. "Other types of externalities," which they do not specify, may be more salient in follower regions (Arora *et al.* 2004: 116).

As discussed in Chapter 6, IT services alone are likely not a sustainable base for Bangalore and for India. In Bangalore, entrepreneurs are beginning to emerge in firms creating software for embedded systems in consumer goods such as cell phones and microwave ovens, transport equipment such as automobiles and aircraft, and industrial process control systems (Parthasarathy and Aoyama 2006).

Birth and life of a digital ecosystem: the example of Grenoble

One of ten "islands of innovation in Europe" (Simmie 2002), Grenoble deserves more than a metaphorical comparison with Silicon Valley. Digital development here is not the result of a recent public initiative, like the science park of Nice–Sophia Antipolis. Rather, it is the outcome of an industrial and entrepreneurial saga begun at the end of the nineteenth century.

The industrial fortune of Grenoble results in great part from the harnessing of the hydraulic power from the surroundings mountains by Aristide Bergès, a local engineer, in 1869. At the turn of the century, hydroelectric power favored the development of chemical, metal, and mechanical engineering industries. "The technical sophistication of the technology required to produce hydroelectric power led to close links between industry and science, for the training of future engineers and for research. Such exchanges sowed the seeds of a new departure, which began to take shape in Grenoble in the 1950s" (Agence d'Etude et de Promotion de l'Isère 2005).

After the Second World War, Grenoble was the birth place of large electrical engineering firms such as Neyrpic (now Alstom) and Merlin-Gerin and Télémécanique, now both consolidated in Schneider Electric, a pillar of Minalogic, the local association of IT industries. After electric industries came electronic. In the 1960s, Serge Kampf founded in Grenoble Cap Gemini, now a world-leading computer services corporation.

However, as in California, public initiative has played since the 1950s a critical role (Druilhe and Garnsey 2000). Because Grenoble already had a good scientific base, abundant electricity, and given its strategic geographic location (far from the Iron Curtain and the Red Army), it was chosen to host French research in nuclear physics, with the creation, in 1956, of the Centre d'Etudes Nucléaires de Grenoble (CENG), a division of the Comissariat à l'Energie Atomique (CEA) (Sternberg 1996). Within the CEA was created in 1967 the LETI (Laboratory of Information Technology Electronics), which has become one of the largest European research centers in micro and nanotechnologies.

The region found its Frederick Terman in the person of the physicist Louis Néel, a Nobel Prize winner, who established his lab in Grenoble during the war. He was the promoter and first director of the CENG, and exerted constant lobbying in favor of Grenoble in the competition between French or European cities for the location of heavy, public research equipment, such as the Franco-German high-flux reactor (Institut Laue Langevin) in 1967, and the European Synchrotron Radiation Facility (ESRF), created by 12 European countries in 1988.

A vibrant entrepreneurial culture has grown from the seedbed of national laboratories and research centers (De Bernardy 1999; Dunford 1991). The Aire de Services et de Transferts de Technologie (ASTEC), an incubator within LETI, was the birthplace of over 80 start-ups. The most spectacular spin-off story is that of SOITEC, which was created within LETI in 1992 by two researchers who patented, developed and exploited the Smart Cut(tm) process, a revolutionary technique used for the transfer of ultra-thin layers of silicon on insulator (SOI). Today, with 1,000 employees, SOITEC is the world leading producer of SOI wafers, with €263 million in sales for the fiscal year ended March 31 2006, according to the company's website. The success of many other start-ups shows that the region has succeeded in shifting from finance based on government funds to venture capital (Brioschi and Cassia 2006; Lawton Smith 2003).

Another landmark of the Grenoble technopolis is Inovallée, formerly the ZIRST, created in 1972 and one of the oldest French science parks. In 2005, it hosted 300 firms and 9,000 jobs (www.inovallee.com). Ten kilometers north of Inovallée is located the massive building of the Crolles2 Alliance: Philips Semiconductors (now NXP), Freescale Semiconductors (formerly Motorola), and ST-Microelectronics have contracted a $3.4 billion, five-year (2002–2007) partnership, which is dedicated to design and production of chips using 30- to 92-nanometer lithography on 300 mm silicon wafers, and SOI technology, according to the companies' websites.

The crown jewel of the Grenoble digital ecosystem is the Minatech, a brand new campus located in the center of the city, inaugurated in April 2006. It was co-founded by the LETI and by the National Polytechnical Institut of Grenoble (INPG), a consortium of nine engineering graduate schools, including the prestigious ENSIMAG, the leading French graduate school in applied mathematics and computing. Built with strong support of local and regional public authorities, Minatech is dedicated to training and research in microelectronics and nanotechnology, and designed to host up to 1,000 students and 2,000 researchers from diverse local institutions. Figure 8.2 presents the ecosystem of Grenoble.

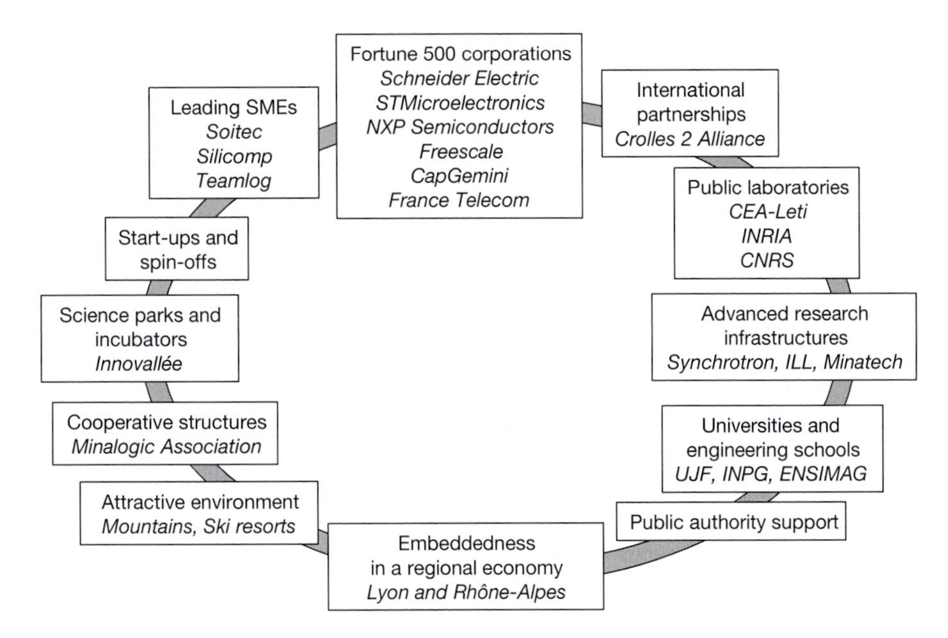

Figure 8.2 ***Grenoble: A local ecosystem of the digital economy***

Source: Author's own research

Such a concentration of innovative activities and high-end educational institutions (more than 50,000 students) is remarkable, given the medium size of the city (500,000 inhabitants in the metropolitan area), and its peripheral location within the economic core and the main transportation corridors of the European Union. However, it is precisely the high degree of specialization and brain concentration in a narrow, cutting-edge sector—although Grenoble has built a presence in biotechnology (Cooke 2004b)—which gives the cluster its international visibility and favor synergies, cooperation, and spin-off phenomena. And the natural environment must not be neglected. With the world's largest ski resorts nearby, it is not surprising that young graduates who are recruited on a global basis (40 percent of ENSIMAG's students are foreigners) seek a job locally. Grenoble is a place where high-tech positions are easy to fill.

Nevertheless, the area suffers from the drawbacks—made worst by the mountainous environment—which often hamper digital clusters: lack of green fields, traffic congestion, and California-like real estate prices. Digital clusters are not only business places, but also living places. A serious problem occurs, therefore, for people who are not IT professionals and do not get "IT salaries"—the majority of the population, as in Silicon Valley. This is a kind of indirect "digital divide," which could threaten the overall prosperity of the cluster if it is not properly treated by policymakers.

Planning digital clusters

Many other places are attempting to rise above the rest in the competition of high-tech centers. Many different types of high-tech formations have evolved, with no clear criteria to distinguish among them (Anttiroiko 2004a). The International Association of

Science Parks (IASP) in mid 2006 had 315 members in 67 countries. Other groups, such as the Association of University Research Parks (AURP) in North America, and the UK Science Park Association (UKSPA), perform similar networking functions.

Macdonald and Deng (2004: 3) conclude that "what little evidence there is does not conclude that science parks offer the optimum location for high technology firms. Indeed, it would seem that the science park offers little advantage at all." The most successful regions, including Silicon Valley, were the product not of planning—but of a serendipity grounded in the benefits of agglomeration economies, externalities, networking, and clustering, with most information flow occurring informally.

Regions in Asia are enthusiastically creating their own versions of Silicon Valley. Many of the more successful ones are by US-educated Asians who can combine their cultural knowledge of their native country with an understanding of the Silicon Valley business culture, community, and institutions. Based on research in Taiwan, India, and China, Saxenian (2006) suggests that these transnational entrepreneurs provide a mechanism for the international diffusion of knowledge and the creation and upgrading of local capabilities. This upgrading is distinct from, but complementary to, global production networks. China has created dozens of "Returning Students Science Parks" to attract foreign-educated Chinese to return home, bringing with them their contacts and cross-cultural knowledge.

China has established a large number of science parks, with the aim of sparking learning both from multinational corporations and from local R&D (Walcott 2003). In most cities, TNCs mainly process and assemble products using imported critical parts plus locally made components with the help of cheap labor and easy access to major ports such as Hong Kong and Shanghai. Shanghai, by contrast, is the best example of a "multinational learning zone" capable of providing a wide range of skills, goods and services. There, TNCs are engaged in manufacturing activities in proximity to Chinese firms, research and development entities, both within and outside designated parks. China touts Hong Kong as an "e-hub" and both Beijing and Shenzhen as "China's Silicon Valley" (Sum 2003; Zhou 2005).

Taiwan's Hsinchu region near Taipei is another area receiving attention as a successful imitator and partner of Silicon Valley (Saxenian 2004; Sum 2003). The return of Taiwanese with educations and business experience elsewhere, including in Silicon Valley, has been a feature of Hsinchu since the 1980s. As a result, firms in the Hsinchu Science-based Industrial Park (HSIP) have developed a large number of global as well as local links (Wang 2005).

Singapore, through its Infocomm Development Authority, has since the mid 1980s followed an IT Development Plan, embodied most recently as iN2015, or Intelligent Nation 2015. The new plan includes advanced ubiquitous broadband technologies, smart sensor technologies, and nanotechnology and biotechnology supported by computing (Corey and Wilson 2006). Singapore's plans have always gone well beyond infrastructure with a goal to become a world-class creative city (Corey 2004). Keivani *et al.* (2003) evaluate Singapore's efforts as "a highly focused economic development strategy and a harmonised, transparent, accountable and supportive institutional milieu." However, Phillips and Yeung (2003: 710) describe the Singapore Science Park as "little more than a form of glorified property development" because of the low level of local linkage and synergy.

Other places trying to imitate Singapore, such as Dubai and Abu Dhabi, fall well short. Dubai is clearly the most ambitious. High-tech projects Dubai Internet City, Dubai Silicon Oasis, Dubiotech (the Dubai Biotechnology and Research Park), and Dubai Media City are among several "cities within the city" mega-projects; others focus on tourism or regional offices of TNCs (Pacione 2005). Whether these projects can be world-class will depend largely on the ability to attract creative talent from abroad. To cite but one additional example, HITEC City is the Hyderabad Information Technology Engineering Consultancy City, a technology township in Hyderabad, India.

A number of examples of innovative agglomerations are included in Table 8.2, based on the work of Anttiroiko (2004a). His compilation omits others, such as Sydney, "Australia's Silicon Valley" (Searle and Pritchard 2005) and Kyoto (Edgington 2007). Some innovative agglomerations have grown from science parks; others have larger ambitions, such as to be known as science cities. The central importance of a research-based university and minimal amounts of manufacturing distinguish science cities from other places (Anttiroiko 2004b). Cyberjaya, part of the Multimedia Super Corridor (MSC), is intended to be exactly that: a science city within the 50-km long MSC (Bunnell 2004). The point to be made is that there is tremendous competition among places to be identified as key nodes in the innovative digital economy (Anttiroiko 2004a). They continue to attract media attention as well as investment (Forrest 2007).

Digital urbanism

In a variety of ways, cities and regions have attempted to empower their citizens with digital technology. Digital urbanism is strategic, aimed at constructing advantage in a competitive world economy. We can see this among some ambitious cities striving to join the list of world cities, including Dubai, Kuala Lumpur, Hong Kong, and Seoul. In these cities, major digital developments—Multimedia Super Corridor in Kuala Lumpur, Dubai Internet City, Hong Kong's Cyberport, and Seoul Digital Media City—represent efforts to construct world-class infrastructures, based on digital technologies.

Reflecting the need for places and people to development aspects of the digital economy *and* the knowledge economy, *intelligent cities* combine real and digital in a local development process.

> An intelligent city is an island (community) of technological innovation integrating real and digital innovation development functions. The functions of an intelligent city relate to the production of knowledge (R&D), technology transfer, funding innovation, new product development services and the production process, technological collaboration activities, and networking.
>
> (Komninos 2002: 201)

The Intelligent Community Forum highlights examples of these activities, making annual awards to the Intelligent Community of the Year and Intelligent Building of the Year, as well as the Intelligent Community Technology and Intelligent Community Visionary of the Year. Table 8.3 lists the awards made thus far to communities, "each a model

Table 8.3 *Intelligent Community Awards*

Year	*Intelligent Community of the Year*	*Intelligent Building of the Year*
2006	Taipei, Taiwan	MaRRS Center, Toronto, Canada
2005	Mitaka (suburban Tokyo), Japan	Ebene Cyber Tower, Mauritius
2004	Glasgow, Scotland, UK	Cyberport, Hong Kong, China
2003	*No awards presented*	
2002	Calgary, Alberta, Canada and Seoul, Korea	International Tech Park, Bangalore, India
2001	New York, New York, US	The Wharf (Holdings), Hong Kong, SAR
2000	LaGrange, Georgia, US	Caracas Teleport, Venezuela and Hong Kong Land, Ltd
1999	Singapore	111 Eighth Avenue, New York, US

Source: Intelligent Community Forum (2006)

for the development of prosperous local economies and vibrant societies in today's hyper-competitive broadband economy," and to buildings.

Table 8.3 confirms that several of the cities seeking world-city status have been successful. Seoul, Singapore, and Taipei have been named an Intelligent Community of the Year, and buildings in Bangalore, Caracas, and Hong Kong have been named as Intelligent Building of the Year.

A digital city is necessarily the combination and overlap of three domains:

- the physical urban area;
- the communities of people associated with that area, and;
- the technological, socioeconomic and ideological possibilities and constraints.

Not only ideologies, but also interpretations of the various metaphors common to IT, greatly influence how citizens perceive and react to a digital city (Couclelis 2004b). Because of these multiple domains, most modern cities have become digital cities in one way or another, reinforced by IT and globalization, telework, and telecommuting, e-government, and virtual diasporas. The greatest change perhaps is the "cyberweek," and the absence of traditional temporal boundaries of work (Laguerre 2005).

The Internet, and its ability to create bottom-up and horizontal networks, led to many attempts to take advantage of these. IT policies in European cities have focused on three areas: infrastructure provision, content development, and enhancing access for underserved groups. Generally, better infrastructure leads to more and better content and services (including e-government services), which leads to more local users, who stimulate demand for infrastructure (van der Meer and van Winden 2003). Provision of broadband has become a high priority recently (van Winden and Woets 2004).

The European Community funded a number of "digital city" initiatives, involving not only websites as repositories of information but also as "civic networks." Overall, "the urbanization of cyberspace was wide but shallow, and . . . few initiatives were designed to encourage and support public discourse and a wider access of citizens and firms" (Aurigi 2005: 77). Most initiatives were primarily for marketing and self-promotion purposes, frequently in English to address foreign audiences. Detailed comparisons of digital cities in Italy and the UK (Aurigi 2005) and in Belgium and the UK (Firmino 2005) suggest there are flaws in even the better examples.

Similar problems were identified by Strover *et al.* (2004) in a study of 36 communities in the US state of Texas. Intended to enhance public access to computers, "providing public access" was commonly done by expansion of computer and Internet availability in schools and libraries. However, the least-used computers were those placed in government (i.e. "public") buildings. The most-used computers were in senior centers, youth clubs, and recreation centers. Other public spaces, such as coffee shops, were not considered. As Firmino (2003: 59) has noted, "Physical space needs to be seen as a fundamental aspect of the whole virtual city phenomenon because its characteristics shape the way citizens access the virtual counterparts of their city."

In Asia, digital cities tend to be top-down, government projects, represented by projects in China, Malaysia, and Singapore. In Europe, local civic networks and their digital cities "have been keeping commercialization at arm's length" (Ishida *et al.* 2003: 196). In the US, many early networks, beginning with Cleveland Free-Net in 1986, were maintained by volunteers, with little commercial or government support. A more recent example is the Blacksburg Electronic Village, one of the oldest and most successful digital cities in the US. It combines civic and commercial purposes and has benefited from the eCorridors Program, led by the local university, Virginia Tech, which has developed plans and a business model to provide broadband across the region (Carroll 2005).

Conclusions

Despite the focus of the digital economy on high-bandwidth fiber-optic links, we have seen in this chapter the importance of modern airports and high-speed trains, because travel remains essential to managing global knowledge found in scattered pockets of technology and local markets. In the end, increased use of the Internet and other IT has not meant the "death of distance"; it merely complements and reinforces face-to-face contacts. This is why cities must continue to promote their local environments and make them attractive to businesses and other organizations.

IT-intensive industries are more clustered than others, and are central in the emergence of public policies focused on building both "digital clusters." The most important processes in clusters and in R&D complexes have little to do with IT. The synergies of innovation result more from human interaction and from perceptive management, as we have shown in earlier chapters. The importance of IT rests on its ability to transcend space—to facilitate distant communication and to facilitate access to pools of knowledge anywhere in the world. The success of a digital place, however, ultimately depends on its intelligence and ability to follow a "high road," knowledge-based development trajectory (Komninos 2004; Malecki 2004b). A comprehensive view takes into account the "the cognitive dimension of agglomeration economies" in a system of "territorial knowledge management" (Cappellin 2003). Unevenness remains across much of the world, however, in the degree to which the resources needed to be involved these activities are available. Chapter 9 addresses some of these issues, which are usually associated with the label *digital divide*.

A disturbing outcome in all digital clusters is the decay or divide in the social conditions found within them. Those who do not work on computers but instead in low-wage service jobs are victims of a dark side to the digital economy noted in Silicon Valley: "many residents risk being left behind if they do not have the knowledge and skills necessary to compete in the rapidly changing global economy" (Joint Venture: Silicon Valley: Network 2006: 5). Soaring prices and low affordability of housing are symptoms of a social and economic divide in Silicon Valley, Grenoble and, even more, in Bangalore.

PERIPHERAL REGIONS AND THE "DIGITAL DIVIDE" 9

> The term "digital divide" refers to the gap between individuals, households, businesses and geographic areas at different socio-economic levels with regard both to their opportunities to access information and communication technologies (ITs) and to their use of the Internet for a wide variety of activities.
>
> (OECD 2001: 5)

Since the emergence of the Internet, the *digital divide* has become both an enormously popular concept and a matter of fact (Dupuy 2007). Great inequalities in IT implementation, uses, and skills actually exist. The digital divide has several dimensions: social, economic, political, and spatial. Poor, less educated people, or minorities, or people living in rural areas and in under-developed countries, or in countries without political freedom, usually show IT indicators lower than those of affluent people, who live in wealthy neighborhoods, preferably in democratic countries.

How to explain the hype surrounding the digital divide (Compaine 2001)? Why are degrees of IT implementation and skills not regarded like ordinary indicators, such as infant mortality, illiteracy rate, the number of cars per inhabitant, and the share of the population with a university degree? After all, the digital divide is not more severe than other forms of economic and technical exclusion. On the contrary: there is evidence that low-income people, communities, and regions are in part digital—see the spread of mobile telephone in regions where paved roads and indoor plumbing are almost nonexistent. "Why is IT sufficiently special that society should support its consumption rather than other goods and services?" (Mariscal 2005: 413).

Actually, there is a digital divide *exceptionalism*, that itself belongs to the mainstream of deterministic, digital utopianism, an ideology which has supported for decades the idea that IT is—or will be—a revolutionary tool for hastening the end of all kinds of inequalities and human plagues. The digital way to regional development has been the subject to an important literature (Baskaran and Muchie 2005; Marshall *et al.* 2003). But the common digital divide rhetoric has inverted the causal link. The lack of IT is not the cause of social and economic under-development phenomena, but rather a consequence of low social and economic status. However, we are witnessing the emergence of a systemic loop. If IT must be recognized as a useful tool for leveraging the improvement of the education system, or the increase of business productivity—actually it must—the lack of IT infrastructure, skills, and uses is likely to reinforce initial inequalities.

The reasonable attitude should be to find an in-between answer. Most initiatives that seek to increase access to IT within less favored areas or communities, such as the "One Laptop Per Child" project we describe more thoroughly below, should be welcomed. But the idea that IT provision (computers, software, telecommunications, Internet access) may in itself enhance health and education—the basic requirements for self-achievement in every society—is misplaced and certainly leads to a waste of money.

In this final chapter, we deal mainly with geographic dimensions of the digital divide. Social, political and economic inequalities—often cumulative—are often entrenched in geographic space. In rural regions of the former "First World," the very question is not merely to get IT capabilities—with a few exceptions, most rural areas in developed countries have access to advanced telecommunication services—but to adapt themselves to the globalized, digital economy. In developing countries, there are enormous inequalities between a small number of digital spots that participate in the global economy, and regions where poverty dominates but where people nevertheless integrate slowly with the digital economy.

The digital economy challenge for rural and industrial communities

The relation between information technologies and the economic development of rural and industrial regions in developed countries has drawn interest from many scholars (Bryden *et al.* 1996; Bryden and Richards 2000; Grimes 1999, 2003b; Malecki 2003; Richardson *et al.* 2000; Richardson and Gillespie 2000a). Since the early 1990s, peripheral areas—including rural regions—have been the field par excellence of "a powerful rhetoric, with effect on policy-makers" (Richardson and Gillespie 2000: 199). Negroponte (1998) forecasted "a complete renaissance of rural living" and even the end of "the need for cities." The *rural penalty* (Hite 1997, Malecki 2003) is caused by the remoteness of people and firms from economic agents, resources and markets. Therefore, a so-called "distance-killing" technology has often been regarded as the tool par excellence for rural economic development.

The "rural penalty" for business

"Rurality" cannot be subjected to a single, objective definition. The Australian "Outback," the Canadian Northern Territories, or the Scottish Hebrides, where people rely on air transportation to break through isolation, have little in common with the countryside of England, where most places are located a mere one-hour drive from main cities. The "rural penalty" varies enormously with the actual accessibility to main cities and the intrinsic desirability of the area, which encompasses a series of features including climate, natural beauty, safety, openness to newcomers, and quality of public services. A high degree of desirability is likely to compound IT-driven economic opportunities. Members of the creative class (Florida 2002a) are more likely to settle in, and to telework from the French Provence, or from rural Utah—an example we develop below—rather than in the monotonous plains of the US Middle West (McGranahan and Wojan 2007). There is a great probability, therefore, that the emergence of the digital economy will favor high-amenity rural areas. At the opposite end of the amenity spectrum, we find small and medium-sized industrial communities whose economic base has relied on primary resource exploitation and/or Weberian, capital-intensive, declining industries. Digital technologies alone offer few benefits to these regions (Polèse and Shearmur 2006).

Producers in rural regions face two intertwined drawbacks: remoteness and low density. Remoteness means difficult access to markets, to exogenous inputs and production factors, and costly, time-consuming business trips. Low density means scarce local resources—above all human resources, and narrow local markets. This is an actual drawback for the provision of transportation and telecommunication services by private, profit-oriented firms. Low density has an additional disadvantage: the small size of rural labor pools generates labor market rigidity and a shortage of talent; the creative "élites" usually wish to live near cultural amenities and high level services, including fine educational institutions (Florida 2002b; McGranahan and Wojan 2007). In a nutshell, the rural penalty for business is the deprivation of most agglomeration economies, traded or untraded externalities presented in Chapter 8, which are the essential constituent of dense business milieus. However, from the manager's point of view, small labor pools present one advantage: rural employees usually show—all things being equal—lower rates of absenteeism, greater loyalty to the firm, and higher productivity.

Rural economies are also imperiled by globalization tendencies. Free trade deregulation, competition from low-cost countries, and the prospective decrease of subsidies in the US and the EU are likely to spur the modernization process of the agro-business sector, which will result in further overall downsizing in employment terms, although at a slower pace. In the US primary sectors (mining, farming, fishing, and forestry), Berman (2005) forecasts a loss of 230,000 jobs from 2004 to 2014, following a loss of 750,000 during the previous decade.

The manufacturing sector faces similar challenges. Isserman (2001) suggests that US rural regions are competitive in manufacturing, warehouses, and distribution. Nevertheless, the overall manufacturing employment remains on a decreasing trend. In the US, the manufacturing sector is forecasted to lose 776,000 jobs between 2004 and 2014, following a 2.7 million loss between 1994 and 2004 (Berman 2005). Challenged by low-cost countries in mass production, rural manufacturing must specialize—as does agriculture—in niche, high value-added segments. However, the future of rural employment, in quantitative terms, is neither in agriculture nor in manufacturing, but in services, which already represent the bulk of both output and employment in most rural regions in developed countries.

The end of the broadband divide?

Since the mid 1990s, the debate on the rural digital divide has been been dominated by the telecommunications issue (Fuentes-Bautista 2001). In the pre-Internet era, universal service regulations achieved a certain degree of equality between telephone users (although rural dwellers and businesses were still penalized by long-distance fees). However, the rapid growth of diverse broadband technologies presented in Chapter 3 (DSL, cable modem, WiFi, and FTTH), which allow permanent Internet connection, have created "islands of inequity" (Grubesic 2003, 2006).

Today, the problem of the rural broadband divide has been partly solved. In France, for example, as of September 2005, 95 percent of the population lived in the footprint of at least one broadband technology. Asymetric digital subscriber line (ADSL) was available in *communes* (the first level of local government) which totaled 93 percent of the country's population (TACTIS and IDATE 2005).

In the US, the increase of broadband deployment since 2000 has been spectacular (Table 9.1). In 2000, low density areas significantly lagged behind. In the 41–67 persons per square mile decile of population density, nearly 20 percent of the population lived in Zip postal codes without any broadband service available. In areas with fewer than six persons per square mile—that is, admittedly, extremely low density, only 43.9 percent of the population had a potential chance to benefit from high-speed connection. Since, the situation has significantly improved. As of June 2006, a very small minority of people still live in Zip codes without any broadband provider present, even in the most sparsely populated areas (Federal Communication Commission 2007).[1] For example, the Delta Regional Authority (2007), which covers several poor southern states in the US, recently found that 15 percent of Zip codes in its region lack a high-speed Internet service provider.

But a fair assessment of the rural broadband issue requires a more thorough analysis. Rural Zip codes often cover a large area, whose extent goes beyond the three miles radius from the telephone central office which usually defines the DSL deployment zone. The same logic applies to numerous French rural *communes*, where isolated settlements cannot benefit from ADSL service. In the end, nobody knows by a reasonable error margin the exact number of people and small businesses who are forced to be content with a dial-up connection.

More importantly, "broadband availability does not lead to its adoption" (Tookey *et al.* 2006: 493). Subscription data are always lower than deployment figures, which measure only a potential. As of July 2005, 28 percent of US households subscribed to any broadband service. But the adoption rate was only 17 percent in rural areas—

Table 9.1 ***Percentage of US population that resides in Zip codes with high-speed service***

Persons per square mile (by deciles)	*June 2000*	*June 2001*	*June 2002*	*June 2003*	*June 2004*	*June 2005*	*June 2006*
More than 3,147	99.7	99.9	99.8	100	99.9	100	99.9
947–3,147	99.4	99.8	99.9	99.9	99.9	99.9	99.9
268–247	98.4	99.5	99.9	99.9	99.9	100	100
118–268	95.9	98.8	99.5	99.7	99.8	99.9	99.8
67–118	90.2	96.8	98.5	99.4	99.6	99.8	99.8
41–67	81.2	93.0	96.3	98.5	99.1	99.4	99.5
25–41	71.4	87.3	92.2	96.9	98.2	99.2	99.4
15–25	59.9	78.4	86.5	93.3	95.6	98.6	98.9
6–15	56.6	74.6	81.9	90.3	93.8	97.7	98.5
Fewer than 6	43.9	60.7	72.6	85.7	91.1	95.1	96.6

Source: Federal Communication Commission 2007, Table 18

defined as areas outside metropolitan statistical areas—against 29 percent in urban areas (GAO 2006). However, the causal link between inferior infrastructure deployment and lesser broadband adoption remains unclear, because IT use is dependent on other interlinked variables, notably purchasing power, education level, and IT use at work, which are likely to be lower, on average, in rural areas.

The deployment of any kind of broadband technology in rural areas does not mean actual equality in terms of quality, price, and speed of connection. In rural regions, few technologies and providers usually compete. In France, for example, as of September 2005, 47 percent of French citizens lived in areas—which include most rural communities—where only one commercial offer was available (TACTIS and IDATE 2005). In these areas, the incumbent telecommunication operator often benefits from a *de facto* monopoly. Even more critical, rural regions usually lag behind metropolitan areas in the perpetual race for more bandwidth. The 200 Kbps threshold considered by US official statistics is as good as obsolete by comparison with upgraded DSL now available in metropolitan areas (up to 20 Mbps), WiMax networks, and the latest fiber-optic technology which is on the way to be deployed in cities throughout Japan, Korea, Western Europe, and North America (see Chapter 3).

Finally, it is clear that a perfect geographic evenness of "teleaccessibility" must not be expected, given the unfavorable balance engendered by, on the one hand, high network deployment costs—a result of great distance and sometimes rugged terrain—and, on the other hand, the low density of the marketable demand. However, the actual problem rural communities face in their process of entering the digital economy is not primarily a question of networks. Leveraging IT implementation for social and economic development of rural communities requires reaching a state of "IT readiness" which is a more arduous task than infrastructure provision (Malecki 2003; Ramirez and Richardson 2005; Tookey *et al.* 2006).

A rural revival?

In many countries, demographers declared the 1990s "the decade of the rural rebound" (Hamrick 2003: 1). This revival has been spurred by a global context of strong economic growth, and by the recognition of rural amenities. In addition to natural demographic growth, rural areas have attracted both new migrants and return migrants. Kotkin (2000) emphasizes the "anti-urban impulse" that leads a growing number of persons and families to flee stress, pollution and sky-high prices that characterize many metropolitan areas. Many authors (Beyers and Lindahl 1996; Kotkin 2000; Moriset 2003a) have identified a certain number of "symbolic analysts" (Reich 1991) or members of the "creative class" (Florida 2002b) among these newcomers. Migrants to rural areas include educated young people, as well as retirees, whether or not they are locally born, returning persons. Stockdale (2006) considers migration as "a pre-requisite for rural economic regeneration." The importance of understanding migration trends is that jobs follow people, rather than the other way around (Vias 1999).

Richardson and Gillespie (2000: 201) identify "two ways for incorporating rural areas in the information society," which show advantages and drawbacks. An "endogenous model" would see the provision of advanced telecommunication infrastructures favoring an uptake by local firms (mostly SMEs) which get better access to the global market. This way of thinking, which Gillespie and Richardson (1996) regard as an "infrastructuralist response," was dominant in the 1990s, notably in programs of the European Union. But the endogenous model could remain pure theory because of a lack of skills and technical awareness among local entrepreneurs. Smallbone and North (2000) surveyed 330 firms in rural England and concluded that they were often late in implementing IT applications, with the risk of further marginalization. The endogenous model is more or less represented by "lone eagles" and "high fliers" (Beyers and Lindahl 1996).

In the "exogenous model," the digital economy is brought to rural areas by external firms in the service sector, including TNCs, which take advantage of advanced telecommunications to exploit remote, albeit cheaper labor pools, without regard for distance. (This dualistic model also applies to developing countries: business process offshoring is a typical application of the exogenous model.) Many authors have questioned the sustainability of this model, which relies on inward investments (Richardson and Gillespie 2000: 203) and often creates few local knowledge and spillover effects. During the last decade, this way materialized notably in call centers locations.

IT-enabled rural service firms, *or* "lone eagles and high fliers"

One way for rural areas to enter the digital economy is the creation of local, small and medium-sized enterprises (SMEs) in the service sector, which use IT to overcome distance constraints and are outward looking. Beyers and Lindahl (1996: 2) described rural firms in producer services as "lone eagles" (individual entrepreneurs) and "high fliers," given they make "at least 40 percent of revenues from outside their local market area."

Indeed, serving clients outside the local market is a prerequisite for IT-enabled service firms to be considered as forming a part of the economic base of a rural territory. A Web designer who works only for local firms belonging to traditional sectors—which are globally stagnant—will have meager growth prospects and does not create directly an inward flow of revenue from outside.

However, this does not mean that traditional industries do not need local producer services. On the contrary: as we wrote above, rural producers must match international standards and find niche markets. Productivity and marketing requirements compel them to implement IT in the whole production cycle (see Chapter 4), at the same level as urban firms. Hence the importance of the local presence of rural producer services in the IT sector.

For diverse reasons, it is difficult to evaluate the statistical importance—and therefore the economic impact—of IT-enabled service firms. Beyers and Lindahl (1996: 9) acknowledge having "not found a good measure of producer service proprietorships in rural areas." While the actual existence of the rural e-business firm is demonstrated through numerous case studies, this kind of activity is not well suited to statistical analysis. Industrial classifications (such as North American Industry Classification System (NAICS) codes) are not well adapted to cover the full range of such services. Therefore, given the very small size and geographical scattering of the firms, this business milieu may be considered as somewhat opaque (Moriset 2003a).

Beyers and Lindahl (1996: 9) estimated in the US "a total of 1.6 million producer services jobs in rural areas," 40 percent of which "are held by Lone Eagles or persons in High Flier businesses." They are right to consider "the multiplier effect through the earnings they bring in from outside the area that are then spent in the local economy," the requirement for widening the economic base of the area. Actually, each job creation in exporting service firms is likely to generate or to save at least two jobs in other sectors, notably in public and consumer services. While this kind of business has a lot in common with telework, these two fields remain partly distinct from each other. First, "electronic home working is overwhelmingly an urban phenomenon" (Richardson and Gillespie 2000: 207). Second, call center and "high flier" employees, despite intensive use of telecommunication at work, are not regarded as teleworkers in a strict sense, because they commute daily from home to their employers' premises.

Rural IT-enabled service firms (including teleworkers) are fairly well known since a series of studies including Gillespie and Richardson (1996), Beyers and Lindahl (1996), Clark (2000), and Moriset (2003a). Among the most important features is perhaps the fact that a vast majority of entrepreneurs are "new or returning residents, who bring enhanced human capital with them" (Malecki 2003: 210). Molini (1997) analyses the migration of "teleprofessionals" to the European Sun Belt, as do Alvstam and Jonsson (2000), surveying Swedish teleworkers in Spain. In addition to their skills and experience, newcomers (including returning people) often bring with them the business network and the portfolio of customers they established while in the city (Beyers and Nelson 2000; Moriset 2003a). This fact is key to success, because the difficulty of reaching new clients outside the locality is perhaps the main drawback of rural business life.

These firms offer a broad spectrum of services that share a common feature: outputs are intangible and may be delivered to clients through telecommunication channels. The scale of operations set apart, these services are similar to those offered by Indian outsourcers: secretarial work, translation, typing, desktop publishing, software writing, Web design, and professional services such as accountancy and architecture.

Nevertheless, rural IT-enabled enterprises suffer from two main drawbacks: the lack of skilled people, and isolation:

- local recruitment is hampered by the scarcity of local talent, and by the reluctance of urban people with IT skills to move in rural areas; and
- rural enterprises suffer from the remoteness from markets, and from the lack of "buzz" and face-to-face meetings opportunities that characterize urban business milieus.

Leveraging local development through IT-enabled enterprises: lessons from Utah Smart Sites

"Utah Smart Site" was a three-year (2002–2004), state-led business development project whose goal was to foster the creation and the growth of IT-based service firms in rural Utah. Local communities were awarded the label "Smart Site" by the Governor's Rural Partnership if they actually hosted one or more "smart businesses" and participating companies were provided with a dedicated facility (publicly or privately owned) with a T1 or higher broadband connection. $3 million from state and federal funds were dedicated to telecommunication infrastructure and technical equipment provision (the program retained ownership of the equipment). In 2004, Utah Smart Site numbered 40 enterprises, totalizing about 750 jobs (Kistner 2004).

We analyzed the database of the program in January 2006, before the website (www.smartsites.utah.gov) was closed. Companies participating in the project may be divided roughly into three categories (we use Beyers and Lindahl's terminology): inward-looking "lone eagles" and small services providers, local "high fliers," and subsidiaries of non-local companies:

1) Enterprises that merely serve the local market, and operate in the less specialized segments, such as Web design, Web hosting and various computer services ranging from medical transcription and desktop publishing. This category comprises the largest number of participants (20 firms have less than five people, 24 out of 40 make less than $250,000 in annual sales). These are the "general practitioners" of rural e-business.
2) Local "high fliers," which reached a regional and even a national market, with fairly sophisticated products and services. Med US is specialized in medical billing. Learnkey Inc. sells e-learning packages and software to about a hundred Fortune 500 companies. Peczuh Printing delivers advanced printing, imaging, and graphics solutions.
3) Back-office and IT-enabled services facilities operated by external firms—usually the largest enterprises. The typical example is Healthaxis, headquartered in Dallas (Texas), which has more than 100 employees in Castle Dale who process imaging, scanning, and data capture for the medical sector.

However, evidence suggests to mitigate the actual impact of the "Utah Smart Site." Many companies were created before the launching of the program (Peczuh Printing was founded in 1962 by the son of a miner). Moreover, while numerous IT-enabled micro-enterprises may be found in "deep" rural Utah (Moab, Price, Tropic), a notable share, including the largest firms, is located on the Salt Lake City–Las Vegas corridor, Interstate 15, notably in Washington County's main city, St George. As a matter of fact, Washington County is booming, from both demographic and economic viewpoints and the phenomenon has little to do with IT-enabled services, but with a rush of both retirees and working families with young children attracted by "the climate, the scenic beauty, the family atmosphere and excellent job opportunities" (St George Area Chamber of Commerce 2006).

Call centers in peripheral areas: "from coal mining to data mining"

The location of call centers in rural communities is the main materialization of an "exogenous model" of IT-enabled economic development. The issue has drawn the attention of several scholars (Arai and Sugizaki 2003; Gillespie and Richardson 2000; Marshall and Richardson 1996; Richardson and Gillespie 2003). Up to now, call centers have shown a strong centralized pattern, which results from internal re-engineering processes within the firms (Bishop *et al.* 2003; Bristow *et al.* 2000). However, overheated labor markets in the main metropolises have led firms to consider peripheral locations with increasing favor. The intrinsic drawbacks of offshoring surveyed in Chapter 6 taken into account, rural places present a series of advantages: cultural proximity with clients, reasonable labor and real estate costs, and generous incentives from public authorities.

Executives who located in small communities unanimously praise their employees' loyalty, punctuality, and work-oriented spirit. Rural agents often have little other employment opportunity. For a given task and salary they get a higher standard of living and social status, and a lower level of stress in their everyday life—by comparison with urban employees. They are more likely to live in their hometown, with positive effects on turnover and, therefore, on experience and productivity. Virtual-Agent Services (VAS), a Chicago-headquartered outsourcer founded in 1999, provides a brilliant demonstration of a rural call center strategy. In the province of New Brunswick, Canada, VAS has implemented a network of 13 facilities dispersed in small towns, working as a single "virtual" call center.

> VAS' unique operating model allows the company to network together a series of fully functional transaction centers using an advanced technical infrastructure. This allows VAS unlimited capacity and reach to an abundant supply of high quality labor while still maintaining the economies of scale associated with having one large center. This capability allows VAS to bring the work to skilled local people without requiring them to travel long distances to obtain work. Creating a wonderful opportunity for those in the area with limited career prospects leads to a dedicated workforce assuring VAS' clients that they are receiving superior service from experienced employees. In an industry where attrition rates average higher than 60%, VAS posts agent turnover rates of less than 15%.
>
> (Virtual-Agent Services 2006)

Call centers have been particularly welcomed in declining industrial towns. The archetypal example of the shift from "coal mining to data mining" is found in Cape Breton-Glace Bay, a town of 19,000 inhabitants in Nova Scotia, Canada. In 2001, Stream, a major outsourcer, opened a 900-employee unit, after more than 1,500 workers had been laid off by the closure of local coal mines. (The second pillar of the local economy had also collapsed with the 1992 moratorium on cod fishing.) Stream received from public authorities a US$8.8 million incentive package and concluded a partnership with Nova Scotia Community College for training purposes. The company has been encouraged by the success of EDS, which had opened two facilities in Port Hawkesbury and Sydney, also in Nova Scotia (Lyne 2001).

A similar process can be seen in France, where a dozen impoverished industrial towns have hosted call centers with 100 seats or more since the end of the 1990s (Moriset and Bonnet 2004). However, call center-based recovery programs, despite being digital, are far from leading to Silicon Valley-fashion developments. The following quotation exposes the hopes that call center growth may raise in communities which have remained apart from the "*régions qui gagnent*" (regions that win) (Benko and Lipietz 1992). It also reflects the limits of such activity for sustainable economic development. Scranton (80,000) is a Northern Appalachian industrial town, located in a region where coal mining employment has undergone an 80 percent decrease since the late 1980s.

> Over the last eight years, the town's technology sector has grown 73 percent, thanks mainly to data and call centers . . . Many Appalachians are unprepared for work in the technology sector, forcing companies to bring mostly jobs requiring low skill levels to these rural communities. Appalachia's emerging technology workforce often gets paid less than the miners who came before them . . . "I don't look at the call centers and customer service centers unfavorably because it got people in front of a keyboard and working with other people . . . It's a springboard."
>
> (King 2001)

This quotation illustrates how the shift from the "old" to the "new" information-based economy works. Call centers produce few "dynamic effects or entrepreneurial spin-offs" that characterize innovative clusters (Richardson *et al.* 2000: 367). Career perspectives for employees remain narrow, and salaries are often lower than those previously paid by manufacturing firms. The problem of "employability" epitomizes the existence of a trans-generational digital divide. In 2003, the CEO of a French outsourcer, which was about to locate in a small industrial town, was asked if he would like to recruit some blue-collar workers made redundant by the recent closure of a factory. The answer was abrupt and clear: "No, but we will recruit their children" (Moriset and Bonnet 2004). The digital divide has materialized in a generational gap. Workers in the declining sectors are often too old, lack keyboarding experience and speaking fluency, all features which prevent them from entering the information economy, even at the lowest end of the qualification spectrum. Their children may have college degrees, but few local job opportunities. For them and the community, call centers are a first step towards the digital economy.

However, as Marshall and Richardson (1999) have pointed out, there are good reasons to cast doubt on the value of call centers as a development tool. Call centers provide mostly low-skilled, low-paid jobs with meager career prospects. The potential for spillover effects is almost nonexistent. Call center development in rural areas could be conceptualized as a transitional phenomenon which contributes to keeping people in the community.

The digital divide in developing countries

> The time has come to move beyond broad discussions of the digital divide . . . We must now set out ways to foster and expand digital opportunities. Those opportunities are immense. Already, in Africa and other developing regions, the rapid spread of mobile telephones and wireless telecommunication has spurred entrepreneurship, and helped small businesses take root, particularly

those run and owned by women. Doctors in remote areas have gained access to medical information on tropical diseases. Students have been able to tap into worldwide databases of books and research. Early warning of natural disasters has improved, and relief workers have been able to provide quicker, better coordinated relief.

(Kofi Annan, former Secretary General of the United Nations, 2005)

A quantitative assessment

The comparison between wealthy and developing countries is the main application field of the digital divide rhetoric. Deep "digital inequalities" between rich and poor countries cannot be denied. The traditional indicator of telecommunications availability is *teledensity*, the number of telephone main lines per 100 inhabitants. Teledensity is correlated with national wealth and level of development, as indicated by gross domestic product (GDP) per capita. This relationship generally shows a strong positive correlation—both teledensity and wealth rise together. At the global level, an increase of US$1,000 in GDP per capita is associated with an increase of 2.4 in teledensity (Kelly 2005).

Table 9.2 shows the teledensity of various regions of the world. High-income countries have teledensities that average 53.7 telephone lines per 100 population, compared to an average teledensity of 3.1 lines per 100 population in low-income countries. Teledensity is a less accurate indicator as mobile telephones have added to individual communications capability. Thus, we see that it is common in Europe and Oceania for the average telephone subscriber to have more than one telephone. Moreover, for some people, a mobile phone substitutes for a traditional telephone at home. High-income countries also have more than 130 telephone subscribers per 100 inhabitants, since many people have one or more mobile phones as well as a landline telephone at home. In low-income countries, less than one-tenth of the population has a telephone.

Table 9.2 *Teledensity: mainlines and mobile cellular, by world region, 2004*

Region	*ITU main lines per 100 inhabitants, 2004*	*Total telephone subscribers per 100 inhabitants*	*Mobile cellular subscribers per 100 inhabitants*
Africa	3.10	11.36	8.97
Americas	33.91	76.09	42.46
Asia	14.48	32.88	18.58
Europe	40.20	105.68	71.30
Oceania	41.07	103.25	62.38
High income	53.72	130.34	76.80
Upper middle income	22.17	61.64	47.77
Lower middle income	18.72	43.26	24.63
Low income	3.08	7.01	4.02
World	18.89	45.54	27.43

Source: Data from ITU (2005)

Infrastructure is among the specialized factors of production that lead to and maintain the competitive advantage of nations (Porter 1990). Technological infrastructure and the technological readiness of nations are central to the annual indices of global status: the *Global Competitiveness Index* of the World Economic Forum (Lopez-Claros *et al.* 2005) and the *World Competitiveness Index* of the Institute for Management Development (IMD 2005), as well as other assessments of technological capability (Archibugi and Coco 2004; Newman *et al.* 2005). Technological infrastructure is among the six "leading edges in competitiveness" (Garelli 2005).

The "networked readiness" index of the World Economic Forum attempts to provide a comprehensive indicator that considers three elements:

- environment, including market, political/regulatory, and infrastructure components;
- readiness of individuals, businesses, and government; and
- usage among individuals, businesses, and government (Dutta and Jain 2006).

Together, 66 variables comprise the index that ranks 115 countries from first-ranked US through last-ranked Ethiopia. In the four years of the index, the US and Finland have consistently been in the top five; Singapore has risen from eighth place to second,

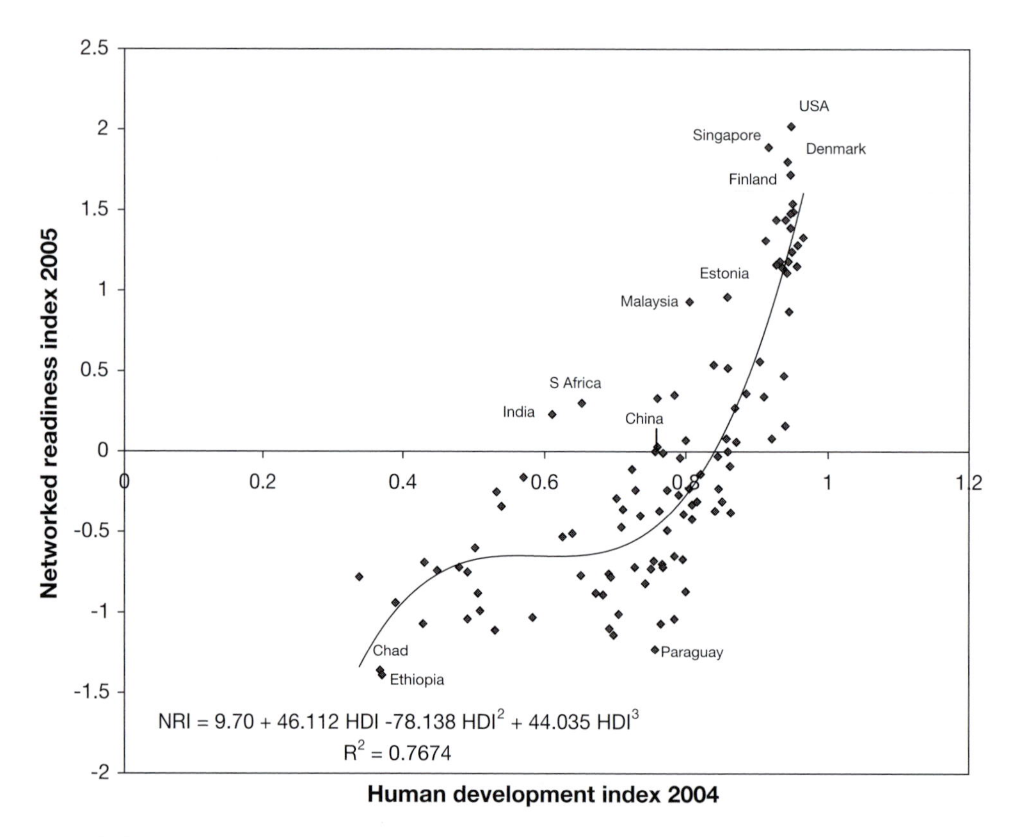

Figure 9.1 ***Networked readiness index increases nonlinearly with the human development index***

Source: Calculated from data in Dutta *et al.* (2006) and UNDP (2006)

and the Nordic countries (Finland, Sweden, Denmark, and Iceland) have consistently been in the top ten, now joined by Canada, Switzerland, and the UK (Dutta and Jain 2006). Figures 9.1 and 9.2 show a clear relationship—if not a causal link—between the networked readiness index and income per capita, or the human development index. However, indicators on a national basis hide the existence of huge inequalities within a given country. While rural areas in wealthy countries, since the beginning of the 2000s, are "catching up" in terms of IT diffusion, the digital divide between urban and rural areas remains dramatically deep in developing countries. Given the poor state of fixed infrastructure, wireless technology is the most promising. Mobile telephony is improving rapidly (see below). Wi-Fi and WiMax might represent the future of rural Internet connectivity (Gunasekaran and Harmantzis 2007).

Regional situations

China

China has made the Internet a priority, including a set of "golden projects" (Dai 2000). The first phase of this was infrastructure, to enable China to make a "jump-start" by deploying a nationwide fiber-optic network. By the end of 1998, a high-speed national

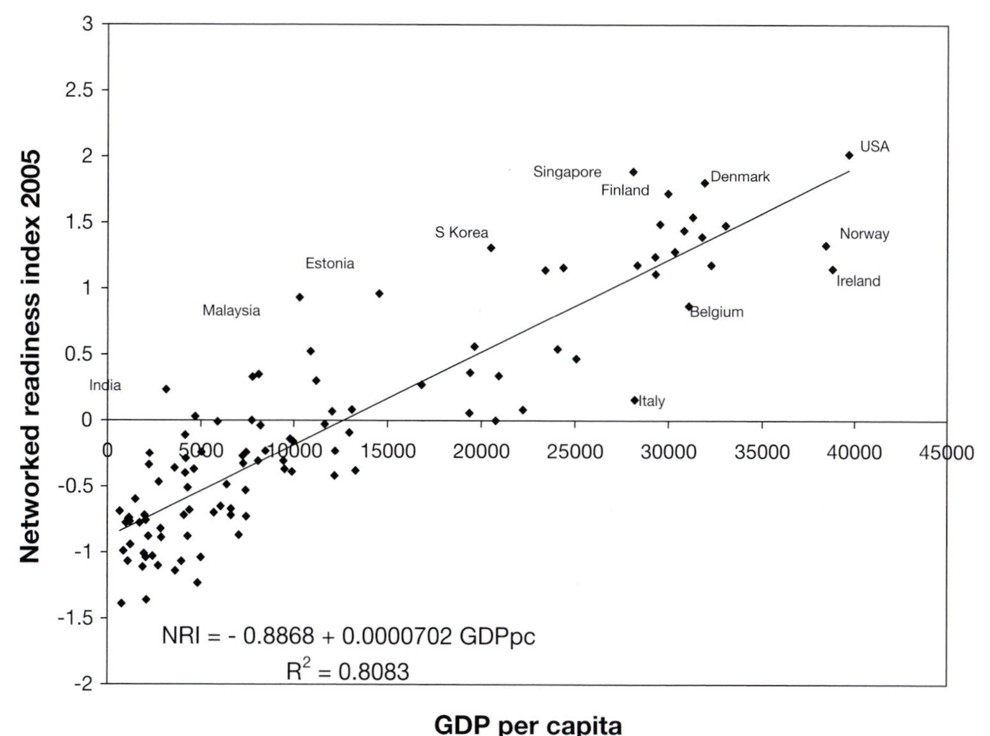

Figure 9.2 ***Networked readiness index increases with national income (GDP per capita)***

Source: Calculated from data in Dutta et al. (2006)

Note: All 105 countries included have populations greater than 1 million population.

grid of info superhighways was already in place, linking together all provinces and major cities. The "Chinese ability to execute by decree rather than consensus building followed by legislative and regulatory reform accelerated the diffusion of the Internet" in China (Press *et al.* 2003: 41). China also was able to mimic aspects of privatization by having state-owned operators compete with each other (Yan and Pitt 2002). Together, these factors jump-started the Chinese Internet, and continue to work in its favor (Dai 2003; Press *et al.* 2003).

The "digital great leap forward" did not include political liberalization. All Internet traffic into and out of sites in China is channeled through three international gateways located in Beijing, Shanghai, and Guangzhou (Dai 2003). This slows down Internet traffic, and it permits a strict regime of filtering of content, screening out pornography, political content, and other material considered unacceptable (Einhorn and Elgin 2006; Rayburn and Conrad 2004; Wacker 2003; Zittrain and Palfrey 2005). Google and Yahoo! have complied with Chinese restrictions, sparking international uproar. Moreover, the Internet industry in China is "an almost exclusively urban phenomenon," reflecting a persistent digital divide between urban and rural China (Zhang 2006b: 103). In 2003, out of 60 million Internet users, only 600,000 were living in the countryside. "Considering that about 62 percent of China's population is still living in the rural areas, this would mean that only about 0.08 percent of rural residents are Internet users, while about 12 percent of China's urban residents use the Internet" (Sun and Wang 2005: 250).

India

India's access to the FLAG (Fibre optic Link Around the Globe) cable in 1996 "provided the kick start in international access that India needed" (Ford 2005: 22). India generally could not match the speed afforded by China's "ability to execute by decree rather than consensus building" but market forces and absence of "Internet Police" in India suggest that it may reduce China's lead (Press *et al.* 2003: 41). Wolcott and Goodman (2003) suggest that "the elephant may finally be learning to dance." Wireless technologies present opportunities for India and other telecom-poor countries to overcome the gap without fixed infrastructure (Dutton *et al.* 2004). A number of innovative projects at Media Lab Asia research lab at IIT Madras are pushing the capabilities of wireless technologies in India to provide "Bits for All": broadband digital connectivity aimed at bringing digitally enabled services to the most remote, rural locations (Media Lab Asia 2006). Stimulated by deregulation and competition, which have resulted "in the cheapest mobile services in the world," India now leads the world in mobile phone subscriber growth (*The Economist* 2006f). Like China, India is characterized by a brutal contrast between Bangalore-like "hotspots" of the digital economy, and poor rural areas where IT indicators tend nearly to zero (Parayil 2006).

Africa

As seen in Chapter 3, African Internet access has lagged far behind that elsewhere in the world. Only about 2.5 percent of African residents are online, compared to the worldwide average of 16 percent. The largest numbers of Internet users exist in Egypt, South Africa,

Morocco, and Kenya. Broadband is becoming widespread in the wealthier parts of South Africa, Morocco and Zimbabwe (TeleGeography 2006b). Africa is heavily dependent on satellites. The main disadvantage of satellite communication is its high cost relative to fiber-optic communication. For example, the average African university has bandwidth equivalent to a broadband residential connection available in Europe, but pays 50 times more for its bandwidth, generally by satellite, than its counterparts at higher education institutions in the rest of the world. The average bandwidth cost per Kbps for a US university is US$0.12, but is US$8.00 in West Africa, US$4.70 in Southern Africa, US$4.38 in East Africa, US$3.18 in Central Africa, and US$0.52 in North Africa (Steiner *et al.* 2005: iii). Several initiatives are underway to provide fiber-optic links to sub-Saharan Africa (Guislain *et al.* 2005; NEPAD e-Africa Commission 2004; Rhoads 2006).

The digital economy, Third-World style

These very low IT indicators seem to compound the idea of an urgent need for enhancement of IT capabilities in the poorest countries. The argument is two-pronged. First, given the rapid emergence of an information economy, the lack of IT infrastructures, skills and uses could prevent under-developed areas from catching up and being integrated in the globalized economy. Second, IT would have *leapfrogging* capabilities. Their *ad hoc* implementation and use would help some regions and communities to enter directly the digital, "post-Fordist" economy. Although the leapfrogging theory has been to a certain degree oversold, the emergence of the digital economy in developing countries shows original, often unexpected features, even in countries in the worst economic, political, and social condition.

Take the paradoxical (although peculiar) example of Somalia (Eastern Africa), a country characterized by almost absolute poverty, rampant civil war, and the absence of any state authorities. A dozen mobile telephone companies have nevertheless flourished in the country. It is precisely because of the absence of government regulation (no taxes and no ministry to corrupt) that they offer the cheapest mobile telecommunication fees in all Africa (*The Economist* 2005f).

Mobile telephony is the most emblematic success of the digital economy in developing countries. In September 2006, India has become the world's leading country for the number of new mobile phone subscribers (*The Economist* 2006f). Even in the poorest countries—notably in Africa—wireless telephone is booming, often showing three-digit growth rates (*The Economist* 2005a). The example of Cameroon (Western Africa) is revealing. In 2005, while the number of fixed lines stagnated around 95,000, 2 million people—out of a population of 15.5 million—had a mobile phone subscription (Nana Nzépa and Tankeu Keutchankeu 2005). Because a large number of people do not have a banking account, a decisive factor of the mobile phone takeoff was the diffusion of pre-paid cards: "in sub-Saharan Africa, more than 90% of the mobile subscribers have a pre-paid subscription" (James and Versteeg 2007). When mobile phone comes to spread in areas where fixed telephone lines have never existed, where electricity at home is almost nonexistent, and where people do not have a banking account, we face a typical case of leapfrogging.

A persistent feature of the digital economy, caused by poverty and the lack of electricity at home, is *technology sharing*. In African cities such as Dakar, Abidjan, and Lagos, the diffusion of the Internet materialized in a dense network of *cybercafés* (Burns 2006; Chéneau-Loquay 2004). Well-advised entrepreneurs, notably in rural areas, have established micro-businesses based on the ownership of a mobile handset: they rent it by the minute, or they become "text interpreters" for illiterate people, by receiving and reading text messages for them (*The Economist* 2005a). Another business is the paid access to a power generator, for refueling the cells of the handsets. In Botswana, 84.8 percent of calls are made through private mobile phone kiosks, 62 percent of the phone owners share their phones with their family, 43.8 percent with their friends, and 20 percent with their neighbors (James and Versteeg 2007). Therefore, the globally low penetration rate of mobile telephones does not reveal the actual number of users.

Bridging the digital divide

Actually, "bridging the digital divide" has been made a priority at the highest level. The World Summit on the Information Society adopted a Plan of Action in December 2003. Its ten targets include:

Box 9.1 Indian fishermen use mobile phones to reduce market inefficiency

In a paper in the *Quarterly Journal of Economics*, R. Jensen (2007) from Harvard University suggests a very intriguing example of how the use of mobile phones may help local economies in developing countries. The story takes place in the fishing industry of Kerala, India. Until recently, when fishermen made a good catch of sardines in a given spot, they used to drop it in the nearest market. Offer often exceeded demand, prices fell, and a part of the production—5 to 8 percent, on average—was wasted. Meanwhile, in other markets along the coast, local fishermen had perhaps not done so well, and prices remained high. But how to know these places and to reach them, given the price of fuel and the high perishability of the product?

Things have deeply changed since 1997, when fishermen began to use mobile phones. They now call while at sea (up to 25 km off the coast), negotiate prices with different nearby markets, and reach directly the most profitable place. The market has become globally more efficient. Waste and price inequities have diminished, with benefits for both producers, whose profits rose by 8 percent, and customers, who saw a 4 percent decrease in prices.

Jensen's analysis is of particular value, because macroeconomic studies often fail to demonstrate a causal link between IT implementation and economic development. It also shows that free information circulation—the main benefit of IT use—is a critical condition of market efficiency.

1) Connect all villages to the Internet and establish community access points.
2) Connect all universities and colleges, secondary and primary schools.
3) Connect all scientific and research institutions.
4) Connect all public libraries, archives, museums, cultural centres and post offices.
5) Connect all health centres and hospitals.
6) Connect all local and central government departments, and establish websites and email addresses for them.
7) Adapt all primary and secondary school curricula to meet the challenges of the Information society, taking into account national circumstances.
8) Ensure that the entire world population has access to television and radio services.
9) Encourage the development of content and put in place technical conditions in order to facilitate the presence and use of all world languages on the Internet.
10) Ensure that more than half the world's inhabitants have personal use of IT (Minges 2006).

In addition to government-led policies, we are witnessing the emergence of "digital philanthropy," with the "One Laptop per Child" (OLPC) project, presented in January 2005 at the World Economic Forum in Davos. The initiative is developed by the eponymous Delaware-based, non-profit organization, founded by a group of scholars from the MIT Media Lab, chaired by Nicholas Negroponte. The key idea of the project is to "bridge" the digital divide by designing, producing, and distributing a cheap laptop computer to children in developing countries.

The first challenge is technical. OLPC endeavors to create a $100 machine, which is likely to be manufactured by Taiwan-based Quanta Computer Inc., the world's leading manufacturer of notebook computers. It has a 500 Mhz processor, a small, energy-saving screen, no hard disk, but a set of USB and wireless connections, which allow Internet access from locations without power (the battery can be charged by means such as a foot pedal). All the installed software is open-source, including Linux OS, according to the OLPC website, www.laptop.org.

OLPC has had to establish partnerships with governments that would like to buy and distribute the computers through the school system. As of the end of 2006, prototypes had been delivered, and several countries had diverse degrees of commitment with the organization: Argentina, Brazil, Cambodia, Costa Rica, Dominican Republic, Egypt, Libya, Nigeria (which has ordered 1 million machines), Tunisia, the US (states of Massachusetts and Maine), and Uruguay. In addition to laptops for 1.2 million school children, Libya has contracted for one server and Internet access to be provided to each school.

Many criticisms have addressed the project's flaws. Some officials, including those of India, pointed out the potential cost of the operation, and considered more urgent needs such as clean water and schools. Schools and libraries are a more cost-effective solution to give children access to basic education, at least in the poorest countries. Some analysts fear laptops could be stolen and resold on the grey market. In the political sphere, opponents complain that OLPC would exploit the poorest nations, or is a misplaced, American vision of developing countries' needs. Other people, including Bill Gates and Craig Barrett (Intel's former chairman) pointed to the technical shortcomings of the laptop. Bill Gates suggested that efforts to connect children should rather put emphasis on an existing network, such as the mobile telephone, rather than on the Internet, which would require a new infrastructure (Surowiecki 2006).

However, Surowiecki brings forward three arguments that mitigate these criticisms:

> 1) philanthropic spending projects might go to projects that do not meet the greatest needs, but still do good;
> 2) poor countries may paradoxically be too poor to afford "not to invest in information and communications technology"; and
> 3) the leapfrogging potential of IT must not be neglected, and developing countries are "surprisingly fertile grounds for new technologies."
>
> (Surowiecki 2006: 56)

"Télécoms sans Frontières" (TSF—"Telecoms without borders") is another kind of non-governmental organization. TSF has the role of opening telecoms centers for the humanitarian community during the first 30 days following a crisis and also to ensure the transition towards more long-term solutions to cover the rest of the emergency phase. With a 24-hour monitoring center in Pau (in the south-west of France) and operational bases in Nicaragua and in Thailand, TSF logistics workers and technical aid crews can intervene anywhere in the world within 24 hours after any catastrophe or conflict and in a matter of minutes set up an operational centre on site. In 2005, TSF deployed in 10 countries to the benefit of more than 250 NGOs and UN agencies.

> The teams, experienced in telecommunications are among the first to arrive on the ground where they open emergency telecommunication centres that help rescue teams from the UN, NGOs and local governments, telephone connections, Internet, video, fax, scanning facilities, and technical assistance in the use of satellite equipment, WiFi and GSM. Rescue teams need reliable systems of communications to ensure logistics, coordination and the evolution of operations.
>
> (TSF 2006)

Conclusion

A persistent flaw in the digital divide discussion is the analysis of the causal effect between the digitalization of peripheral economies and their overall economic development. In their study of IT uses by rural businesses in the province of Jiangsu (north-east of Shangai, China), Sun and Wang (2005: 257) found a strong positive correlation between firms' degree of IT implementation and their performance. However, they acknowledge that this relationship "could indicate the positive impacts of enterprise's general competence and professionalism, rather than the causal effects of Internet on raising productivity, profitability, and innovation . . . It is not easy to tell the exact cause–consequence relationship between enterprise performance and Internet access."

Technological determinism has indeed contaminated the discourse about the digital divide. Information technology "is only one piece in the more complex puzzle of rural development . . . the issue is wrapped up in human capital" (Malecki 2003: 212). The lesson we learn from mobile phone success stories is that the ingredient peripheral regions in emerging countries need first is civil peace and free entrepreneurship, coupled with educational efforts. Given this objective is met, individuals and communities are prompt to adopt new technologies, often in an unexpectedly innovative way. The emphasis on human capital improvement is equally relevant in rural regions of developed countries, or in former industrial districts. Having said that, public initiatives in the field of infrastructure provision or public services digitalization remain welcome, especially when they come from local authorities, which are more able to tailor projects adapted to actual needs and local degree of "IT-readiness."

EPILOGUE:
The distortion of economic space in a "postmodern" era

Stretched and distorted, rather than fractured?

The key feature of the digital economy is the ability to squeeze the hierarchy of scales which has been set up by the traditional models of economic geography since the time of Von Thünen, Christaller, and Weber. Yeung (2002) suggests the "switchability" of scales. We have opened this book on the example of Digital Divide Data, a small firm located in Laos and Cambodia, which processes transcription work for American firms. The output of Cambodian employees—which is in digital form—is not transported to the nearest port or market-town, but is directly and instantaneously routed to US clients in Boston or New York.

This kind of tunnel effect is typical of the digital economy. Like small towns bypassed by high-speed railways, places, regions, and people who live there are often bypassed by the networks that transport goods and services from one hotspot of digital production to another one. However, it is simplistic to think that those places—for example, Grenoble, Stockholm, or Taipei—are isolated from their environment, which provides them goods and services as basic as food, energy, and support services.

Concomitant centripetal and centrifugal forces have great organizational and spatial impacts. The digitalization of production has favored the vertical disintegration of the value chain, simultaneously with horizontal re-integration processes illustrated by the emergence of large outsourcers, in manufacturing as well as in services. The outcome of this evolution is the emergence of localized "horizontal" clusters, such as London in finance or Grenoble in nanotechnologies, but also the dispersal of activities, seen in telework practices and offshoring processes (although offshored activities are often re-clustered in the countries where they are relocated).

Changes in the hierarchy of scales are concomitant with a recombining of organizational hierarchies, through phenomena of disintermediation and re-intermediation. This is typically the case in electronic commerce operations. If people, living in a peripheral place, want to buy a book or a travel ticket, they no longer need to drive to the nearest city. A single website can deliver directly to their homes the goods or services which were previously provided only by a dense network of bookstores and travel agencies. However, value chain unbundling and multichannel operations have rendered the spatial scheme even more complex: a part of sales are made by telephone, and the related call centers may be located almost everywhere. Sheppard (2002) describes a topological economic space. Indeed, rather than the terms splintering or fracturing (Wheeler *et al.* 2000), which suggest the existence of impenetrable barriers and empty spaces, the geography of the contemporary economic space might be better described by *stretching* and *distortion*.

These words also give justice to the *time* dimension, fundamental in the analysis of the digital economy. Time is at the core of the management of network-centric and customer-centric organizations. We have seen throughout this book several materializations of apparent "space/time compression." When people dial a toll-free number, the call may be instantaneously re-routed to Africa or India. The process is invisible and lasts only milliseconds. However, the "collapsing of space and time" (Brunn and Leinbach 1991) remains a metaphor. The geographic imperatives of telecommunications accessibility demonstrate that euclidean distance has not disappeared. But speed has increased. Today's capitalism has become "smart and fast" (Thrift 2000). E-business participates in the rise in speed and complexity of society (Hassan 2003).

Convergences and divergences

The distortion of the economic space created by value chain unbundling, and by the dual logic of concentration and dispersion, results in both convergences and divergences of individuals, communities, and regions. Inequities are subjected to a geographical shift in terms of location and scale.

On the one hand, we see the emergence of a worldwide class of IT workers, who share similar features without regard to geography. The end of the dualism between the old, rich "First World" and the former poor "Third World" is epitomized by the emergence of Indian IT professionals, whose standard of living resembles that of their American or European counterparts. On the other hand, the rise of the digital economy has generated self-reinforcing inequalities between the core and the periphery, in every country, developed or developing. Less creative, mostly non-IT workers, face stagnating salaries. The well-paid, blue-collar aristocracy—such as the members of labor unions in the US—is shrinking, while IT jobs situated at the lowest end of the skill and salary spectrum, such as telemarketers, do not pay more than jobs of Wal-Mart cashiers, because low cost-based offshoring generates a levelling-down effect.

This two-pronged phenomenon of convergence on a large scale and divergence on the local scale creates micro-scale divides that may be observed in Bangalore as well as in London, Silicon Valley, or Grenoble, where salary inequalities between "players" and "non-players" have become outrageous and reinforce socio-spatial exclusions. It must be recognized that this trend is also the outcome of the shifting of power between labor and capital that has followed the entering of Asian countries into the free trade economy.

IT drives globalization?

Admittedly, globalization is a recurrent theme throughout the present book. Information technology is a major driver of globalization-like tendencies. This is obvious in the field of business services that can be processed "offshore." It is also true in the production and commerce of tangible goods, from textiles and apparel to aircraft. We have seen how IT-enabled methods of production and supply chains allow the unbundling of the value chain, and the emergence of a much more complex international division of labor.

Because of size limitations, we have neglected some sectors of the digital economy which play a great role in globalization trends. The media and entertainment industry, notably, orchestrates a worldwide flow of digitalized information which nurtures the myth of the global village popularized by McLuhan (1962). When the great tsunami of December 26, 2004 devastated numerous beach resorts in Thailand, it is the globalization of tourism, and the digitalization of the information chain which made the catastrophe a planetary event almost instantaneously. The fact that several tourists on the shore had digital cameras made it possible for TV networks to broadcast the event in their evening news, almost "live." The terrible toll of death in "non-globalized," "non-digitalized" areas around the Indian Ocean was discovered hours and days later.

Actually, electronic mass media have made the world the stage of a 24-hour show. But it is a simulacra of the real world. Most of the empirical elements in this book support some fundamental reservations about a simplistic view of globalization, a concept which has been oversold.

Globalization is a facile word—and we have used it—to describe the increase in intensity and geographic span of interlinkages and integration processes in different fields: economic, but also cultural, political, and environmental. From a theoretical viewpoint, it is usually misused. Business organizations and structures—such as "global production networks" or "global companies"—are almost never truly global. Wal-Mart rules over America, but has failed to develop in Europe. Despite the worldwide spread of McDonald's-like restaurants—a vaunted feature of globalization, the French *sandwich-baguette*, the Chinese noodle soup, and other local meals still reign supreme over the fast food market.

The digital economy shows similarly the remaining importance of local idiosyncrasies in business organization. E-finance is not globalized yet. Global e-retailers, such as Amazon.com, are challenged by incumbent "brick-and-click" vendors whose market is local. E-grocers have local footprints. The emergence of so-called globalized value chains, orchestrated by so-called global companies such as Dell or Li & Fung, is not exclusive of the success of locally anchored firms, such as Inditex Group (the owner of Zara), which has retained half of the production in Spain, thanks to advanced supply chain management. The telecommunications industry still features traits of localism. For example, mobile telephone markets remain in large part national, because of the opacity and high prices of roaming.[1] Last, the Internet, the World Wide Web, and Google, which are popular epitomes of globalization, are not global in the way that over 80 percent of humanity still live outside their reach (Internet World Stats 2007).

The second essential reservation, is about the opposition between global and local (Cox 1997). All things are local, because they are located at a certain point on earth. One more time, the concept of global supply chain or global firm is an easy word: every office, every factory, every seaport and airport, is located in a certain place, a certain region, a certain country, whose specific features play a decisive role in the creation of a given piece of value-added, either tangible or intangible. The power of IT implementation in business is precisely to help producers to benefit from the diversity of places and business environments throughout the world, thanks to its coordination capacity.

To overcome this contradiction, some authors have suggested the word *glocal* (Swyngedouw 1997) that describes the feature of being simultaneously global and local. "Think globally, act locally" has become a catch phrase in business milieus. Actually, the nexus of contemporary management is to deal with large scale—sometimes worldwide-scale—operations, without neglecting the local dimension which is at the center, for the most part, of both employees' and customers' lives.

A glance at the future

The ubiquity and mobility of teleaccess, the commoditization of versatile digital devices, the pervasiveness of chips, processors and digital sensors (Cukier 2007), are partly materializing the evolution predicted by "techno-guru" N. Negroponte (1995). The world's society is "Being Digital." In the medium term, people themselves might become digital, with the embeddness of digital sensors and accessories in bodies and brains.

However, it is doubtful that the predictions about the fundamentals of society's temporal and spatial organization—end of social synchronism and end of the city—will materialize. E-business trends show that the willingness of people for travel and face-to-face meetings continues to increase. Telework will not lead soon to the end of commuting, because most people do not like isolation, and because the workplace is a fundamental element of sociability. For similar reasons, electronic shopping will not cause the "fall of the mall" (Wilson 2000). Main Street is subjected to a normal phenomenon of "churning" within commercial activities: from place to place, McDonald's and cell phone retailers have replaced bookstores and travel agencies.

Therefore, it is unlikely—at least in the medium term—that the digitalization of the economy will result in a complete upheaval of urban life and landscape, as foreseen by utopianist thinkers. On the contrary, information technology may help current economic and spatial trends to become more sustainable from both energy and environmental points of view. Two sectors appear as very promising fields for IT-enabled, sustainable growth: transportation and the architecture and building industry (transportation and buildings' HVAC are the two biggest energy-consuming and carbon dioxide emitting sectors).

In the years to come, digital technology will play an increasing role in the implementation of intelligent transportation systems (ITS) that will be more comfortable, safe, energy-efficient, and green. ITS combines intelligent infrastructure (freeway management, electronic payment, traveller information, emergency management) with intelligent vehicles (collision avoidance and driver assistance systems) (ITS America 2006).

Green, intelligent architecture and planning also offer promising prospects. In America, buildings account for 65 percent of electricity consumption, 36 percent of total energy use and 30 percent of greenhouse gas emissions. Green buildings, such as the Condé Nast Building at 4 Times Square in New York, may reduce their environmental impact by 40 percent, thanks to special glass windows, natural-gas-powered fuel cells, photovoltaic panels, electronic, and automated control of lighting and blinds (*The Economist*

2004). Buildings are becoming green by design: digital architecture uses three-dimensional computer models and sophisticated analytical tools. IT implementation may benefit planning and building on the whole city scale. On the island of Chongming, near Shanghai, local authorities are building a new, "low-carbon" eco-city (*The Economist* 2006b). However, given the enormous, urgent needs created by urban growth in emerging countries, the advent of the green "digital" city forecast by cybergurus is a long-term vision.

A "postmodern" Digiworld?

The spread of digital technology over the society features numerous elements which might be subjected to post modern or hypermodern interpretations.

Information technology enables post-Fordist methods of production and "time–space compression" phenomena which characterize the new forms of accumulation of "late capitalism" (Harvey 1989). The emergence of virtualness may be thought of as a typical postmodern feature. The virtual dimension that is inherent to many e-business operations is backed by simulation-based practices, whose aim is to transcend both cultural and physical distances. The use of aliases by African and Indian tele-agents is "a peculiar form of cyber-reality" (Bryson *et al.* 2003: 127). E-commerce and call center operations may be regarded as "simulacra and simulation" (Baudrillard 1985), which tend to precede, or even to supersede reality, in the "apotheosis of immediacy" (Nunes 1995). Also typically postmodern is the conceptualization of the "cyberspace," a word coined by William Gibson in the novel *Neuromancer* (1984) and which has been used widely by IT researchers (Dodge and Kitchin 2000; Janelle and Hodge 2000).

Digital technology is the key driver of the emergence of an over-mediatized society which features an increasing level of "panopticism" (Foucault 1975). Call center agents are recorded. Calls and Internet queries feed customer databases, which are exploited to target teleprospecting. These elements complement digital public surveillance, RFID, and computer cookies (Elmer 2004; Graham and Wood 2003). Facing these developments, we should recall Orwell's *1984* (1948), as well as the "soft totalitarianism" of Huxley's *Brave New World* (1938).

Undoubtedly, the rise of digital technology generates as much fear as hope. The forthcoming association of computer science, nanotechnology and biotechnology will pose serious ethical problems. The old ideologies, the faith in progress, and the long-established structures of society—the family, the enterprise, the school, the local community—are challenged by unstable structures, unpredictable, threatening dynamics, and uncontrolled forms of knowledge, many of which are enabled by computers and digital networks: the virtual worlds of electronic mass media, the Web and the blogsphere, e-commerce, offshoring, and telework. Following Lyotard (1979), we could say that the latest advances in digital technologies have given the "great narratives" the *coup de grâce*—unless we interpret "Digiworld" itself as the latest great narrative?

NOTES

1 Introduction: the digital economy and the splintering of economic space

1 Gordon Moore, a former Chairman of Intel (the leading producer of micro processors), made in 1965 the observation that the number of transistors in an integrated circuit doubled every 24 months (it is often quoted as every 18 months). Since then, "Moore's law" has materialized in an exponential increase of computing power of a single processor, which has made possible miniaturization and a wide range of applications, in wireless telephony as well as cars and household appliances.

3 Where local meets global: the rise of the digital network

1 Dense wavelength division multiplexing (DWDM) works by combining and transmitting multiple signals simultaneously at different wavelengths on the same fiber. Currently, because of DWDM, single fibers have been able to transmit data at speeds up to 400 Gbps. This means that DWDM-based networks also are able to carry different types of traffic at different speeds over the same optical channel (Jupitermedia 2006).
2 In part, this was a result of the Japanese Ministry of Communications having determined that NTT had already amortized the cost of its local "last-mile" copper network (Fransman 2006b: 34).

4 Digital production and the business organizations

1 ERP is the key platform for various applications: product planning, purchasing, inventory management, order tracking, and finance.
2 Data in this section come mainly from the websites of AB Volvo-owned companies.
3 However, quality problems do not systematically require an inspector being sent from Hong Kong: high resolution digital photos are taken on-site by Li & Fung representatives, then sent by e-mail to quality managers.

5 The multiscale geographies of electronic commerce and electronic finance

1 Since April 4, 2007, NYSE Group is merged with Euronext (the owner of several European stock exchanges including Paris and Amsterdam) in a new entity named NYSE Euronext.
2 Amazon trades from seven websites, which are dedicated to the US, Canada, the UK, Germany, France, Japan, and China.

3 After a period of reorganization and consolidation, the four ERSs have been founded as independent companies or autonomous subsidiaries. Amadeus is controlled by Wam Acquisition SA, itself controlled by three airlines (Air France, Iberia, Lufthansa) and two private equity firms, BC Partners and Cinven. Sabre Travel Networks is the reservation arm of Sabre Holdings Corporation, Worldspan is a subsidiary of Worldspan Technologies Inc., and Galileo is a fully owned subsidiary of Cendant, a New York-based travel and real estate corporation.

4 Yield management seeks to optimize the relation between offer (seat capacity) and actual demand.

5 Expedia was founded by Microsoft in 1996, then sold to Barry Diller's IAC (InterActive Corp.), according to *The Economist* (2005b).

6 We may regard planned business meetings, or the financial press, as examples of formal information channels, while gossips, tips, and rumors traded in cafeterias, restaurants, or even elevators, may be regarded as unconventional channels.

6 Splintering the economic space: the offshoring of corporate services

1 Fifty million US citizens have already subscribed to the federal "do-not-call" list, which bans unsolicited telephone messages from televendors (www.donotcall.org).

2 However, there is a bias in such a measurement, because headquarters location does not suggest the size of business units.

3 The seven Indian Institutes of Technology (IIT) are a kind of Ivy League whose founding began immediately after independence in 1947. Each year, 300,000 students apply to the Joint Entrance Examination, for 5,500 seats.

7 Telework/telecommunicating: tome and space flexibilities in work and business orgaanization

1 Thierry Breton is a former CEO of France Télécom and Ministre de l'Economie et des Finances.

2 For greater convenience, we keep the acronym VMT, even if measurement is in kilometers.

3 Given the enormous margin of error in telework and commuting length estimations, the transformation of 1000 liters of gasoline to one ton is a benign approximation (the volumetric mass of gasoline is inferior to that of water).

9 Peripheral regions and the "digital divide"

1 In this paragraph, high-speed or broadband is defined as "any connection that exceeds 200 Kbps in at least one direction" (Federal Communication Commission 2007).

Epilogue

1 When mobile phone users leave a country, they usually leave the footprint of the home operator. Therefore, inbound and outbound calls must be transferred to another network in the destination country. In the EU, a new regulation voted by the European Parliament on 23 May 2007 seeks to reduce roaming charges by 70 percent.

BIBLIOGRAPHY

Abbate, J. (1999) *Inventing the Internet*, Cambridge, MA: MIT Press.

Abboud, L. (2006) "How France became a leader in offering faster broadband," *The Wall Street Journal*, 28 March: A1.

Abler, R., Janelle, D., Philbrick, A. and Sommer, J. (1975) "Introduction: the study of spatial futures," in R. Abler, D. Janelle, A. Philbrick, and J. Sommer (eds) *Human Geography in a Shrinking World*, North Scituate, MA: Duxbury, 3–16.

Accenture (2006) "The geography of high performance," www.accenture.com/Global/About_Accenture/Locations/GeographyHighPerformance.htm (accessed 14 December 2006).

Afuah, A. and Tucci, C.L. (2003) "A model of the Internet as creative destroyer," *IEEE Transactions on Engineering Management*, 50: 395–402.

Agence d'Etude et de Promotion de l'Isère (2005) *From hydraulic power to information and communications technology*, http://english.grenoble-isere.com/378-grenoble.htm (accessed 20 December 2006).

Agentrics (2005) "GNX and WWRE complete merger transaction to form Agentrics," press release, 14 November, Alexandria (VA) and Chicago (IL): Agentrics LLC.

Agnes, P. (2000) "The 'end of geography' in financial services? Local embeddedness and territorialization in the interest rate swaps industry," *Economic Geography*, 76: 347–66.

Alcaly, R. (2003) *The New Economy*, New York: Farrar, Straus & Giroux.

Alfasi, N. and Portugali, J. (2004) "Planning just-in-time versus planning just-in-case," *Cities*, 21: 29–39.

Allenby, B. and Roitz, J. (2003) "Implementing the knowledge economy: the theory and practice of telework," working paper, The Batten Institute, Darden Graduate School of Business, Charlottesville: University of Virginia, www.telcoa.org/id146.htm (accessed 11 May 2006).

Alvstam, C. and Jonsson, A. (2000) "On promoting regional economic growth in industrial peripheries. hyper-footloose business services—the case of Swedish teleworkers in Spain," *Netcom*, 14: 5–25.

Amazon.com, Inc. (2005) *Form 10-K Annual Report for the fiscal year ended December 31, 2004*, Washington, DC: US Securities and Exchange Commission.

Amazon.com, Inc. (2006) *Form 10-K Annual Report for the fiscal year ended December 31, 2005*, Washington, DC: US Securities and Exchange Commission.

Amazon Jersey Ltd. (2007) "Buying from Amazon Jersey," www.amazon.co.uk/gp/help/customer/display.html?nodeId=13378011 (accessed 16 April 2007).

Amin, A. and Cohendet, P. (2004) *Architectures of Knowledge: firms, capabilities, and communities*, Oxford: Oxford University Press.

Amirahmadi, H. and Wallace, C. (1995) "Information technology, the organization of production and regional development," *Environment and Planning A*, 27: 1745–75.

Andal-Ancion, A., Cartwright, P.A. and Yip, G.S. (2003) "The digital transformation of traditional business," *MIT Sloan Management Review*, 44 (4): 34–41.

Anderson, C. (2006) *The Long Tail: why the future of business is selling less of more*, New York: Hyperion.

Anderson, W.P., Chatterjee, L. and Lakshmanan, T.R. (2003) "E-commerce, transportation, and economic geography," *Growth and Change*, 34: 415–32.

Andersson, A.E. (1985) "Creativity and regional development," *Papers of the Regional Science Association*, 56: 5–20.

Annan, K. (2005) *Statement by H.E. Mr. Kofi Annan, Secretary-General of the United Nations*, Second phase of the World Summit for the Information Society, 16 November, www.itu.int/wsis/tunis/statements/docs/io-un-opening/1.html (accessed 27 December 2006).

Anton, J., Chatterley, J.R., Trobaugh, J.J., and Seeley, M.W. (2002) *Improving Call Center Performance through Optimized Site Selection*, White paper, Purdue Research Foundation, Santa Maria (California): BenchmarkPortal, Inc., www.answerstat.com/papers/4/03.pdf (accessed 17 October 2006).

Antonelli, C. (2003) "The digital divide: understanding the economics of new information and communication technology in the global economy," *Information Economics and Policy*, 15: 173–99.

Anttiroiko, A.-V. (2004a) "Editorial: global competition of high-tech centres," *International Journal of Technology Management*, 28: 289–323.

Anttiroiko, A.-V. (2004b) "Science cities: their characteristics and future challenges," *International Journal of Technology Management*, 28: 395–418.

Aoyama, Y. (2001) "The information society, Japanese style: corner stores as hubs for e-commerce access," in T.R. Leinbach and S.D. Brunn (eds) *Worlds of Electronic Commerce: economic, geographical and social dimensions*, New York: John Wiley, 109–28.

Aoyama, Y., Ratick, S.J. and Schwarz, G. (2005) "Modeling the impact of business-to-business electronic commerce on the organization of the logistics industry," *Geographical Analysis*, 37: 46–68.

Aoyama, Y., Ratick, S.J. and Schwarz, G. (2006) "Organizational dynamics of the US logistics industry: an economic geography perspective," *The Professional Geographer*, 58: 327–40.

APL Logistics (2005) "APL Logistics announces site for Dell's Winston-Salem logistics center," press release, August 4, Oakland (California): APL Logistics.

Arai, Y. and Sugizaki, K. (2003) "Concentration of call centers in peripheral areas: cases in Japan," *Netcom*, 17: 187–202.

Archibugi, D. and Coco, A. (2004) "A new indicator of technological capabilities for developed and developing countries (ArCo)," *World Development*, 32: 629–54.

Arnold, E. and Guy, K. (1989) "Policy options for promoting growth through information technology," in OECD, *Information Technology and New Growth Opportunities*, Paris: Organisation for Economic Co-operation and Development, 133–201.

Arnold, S.E. (2005) *The Google Legacy: how Google's Internet search is transforming applications software*, Tetbury, England: Infonortics, www.infonortics.com/publications/google/technology.pdf (accessed 11 June 2006).

Aron, R. and Singh, J.V. (2005) "Getting offshoring right," *Harvard Business Review*, 83 (12): 135–43.

Arora, A. and Athreye, S. (2002) "The software industry and India's economic development," *Information Economics and Policy*, 14: 253–73.

Arora, A., Gambardella, A. and Torrisi, S. (2004) "In the footsteps of Silicon Valley? Indian and Irish software in the international division of labor," in T. Bresnahan and A. Gambardella (eds) *Building High-Tech Clusters: Silicon Valley and beyond*, Cambridge: Cambridge University Press, 78–120.

Arora, A., Arunchalam, V.S., Asundi, J. and Fernandes, R. (2001) "The Indian software services industry," *Research Policy*, 30: 1267–87.

Arthur W.B. (1996) "Increasing returns and the new world of business," *Harvard Business Review*, 74 (4): 100–9.

Ascher, F. (2000) *Ces événements nous dépassent, feignont d'en être les organisateurs. Essai sur la société contemporaraine*, La Tour d'Aigues: Editions de l'Aube.

Asheim, B.T. and Gertler, M.S. (2006) "The geography of innovation: regional innovation systems," in J. Fagerberg, D.C. Mowery and R.R. Nelson (eds), *The Oxford Handbook of Innovation*, Oxford: Oxford University Press: 291–317.

Asheim, B., Cooke, P. and Martin, R. (2006) "The rise of the cluster concept," in B. Asheim, P. Cooke and R. Martin (eds), *Clusters and Regional Development: critical reflections and explorations*, London: Routledge, 1–29.

Aspray, W., Mayadas, F. and Vardi, M.Y. (eds) (2006) *Globalization and Offshoring of Software*, New York: Association for Computing Machinery, www.acm.org/globalizationreport/pdf/fullfinal.pdf (accessed 22 June 2006).

Athreye, S.S. (2005) "The Indian software industry and its evolving service capability," *Industrial and Corporate Change*, 14: 393–418.

Atkinson, R.D. (2004) *The Past and Future of America's Economy: long waves of innovation that power cycles of growth*, Cheltenham: Edward Elgar.

Atkinson, R.D. and Correa, D.K. (2007) *The 2007 State New Economy Index: benchmarking economic transformation in the states*, Washington, DC: Information Technology and Innovation Foundation.

Atkinson, R.D. and McKay, A.S. (2007) *Digital Prosperity: understanding the economic benefits of the information technology revolution*, Washington, DC: Information Technology and Innovation Foundation.

Aubuchon, V. (2006) Vaughns Google data centers chart, www.vaughns-1-pagers.com/internet/google-data-centers.htm (accessed 11 June 2006).

Audirac, I. (2003) "Information-age landscapes outside the developed world," *Journal of the American Planning Association*, 69: 16–32.

Audirac, I. (2005) "Information technology and urban form: challenges to smart growth," *International Regional Science Review*, 28: 119–45.

Aurigi, A. (2005) *Making the Digital City: the early shaping of urban Internet space*, Aldershot: Ashgate.

Bahrami, H. and Evans, S. (1995) "Flexible re-cycling and high-technology entrepreneurship," *California Management Review*, 37 (3): 62–89.

Baily, M.N. and Farrell, D. (2004) "Exploding the myths of offshoring," *The McKinsey Quarterly*, June, www.mckinseyquarterly.com/article_page.aspx?ar=1453 (accessed 22 December 2006).

Bakis, H. (1987) "Telecommunications and the global firm," in F.E.I. Hamilton (ed.) *Industrial Change in Advanced Economies*, London: Croom Helm, 130–60.

Baldwin, R. (2006) *The Great Unbundling*, Helsinki: Economic Council of Finland.

Baljko Shah, J. and Serant, C. (2002) "Microsoft's Xbox sets supply chain standard," www.agile.com/news/2002/ebn_031102.pdf (accessed 3 November 2006).

Baran, N. (1998) "The privatization of telecommunication," in R.W. McChesney, E.M. Wood and J.B. Foster (eds) *Capitalism and the Information Age: the political economy of the global communication revolution*, New York: Monthly Review Press: 123–33.

Bardhan, A.D. and Kroll, C.A. (2003) "The new wave of outsourcing," Fisher Centre Research Report No. 1103, Berkeley: University of California, www.repositories.cdlib.org/iber/fcreue/reports/1103 (accessed 27 June 2005).

Barnes & Noble Inc. (2006) *Form 10-K Annual Report for the fiscal year ended January 28, 2006*, Washington, DC: US Securities and Exchange Commission.

Bartlett, C.A. and Ghoshal, S. (1990) "Matrix management: not a structure, a frame of mind," *Harvard Business Review*, 68 (4): 138–45.

Baskaran, A. and Muchie, M. (eds) (2005) *Bridging the digital divide: innovation systems for ICT in Brazil, China, India, Thailand and Southern Africa*, London: Adonis & Abbey Publishers.

Bathelt, H. (2001) "Regional competence and economic recovery: divergent growth paths in Boston's high technology economy," *Entrepreneurship and Regional Development*, 13: 287–314.

Bathelt, H., Malmberg, A. and Maskell, P. (2004) "Clusters and knowledge: local buzz, global pipelines and the process of knowledge creation," *Progress in Human Geography* 28, 31–56.

Batt, R., Doellgast, V., Kwon, H., Nopany, M., Nopany, P., and da Costa, A. (2005) "The Indian call centre industry: national benchmarking report strategy, H.R. practices, & performance," *Working Paper 05—07*, School of Industrial and Labor Relations, Center for Advanced Human Resource Studies, Ithaca, NY: Cornell University.

Batty, M. (2006) "Public sector information: chains of value added," *Environment and Planning B: Planning and Design*, 33: 163–4.

Baudrillard, J. (1985) *Simulacres et simulation*, Paris: Galilée.

Baumol, W.J. (1967) "Macroeconomics of unbalanced growth: the anatomy of urban crisis," *American Economic Review*, 57 (3): 415–26.

Bauer, J.M. (2006) "Broadband in the United States," in M. Fransman (ed.) *Global Broadband Battles: why the US and Europe lag while Asia leads*, Stanford: Stanford University Press, H.HH 133–63.

Bax, A. (2005) "The rise of regional submarine networks," *Submarine Telecoms Forum*, 20: 19–20.

BBC News (2001) "Doctors claim world first in telesurgery" www.news.bbc.co.uk/1/hi/sci/tech/1552211.stm (accessed 9 November 2006).

Bell, D. (1973) *The Coming of Post-Industrial Society: a venture in social forecasting*, New York: Basic Books.

Bender, S. and Fish, A. (2000) "The transfer of knowledge and the retention of expertise: the continuing need for global assignments," *Journal of Knowledge Management*, 4: 125–37.

Benetton Group (2003) "Benetton's position on RFID technology," www.press.benettongroup.com/ben_en/about/facts/fact3 (accessed 17 November 2006).

Benghozi, P.-J. and Licoppe, C. (2003) "Technological national learning in France: from Minitel to Internet," in B. Kogut (ed.) *The Global Internet Economy*, Cambridge, MA: MIT Press, 153–90.

Benhamou, E., Engel, D., Johri, S., North, L., Phillips, S. and Elston, S (1998) *Smart Valley Telecommuting Guide*, Palo Alto, CA: Smart Valley Inc., www.cisco.com/warp/public/779/smbiz/netsolutions/find/telecommuting/98guide.pdf (accessed 11 May 2006).

Beniger, J.R. (1986) *The Control Revolution: technological and economic origins of the information society*, Cambridge, MA: Harvard University Press.

Benko, G. and Lipietz, A. (eds) (1992) *Les Régions qui Gagnent. Districts et réseaux: les nouveaux paradigmes de la géographie économique*, Paris: Presses Universitaires de France.

Benner, C. (2006) "South Africa on-call": information technology and labour restructuring in South African call centres," *Regional Studies*, 40: 1025–40.

Berman, J.M. (2005) "Industry output and employment projections to 2014," *Monthly Labor Review*, 128 (11): 45–69.

Best Workplaces for Commuters (2006) *BWC National Benefits and Related Facts*, Washington DC: US Department of Transportation and US Environmental Protection Agency, June, www.bestworkplacesforcommuters.gov/pdf/bwc-reference-facts-june-06.pdf (accessed 4 October 2006).

Beveridge, C. (2003) "The new workplace. How to accommodate increased mobility—and save money in the process," Sun Microsystems Inc., January 14, www.sun.com/2003–0114/feature/index.html (accessed 19 October 2006).

Beveridge, T. and Perks, C. (2000) "Blueprint for the Flexible Enterprise," *Intelligent Enterprise*, 1 March, 3 (4), www.intelligententerprise.com/000301/feat3.jhtml (accessed 3 November 2006).

Beyers, W.B. and Lindahl, D.P. (1996) "Lone eagles and high fliers in rural producer services," *Rural Development Perspectives*, 11 (3): 2–10.

Beyers, W.B. and Nelson, P.B. (2000) "Contemporary development forces in the non metropolitan West: new insights from rapidly growing communities," *Journal of Rural Studies*, 16: 459–74.

Bhagowati, G. (2005) "Why Indian suppliers are moving to tier 2 cities," Dallas: Everest Partners LP, www.outsourcing-offshore.com/tier2.html (accessed 3 November 2006).

Biederman, D. (2005) "Growth business: the fourth-party logistics industry, whatever that is, appears to be doing quite well," *The Journal of Commerce*, 6 (23) (6 June): 28–31.

Birkinshaw, J. (2000) "Network relationships inside and outside the firm, and the development of capabilities," in J. Birkinshaw and P. Hagström (eds) *The Flexible Firm: capability management in network organizations*, Oxford: Oxford University Press, 4–17.

Biseul, X. (2004) "Les salariés franciliens d'IBM expérimentent le bureau 'on demand'," *01 Informatique*, 13 October, www.01net.com/article/253383.html (accessed 19 December 2006).

Bishop, J. and Walker, J. (2006) "There is no safety in numbers . . . (James Thurber): the security issues of multiple cable landings," *Submarine Telecoms Forum*, 25: 16–18.

Bishop, P., Gripaios, P. and Bristow G. (2003) "Determinants of call centre location: some evidence for UK urban areas," *Urban Studies*, 40: 2751–68.

Blanc, G. (1998) *Panorama des télécentres dans le monde*, Paris: Eurotechnopolis Institut, www.eurotechnopolis.com/fr/bookstore/telecent1.html (accessed 3 November 2006).

Blanc, H. and Sierra, C. (1999) "The internationalisation of R&D by multinationals: a trade- off between external and internal proximity," *Cambridge Journal of Economics*, 23: 187–206.

Blinder, A.S. (2006) "Offshoring: the next industrial revolution?," *Foreign Affairs*, 85 (2) March/April: 113–28.

Bloom, N. and Van Reenen, J. (2006) "Measuring and explaining management practices across firms and countries," mimeo, Center for Economic Performance, London School of Economics, http://cep.lse.ac.uk/textonly/people/bloom/papers/BloomVanReenen2.pdf (accessed 5 December 2006).

Boeing (2007) "787 Dreamliner international team facts," www.boeing.com/commercial/787family/dev_team.html (accessed 29 April 2007).

Bohn, R.E. (1994) "Measuring and managing technological knowledge," *Sloan Management Review*, 36 (1): 61–73.

Boschma, R. (1999) "The rise of clusters of innovative industries in Belgium during the industrial epoch," *Research Policy*, 28: 853–71.

Boutellier, R., Gassmann, O. and von Zedtwitz, M. (2000) *Managing Global Innovation*, Berlin: Springer.

Boutin, P. (2006) "A grand unified theory of YouTube and MySpace," *slate.com*, 28 April, www.slate.com/id/2140635/ (accessed 12 June 2006).

Bowonder, B. (2001) "Globalisation of R&D: the Indian experience and implications for developing countries," *Interdisciplinary Science Reviews*, 26 (3): 191–203.

Boyer, R. (2004) *The Future of Economic Growth: as new becomes old*, Cheltenham: Edward Elgar [translated from the French, *Croissance début de siècle: de l'octet au gène*, Paris: Albin Michel, 2002].

Bradsher, K. (2006) "The next industrial giant is . . . India?," *The New York Times*, 1 September: A1.

Brafman, N. (2006) "France Télécom lance l'Internet à très haut débit," *Le Monde*, 17 December.

Breen, B. (2004) "Living in Dell time," *Fast Company*, November, 88: 86–97, www.fastcompany.com/magazine/88/dell.html (accessed 3 November 2005).

Breschi, A. and Malerba, F. (2001) "The geography of innovation and economic clusterings: some introductory notes," *Industrial and orporate change*, 10: 817–33

Bresnahan, T., Gambardella, A. and Saxenian, A. (2001) " 'Old economy' inputs for 'new economy' outcomes: cluster formation in the new Silicon Valleys," *Industrial and Corporate Change*, 10: 835–60.

Bresnahan, T.F. and Trajtenberg, M. (1995) "General purpose technologies: 'engines of growth'?" *Journal of Econometrics*, 65: 83–108.

Breton, T. (1995) *Le télétravail en France*, Paris: La Documentation française.

Brioschi, M.S. and Cassia, L. (2006) "Common trajectories of regional competitiveness in the knowledge economy: a European investigation," *Industry and Higher Education*, 20: 387–401.

Briscoe, B., Odlyzko, A. and Tilly, B. (2006) "Metcalfe's Law is wrong," *IEEE Spectrum*, 43 (7): 34–9.

Bristow, G., Gripaios, P. and Munday, M. (2000) "Call centre growth and location: corporate strategy and the spatial division of labour," *Environment and Planning A*, 32: 519–38.

British Bankers Association (2004) "Credit derivatives market predicted to reach $8.2 trillion by 2006," *press release*, 22 September, www.bba.org.uk/bba/jsp/polopoly.jsp?d=441&a=4492 (accessed 10 September 2005).

Brittain-Catlin, W. (2005) *Offshore: the dark side of the global economy*, New York: Farrar, Straus & Giroux.

Brock, G.W. (2003) *The Second Information Revolution*, Cambridge, MA: Harvard University Press.

Brown, D. and Wilson, S. (2005) *The Black Book of Outsourcing: how to manage the changes, challenges, and opportunities*, New York: John Wiley.

Brown, J.S. and Hagel, J. (2003) "Does IT matter?," *Harvard Business Review*, 81 (7), 109–12.

Brunn, S.D. and Leinbach, T.R. (eds) (1991) *Collapsing Space and Time: geographic aspects of communication and information*, New York: HarperCollins.

Bruton G.D., Fried, V.H. and Manigart, S. (2005) "Institutional influences on the worldwide expansion of venture capital," *Entrepreneurship Theory and Practice*, 29: 737–60.

Bryden, J. and Richards, C. (2000) "Information technology and rural development in the Scottish Highlands and Islands: a preliminary review of the issues and evidence," *Géocarrefour*, 75: 71–6.

Bryden, J., Rennie, F. and Fuller, A.M. (1996) *Implications of the Information Highway for Rural Development and Education*, Oxford: The Arkleton Trust.

Brynjolfsson, E. (1993) "The productivity paradox of information technology," *Communications of the ACM*, 36 (12): 67–77.

Brynjolfsson, E. and Hitt, L. (1996) "Paradox lost? Firm-level evidence on the returns to information systems spending," *Management Science*, 42: 541–58.

Brynjolfsson, E. and Hitt, L.M. (2000) "Beyond computation: information technology, organizational transformation and business performance," *Journal of Economic Perspectives*, 14 (4): 23–48.

Brynjolfsson, E. and Hitt, L.M. (2003) "Computing productivity: firm-level evidence," *Review of Economics and Statistics*, 85: 793–808.

Brynjolfsson, E. and Hitt, L.M. (2005) "Intangible assets and the economic impact of computers," in W.H. Dutton, B. Kahin, R. O'Callaghan and A.W. Wyckoff (eds) *Transforming Enterprise: the economic and social implications of information technology*, Cambridge, MA: MIT Press, 21–48.

Brynjolfsson, E., Malone, T.W., Gurbaxani, V. and Kambil, A. (1994) "Does information technology lead to smaller firms?," *Management Science*, 40: 1628–44.

Bryson, J.R., Daniels, P.W. and Warf, B. (2003) *Service Worlds: people, organizations, technologies*, London: Routledge.

Bunnell, T. (2004) *Malaysia, Modernity and the Multimedia Super Corridor: a critical geography of intelligent landscapes*, London: RoutledgeCurzon.

Bureau of Labor Statistics (2005) *Work at Home in 2004*, Washington DC: US Department of Labor, 22 September, ftp://ftp.bls.gov/pub/news.release/History/homey.09222005.news (accessed 10 May 2006).

Burns, E. (2006) "Web usage climbs in Africa," London: Incisive Media. www.clickz.com/stats/sectors/geographics/article.php/3603526 (accessed 10 June 2006).

Burrows, P. (2006a) "Stopping the sprawl at HP," *Business Week*, 29 May: 54–6, www.businessweek.com/technology/content/may2006/tc20060517_717233.htm?campaign_id=rss_tech (accessed 10 June 2006).

Burrows, P. (2006b) "Servers as high as an elephant's eye," *Business Week*, 12 June: 73–4, www.businessweek.com/magazine/content/06_24/b3988087.htm (accessed 10 June 2006).

Burton-Jones, A. (1999) *Knowledge Capitalism: business, work, and learning in the new economy*, London: Routledge.

Byrne, J.A., Brandt, R. and Port, O. (1993) "The virtual corporation," *Business Week*, 8 February: 36–41.

CacheLogic (2005) "First-ever, real-time traffic analysis of file formats crossing peer-to-peer released by CacheLogic" (press release 9 August 2005), Cambridge (UK): CacheLogic, www.cachelogic.com/home/pages/research/filetypestudy.php (accessed 20 September 2006).

Cairncross, F. (2001) *The Death of Distance: how the communications revolution is changing our lives*, Boston: Harvard Business School Press.

Cappellin, R. (2003) "Territorial knowledge management: towards a metrics of the cognitive dimension of agglomeration economies," *International Journal of Technology Management*, 26: 303–25.

Carlsson, B. (2004) "The digital economy: what is new and what is not?," *Structural Change and Economic Dynamics*, 15: 245–64.

Carr, N.G. (2003) "IT doesn't matter," *Harvard Business Review*, 81 (5): 41–9.

Carroll, J.M. (2005) "The Blacksburg Electronic Village: a study in community computing," in P. van den Besselaar and S. Koizumi (eds) *Digital Cities 2003*, Berlin: Springer-Verlag, 43–65.

Castellacci, F. (2006) "Innovation, diffusion and catching up in the fifth long wave," *Futures*, 38: 841–63.

Castells, M. (1989) *The Informational City: information technology, economic restructuring and the urban-regional process*, Oxford: Blackwell.

Castells, M. (1996; 2nd edn 2000) *The Rise of the Network Society*, Oxford: Blackwell.

Castells, M. (2001) *The Internet Galaxy: reflections on the Internet, business, and society*, Oxford: Oxford University Press.

Castells, M. and Hall, P. (1994) *Technopoles of the World: the making of 21st century industrial complexes*, London: Routledge.

Cava-Ferreruela, I. and Alabau-Muñoz, A. (2006) "Broadband policy assessment: a cross-national empirical analysis," *Telecommunications Policy*, 30: 445–63.

Cawley, R. (2003) "The European Union and world telecommunications markets," in G. Madden (ed.) *World Telecommunications Markets* (*The International Handbook of Telecommunications Economics*, vol. III), Cheltenham: Edward Elgar, 153–72.

Charlès, B. (2006) "Looking to the future of the manufacturing world in 3D," Dassault Systèmes: message from the President & CEO, www.3ds.com/corporate/about-us/message-from-the-president/ (accessed 29 April 2007).

Charles, D.R. (1996) "Information technology and production systems," in P.W. Daniels and W.F. Lever (eds) *The Global Economy in Transition*, London: Longman, 83–102.

Chatterjee, S. (2005) "Dell opens third Indian call center," *The Austin American Statesman*, 22 March, www.statesman.com/business/content/business/stories/03/22dell.html (accessed 8 September 2005).

Chéneau-Loquay, A. (ed.) (2004) *Mondialisation et Technologie de la Communication en Afrique*, Paris: Karthala.

Chesbrough, H.W. (2003) *Open Innovation: the new imperative for creating and profiting from technology*, Boston: Harvard Business School Press.

Chesbrough, H.W. and Teece, D.J. (1996) "Organizing for innovation: when is virtual virtuous?" *Harvard Business Review*, 74 (1): 65–73.

Cisco Systems (2002) "Internetworking basics," in *Internetworking Technology Handbook*, Santa Clara, (CA): Cisco Systems, www.cisco.com/univercd/cc/td/doc/cisintwk/ito_doc/introint.htm (accessed 8 June 2006).

Cisco Systems (2003a) *The Bridge—Connecting Business and Technology Strategies*, San Jose, CA: Cisco Systems, Inc., www.cisco.com/en/US/about/ac79/wp/bridge.html (accessed 22 December 2006).

Cisco Systems (2003b) *Mountain America Credit Union*, Success Story, San Jose, CA: Cisco Systems, Inc., www.cisco.com/warp/public/cc/pd/sqsw/sq/prodlit/mtass_ss.pdf (accessed 24 August 2005).

Clark, G.L. (2002) "London in the European financial services industry: locational advantage and product complementarities" *Journal of Economic Geography*, 2: 433–53.

Clark, G.L. and O'Connor, K. (1997) "The informational content of financial products and the spatial structure of the global finance industry." In K. Cox (ed.) *Spaces of globalization: reasserting the power of the local*, New York: Guilford, 89–114.

Clark, G.L. and Wójcik, D. (2007) *The Geography of Finance: corporate governance in a global marketplace*, Oxford: Oxford University Press.

Clark, M.A. (2000) *Teleworking in the Countryside: home-based working in the information society*, Aldershot: Ashgate.

CMA CGM (2006) "CMA CGM and IBM establish Joint Venture," press release, October 30, www.cma-cgm.com/AboutUs/PressRoom/PressReleaseDetail.aspx?Id=4706 (accessed 22 November 2006).

Coase, R.H. (1937) "The nature of the firm," *Economica*, 4 (16): 386–405.

Cobb, S. (2003) "Offshore financial services and the Internet: creating confidence in the use of cyberspace?" *Growth and Change* 34: 244–59.

Coffey, W.J. and Bailly, A.S. (1992) "Producer services and systems of flexible production," *Urban Studies*, 29: 857–68.

Cohen, S.S., DeLong, J.B. and Zysman, J. (2000) *Tools for Thought: what is new and important about the 'E-conomy'*, BRIE Working Paper 138, Berkeley, CA: Berkeley Roundtable on the International Economy.

Compaine, B.M. (ed.) (2001) *The Digital Divide: facing a crisis or creating a myth?*, Cambridge, MA: MIT Press.

Conlin, M. (2006) "Call centers in the rec room," *Business Week*, 23 January: 76–9.

Constable, G. and Sommerville, B. (2003) *A Century of Innovation: twenty engineering achievements that transformed our lives*, Washington, DC: Joseph Henry Press; also www.greatachievements.org/ (accessed 21 May 2007).

Cooke, P. (2002) *Knowledge Economies: clusters, learning and cooperative advantage*, London: Routledge.

Cooke, P. (2004a) "The molecular biology revolution and the rise of bioscience megacentres in North America and Europe," *Environment and Planning C: Government and Policy*, 22: 161–77.

Cooke, P. (2004b) "The accelerating evolution of biotechnology clusters," *European Planning Studies*, 12: 915–20.

Cooke, P. (2005) "Regionally asymmetric knowledge capabilities and open innovation: exploring 'Globalisation 2'—a new model of industry organization," *Research Policy*, 34: 1128–49.

Cooke, P. and Leydesdorff, L. (2006) "Regional development in the knowledge-based economy: the construction of advantage," *Journal of Technology Transfer*, 31: 5–15.

Cooke, P. and Morgan, K. (1998) *The Associational Economy: firms, regions, and innovation*, Oxford: Oxford University Press.

Cooke, P., Davies, C. and Wilson, R. (2002) "Innovation advantages of cities: from knowledge to equity in five basic steps," *European Planning Studies*, 10: 233–50.

Corbridge, S., Martin, R. and Thrift, N. (eds) (1994) *Money, Power and Space*, Oxford: Blackwell.

Corey, K.E. (2004) "Moving people, goods, and information in Singapore: intelligent corridors," in R. Hanley (ed.), *Moving People, Goods, and Information in the 21st century: the cutting-edge infrastructures of networked cities*, London: Routledge, 293–324.

Corey, K.E. and Wilson, M.I. (2006) *Urban and Regional Technology Planning: planning practice in the global knowledge economy*, London: Routledge.

Corona, L., Doutriaux, J. and Mian, A.A. (2006) *Building Knowledge Regions in North America*. Cheltenham: Edward Elgar.

Cortada, J.W. (2004) *The Digital Hand: how computers changed the work of American manufacturing, transportation and retail industries*, Oxford: Oxford University Press.

Cortright, J. and Mayer, H. (2002) *Signs of Life: the growth of biotechnology centers in the US*, Washington, DC: Brookings Institution, www.brookings.edu/es/urban/publications/biotech.pdf (accessed 25 October 2006).

Couclelis, H. (2004a) "Pizza over the Internet: e-commerce, the fragmentation of activity and the tyranny of the region," *Entrepreneurship & Regional Development*, 16: 41–54.

Couclelis, H. (2004b) "The construction of the digital city," *Environment and Planning B: Planning and Design*, 31: 5–19.

Council on Competitiveness (2005) *2005 National Innovation Survey*, Washington, DC: Council on Competitiveness, www.innovateamerica.org/download/2005_National_Innovation_Survey.pdf (accessed 23 December 2006).

Covisint (2006) "Customers: case studies," Detroit: Compuware Corp., www.covisint.com/customers/casestudy (accessed 12 December 2006).

Cox, K.R (ed.) (1997) *Spaces of Globalization: reasserting the power of the local*, New York: Guilford Press.

Cox, W.M. and Alm, R. (1997) "The economy at light speed: technology and growth in the information age—and beyond," in *Federal Reserve Bank of Dallas Annual Report 1996*, Dallas: Federal Reserve Bank of Dallas, 2–17, www.dallasfed.org/fed/annual/1999p/ar96.pdf (accessed 23 December 2006).

Coyle, D. (1998) *The Weightless World: strategies for managing the digital economy*, Cambridge, MA: MIT Press.

Crandall, R.W. (2005) "Broadband communications," in S.K. Majumdar, I. Vogelsang and M.E. Cave (eds) *Handbook of Telecommunications Economics*, vol. 2: *Technology evolution and the Internet*, Amsterdam: Elsevier, 155–90.

Cross, R.G. and Dixit, A. (2005) "Customer-centric pricing: the surprising secret for profitability," *Business Horizons*, 48: 483–91.

Cukier, K.N. (1999) "Bandwidth colonialism? The implications of Internet infrastructure on international e-commerce," *INET Proceedings 1999*, www.isoc.org/inet99/proceedings/1e/1e_2.htm (accessed 4 June 2006).

Cukier, K. (2005) "A market for ideas: a survey of patents and technology," *The Economist*, 22 October.

Cukier, K. (2007) "A world of connections: a special report on telecoms," *The Economist*, 28 April.

Currah, A. (2002) "Behind the web store: the organisational and spatial evolution of multichannel retailing in Toronto," *Environment and Planning A*, 34: 1411–41.

Currah, A. (2006) "Hollywood versus the Internet: the media and entertainment industries in a digital and networked economy," *Journal of Economic Geography*, 6: 439–68.

Cyert, R.M. andd Mowery, D.C. (1987) *Technology and Employment: innovation and growth in the US economy*, Washington DC: National Academy Press.

Dai, X. (2000) *The Digital Revolution and Governance*, Aldershot: Ashgate.

Dai, X. (2003) "ICTs in China's development strategy," in C.R. Hughes and G. Wacker (eds) *China and the Internet: politics of the digital leap forward*, London: RoutledgeCurzon, 8–29.

Darby, M.R. and Zucker, L.G. (2003) "Grilichesian breakthroughs: inventions of methods of inventing and firm entry in nanotechnology," NBER Working Paper No. 9825, Cambridge, MA: National Bureau of Economic Research, http://papers.nber.org/papers/w9825.pdf (accessed 25 October 2006).

Davenport, T.H. (1997) *Information Ecology: mastering the information and knowledge environment*, Oxford: Oxford University Press.

Daveri, F. (2002) "The new economy in Europe, 1992–2001," *Oxford Review of Economic Policy*, 18: 345–62.

Daveri, F. and Silva, O. (2004) "Not only Nokia: what Finland tells us about *new economy* growth," *Economic Policy* 19 (38): 117–63.

David, P.A. (1991) "Computer and dynamo: the modern productivity paradox in a not-too-distant mirror," in OECD, *Technology and Productivity: the challenge for economic policy*, Paris: Organisation for Economic Co-operation and Development, 315–47.

Davidow, W.H. and Malone, M.S. (1993) *The Virtual Corporation: structuring and revitalizing the corporation for the 21st century*, New York: Harper.

Dawson, K. (2001) *Siliconia*, www.tbtf.com/siliconia.html (accessed 23 December 2006).

Day, G.S. (1981) "The product life cycle: analysis and applications issues," *Journal of Marketing*, 45 (4): 60–7.

Day, G.S., Fein, A.J. and Ruppersberger, G. (2003) "Shakeouts in digital markets: lessons from B2B exchanges," *California Management Review*, 45 (2): 131–50.

De Bernardy, M. (1999) "Reactive and proactive local territory: co-operation and community in Grenoble," *Regional Studies*, 33: 343–52.

Dedrick, J. and Kraemer, K.L. (2005) "The impacts of IT on firm and industry structure: the personal computer industry," *California Management Review*, 47 (3): 122–42.

DeFanti, T., de Laat, C., Mambretti, J., Neggers, K. and St. Arnaud, B. (2003) "TransLight: a global scale LambdaGrid for e-science," *Communications of the ACM*, 46 (11): 35–41.

Dell Inc. (2006) *Form 10-K Annual Report for the fiscal year ended February 3, 2006*, Washington, DC: US Securities and Exchange Commission.

Deloitte Research (2003) *Mastering Complexity in Global Manufacturing: powering profits and growth through value chain synchronization*, London: Deloitte & Touche LLP, www.deloitte.com/dtt/cda/doc/content/global_mfg_MasteringComplexity_10_2003.pdf (accessed 5 December 2005).

Delta Regional Authority (2007) *iDelta: Information Technology in the Delta*, Clarksdale (Mississippi): Delta Regional Authority.

De Meyer, A. (1993) "Management of an international network of industrial R&D laboratories," *R&D Management*, 23: 109–20.

den Hertog, P., Broersma, L. and van Ark, B. (2003) "On the soft side of innovation: services innovation and its policy implications," *De Economist*, 151: 433–52.

Derudder, B. and Witlox, F. (2005) "An appraisal of the use of airline data in assessing the world city network: a research note on data," *Urban Studies*, 42: 2371–88.

Desautez, L. (2005) "Egg: de Zebank à 'Ze end," *Le Journal du Net*, February, www.journaldunet.com/dossiers/egg/ (accessed 12 December 2006).

De Young, R., Klier, T. and McMillen, D.P. (2004) "The changing geography of the US banking industry," *The Industrial Geographer*, 2: 29–48.

Dicken, P. (2007) *Global Shift: mapping the changing contours of the world economy*, 5th edn, London: Sage.

Dixon, T. and Marston, A. (2002) "U.K. retail real estate and the effects of online shopping," *Journal of Urban Technology*, 9 (3): 19–47.

Dodge, M. (2001) "Finding the source of Amazon.com," in T.R. Leinbach and S.D. Brunn (eds) *Worlds of Electronic Commerce: economic, geographical and social dimensions*, New York: John Wiley, 167–80.

Dodge, M. and Kitchin, R. (2000) *Mapping Cyberspace*, London: Routledge.

Dodgson, M., Gann, D. and Salter, A. (2006) "The role of technology in the shift towards open innovation: the case of Procter & Gamble," *R&D Management*, 36: 333–46.

Donaghu, M.T. and Barff, R. (1990) "Nike just did it: international subcontracting and flexibility in athletic footware production," *Regional Studies*, 24: 537–52.

Dossani, R. and Kenney, M. (2003) "Lift and shift': moving the back office to India," *Information Technologies and International Development*, 1: 21–37.

Dow Jones & Company, Inc. (2007) Advertisement, *The Wall Street Journal*, 9 April: A8.

Doz, Y., Santos, J. and Williamson, P. (2001) *From Global to Metanational: how companies win in the knowledge economy*, Boston, MA: Harvard Business School Press.

Doz, Y., Wilson, K., Veldhoen, S., Goldbrunner, T. and Altman, G. (2006) *Innovation: Is global the way forward? Survey results*. Fontainebleau: INSEAD and Booz Allen Hamilton, www.strategy-business.com/media/file/global_innovation.pdf (accessed 24 October 2006).

Drennan, M.P. (2002) *The Information Economy and American Cities*, Baltimore, MD: Johns Hopkins University Press.

Drucker, P.F. (1999) "Beyond the information revolution," *The Atlantic Monthly*, 284 (4): 47–70.

Druilhe, C. and Garnsey, E. (2000) "Emergence and growth of high-tech activity in Cambridge and Grenoble," *Entrepreneurship and Regional Development*, 12: 163–77.

Dumout, E. (2004) "La délocalisation des centres d'appel fait le succès de Webhelp," *ZDNet France*, January 28, www.zdnet.fr/actualites/internet/0,39020774,39139405,00.htm (accessed 22 December 2006).

Dunford, M. (1991) "Industrial trajectories and social relations in areas of new industrial growth," in G. Benko and M. Dunford (eds) *Industrial Change and Regional Development*, London: Belhaven: 51–82.

Dunn, D. (2006) "The best and worst cities for data centers," *InformationWeek*, 23 October, www.informationweek.com/news/showArticle.jhtml?articleID=193401105 (accessed 7 December 2006).

Dupuy, G. (2002) *Internet: geographie d'un réseau*, Paris: Éditions Éllipses.

Dupuy, G. (2007) *La fracture numérique*, Paris: Éditions Ellipses.

Duranton, G. (1999) "Distance, land, and proximity: economic analysis and the evolution of cities," *Environment and Planning A*, 31: 2169–88.

Dutta, S. and Jain, A. (2006) "Networked readiness and the benchmarking of ICT competitiveness," in S. Dutta, A. Lopez-Claros and I. Mia (eds) *The Global Information Technology Report 2005–2006: leveraging ICT for development*, London: Palgrave Macmillan, 3–24.

Dutton, W.H., Gillett, S.E., McKnight, L.W. and Peltu, M. (2004) "Bridging broadband Internet divides: reconfiguring access to enhance communicative power," *Journal of Information Technology*, 19: 28–38.

Dutton, W.H., Kahin, B., O'Callaghan, R. and Wyckoff, A.W. (2005) (eds) *Transforming Enterprise: the economic and social implications of information technology*, Cambridge, MA: MIT Press.

Earl, M. (2001) "Knowledge management strategies: towards a taxonomy," *Journal of Management Information Systems*, 18: 215–33.

Edgington, D.W. (2007) "The Kyoto Research Park and innovation in Japanese cities," paper presented at the annual meeting of the Association of American Geographers, San Francisco, CA, April.

Edquist, H. (2005) "The Swedish ICT miracle—myth or reality?," *Information Economics and Policy*, 17: 275–301.

Edwards, C. and Ihlwan, M. (2006) "Upward mobility," *Business Week*, 4 December: 69–82.

Egg plc (2005) *Annual Report 2004*, London: Egg plc. www.investis.com/egg/annual04/pdf/egg_ar2004.pdf (accessed 12 December 2006).

Ehrle, D. (2006) *WebSphere Everyplace Connection Manager*, Zürich: IBM Switzerland Software Group, 26 September, www-5.ibm.com/ch/events/ip-communication/pdf/WECM_IPC.pdf (accessed 19 October 2006).

Einhorn, B. and Elgin, B. (2006) "The great firewall of China," *Business Week*, 23 January: 32–4.

Eliasson, G. (2005) "The nature of economic change and management in a new knowledge based information economy," *Information Economics and Policy*, 17: 428–56.

Elmer, G. (2004) *Profiling Machines: mapping the personal information economy*, Cambridge, MA: MIT Press.

Empirica (2003) *e-Business W@tch—the European e-Business report 2003 edition*, European Commission, Enterprise and Industry Directorate General, Luxembourg: office for official publications of the European Communities, www.ebusiness-watch.org/resourses/document/E-Business-2003.pdf (accessed 9 August 2007)

Empirica (2005) *e-Business W@tch—The European e-Business Report 2005 edition*, European Commission, Enterprise & Industry Directorate General, Luxembourg: Office for official publications of the European Communities, www.ebusiness-watch.org/resources/documents/eBusiness-Report-2005.pdf (accessed 30 November 2006).

Engardio, P., Bernstein, A. and Kripalani, M. (2003) "Is your job next?" *Business Week*, 3 February, 50–60. www.businessweek.com/magazine/toc/03_05/B3818magazine.htm (accessed 22 December 2006).

Engelbrecht, H.-J. (2003) "Data issues in the new economy," in D.C. Jones (ed.) *New Economy Handbook*, Amsterdam: Elsevier, 57–76.

ENX Association (2006) "What is the ENX?" www.enxo.com/lang/en/_basics_01.php (accessed 25 November 2006).

E.ON (2006) "Powergen withdraws call handling from India," press release, June 15, Coventry (UK): E.ON UK plc.

Etzkowitz, H. and Klofsten, M. (2005) "The innovating region: towards a theory of knowledge-based regional development," *R&D Management*, 35: 243–55.

European Commission (2005a) *Information Society Benchmarking Report*, Brussels: Commission of the European Communities, www.europa.eu.int/information_society/eeurope/i2010/docs/benchmarking/051222%20Final%20Benchmarking%20Report.pdf (accessed 26 September 2006).

European Commission (2005b) *Digital Divide Forum Report: broadband access and public support in under-served areas*, Brussels: Commission of the European Communities, www.europa.eu.int/information_society/eeurope/i2010/docs/implementation/ddf_report_final.pdf (accessed 26 September 2006).

European Commission (2006) *Bridging the Broadband Gap* (COM (2006) 129 FINAL), Brussels: Commission of the European Communities, www.europa.eu.int/information_society/eeurope/i2010/docs/digital_divide/communication/com_bridging_bb_gap_2006_en.pdf (accessed 11 June 2006).

European Trade Union Confederation (2002) *Framework Agreement on Telework*, Brussels, July 16, www.ec.europa.eu/employment_social/news/2002/jul/telework_en.pdf (accessed 19 December 2006).

Evans, P. and Wurster, T.S. (1999) "Getting real about virtual commerce," *Harvard Business Review*, 77 (6): 84–94.

Evans, P. and Wurster, T.S. (2000) *Blown to Bits: how the new economics of information transforms strategy*, Boston, MA: Harvard Business School Press.

Falk, M. (2005) "ICT-linked firm reorganisation and productivity gains," *Technovation*, 25: 1229–50.

Farrell, D., Laboissière, M., Rosenfeld, J., Stürze, S. and Umezawa, F. (2005b) *The Emerging Global Labor Market: Part II—the supply of offshore talent in services*, San Francisco: McKinsey Global Institute, www.mckinsey.com/mgi/reports/pdfs/emerginggloballabormarket/part2/MGI_supply_fullreport.pdf (accessed 22 December 2006).

Farrell, D., Laboissière, M., Pascal, R., Rosenfeld, J., de Segundo, C. and Stürze, S. (2005a) *The Emerging Global Labor Market: Part I—the demand for offshore talent in services*, San Francisco, CA: McKinsey Global Institute, www.mckinsey.com/mgi/reports/pdfs/emerginggloballabormarket/part1/MGI_demand_fullreport.pdf (accessed 22 December 2006).

Farrell, D., Laboissière, M., Pascal, R., de Segundo, C., Rosenfeld, J., Stürze, S. and Umezawa, F. (2005c) *The Emerging Global Labor Market: Part III—how supply and demand for offshore talent meet*, San Francisco, CA: McKinsey Global Institute, www.mckinsey.com/mgi/reports/pdfs/emerging globallabormarket/part3/MGI_supplyanddemand_fullreport.pdf (accessed 23 December 2006).

Federal Communication Commission (2007) *High-Speed Services for Internet Access: status as of June 30, 2006*, Washington, DC: Wireline Competition Bureau, January, www.hraunfoss.fcc.gov/edocs_public/attachmatch/DOC-270128A1.pdf (accessed 11 May 2007).

Federal Highway Administration (2005) "Annual vehicle distance traveled in kilometers—2004, by highway category and vehicle type," *Highway Statistics 2004*, Washington DC: US Department of Transportation, www.fhwa.dot.gov/policy/ohim/hs04/htm/vm1m.htm (accessed 4 October 2006).

FedEx Corporation (2005) *Form 10-K Annual Report for the fiscal year ended May 31, 2005*, Washington, DC: US Securities and Exchange Commission.

FedEx Corporation (2006) "About FedEx: Technology," www.fedex.com/us/about/unitedstates/technology (accessed 1 December 2006).

Feitzinger, E. and Lee, H.L. (1997) "Mass customization at Hewlett-Packard: the power of postponement," *Harvard Business Review*, 71 (1): 116–21.

Feldman, M.P. and Florida, R. (1994) "The geographic sources of innovation: technological infrastructure and product innovation in the United States," *Annals of the Association of American Geographers*, 84: 210–29.

Feldman, M.P., Francis, J. and Bercovitz, J. (2005) "Creating a cluster while building a firm: entrepreneurs and the formation of industrial clusters," *Regional Studies*, 39: 129–41.

Ferguson, C.H. (2004) *The Broadband Problem: anatomy of a market failure and a policy dilemma*, Washington, DC: Brookings Institution.

FiberNet Telecom Group (2005) *60 Hudson Street Meet-Me Room*, www.ftgx.com/services/services-meet-me/ (accessed 9 June 2006).

Fields, G. (2006) "Innovation, time, and territory: space and the business organization of Dell Computer," *Economic Geography*, 82: 119–46.

Finnie, G. (1998) "Wired cities," *CommunicationsWeek International*, 18 May: 19–23.

Firmino, R.J. (2003) "Not just portals': virtual cities as complex sociotechnical phenomena," *Journal of Urban Technology*, 10 (3): 41–62.

Firmino, R.J. (2005) "Planning the unplannable: how local authorities integrate urban and ICT policy making," *Journal of Urban Technology*, 12 (2): 49–69.

Florida, R. (1997) "The globalization of R&D: results of a survey of foreign-affiliated R&D laboratories in the US," *Research Policy*, 26: 85–103.

Florida, R. (2002a) *The Rise of the Creative Class . . . and how it's transforming work, leisure, community, & everyday life*, New York: Basic Books.

Florida, R. (2002b) "The economic geography of talent," *Annals of the Association of American Geographers*, 92: 743–55.

Florida, R. (2005) *Cities and the Creative Class*, London: Routledge.

Foray, D. (2004) *Economics of Knowledge*, Cambridge, MA: MIT Press [first published in French in 2000 as *L'économie de la connaissance* by La Découverte Repères].

Foray, D. (2005) "New models of innovation and the role of information technologies in the knowledge economy," in W.H. Dutton, B. Kahin, R. O'Callaghan and A.W. Wyckoff (eds) *Transforming Enterprise: the economic and social implications of information technology*, Cambridge, MA: MIT Press, 113–29.

Forbes.com (2007) "Matsushita Electric to allow 30,000 white-collar staff to work from home," 28 March, www.forbes.com/technology/feeds/afx/2007/03/28/afx3557880.html (accessed 21 April 2007).

Ford, P. (2005) "Digital India: the underwater connection," *Submarine Telecoms Forum*, 21: 21–3.

Forester, T. (1988) "The myth of the electronic cottage," *Futures*, 20: 227–40.

Forrest, B. (2007) "Silicon Siberia," *Fortune*, 155 (5) 2 April: 33–36.

Foucault, M. (1975) *Surveiller et punir*, Paris: Gallimard.

Franke, U. (ed.) (2002) *Managing Virtual Web Organizations in the 21st Century: issues and challenges*, Hershey, PA: Idea Group Publishing.

Fransman, M. (2002) *Telecoms in the Internet Age: from boom to bust to . . . ?*, Oxford: Oxford University Press.

Fransman, M. (ed.) (2006a) *Global Broadband Battles: why the US and Europe lag while Asia leads*, Stanford, CA: Stanford University Press.

Fransman, M. (2006b) "Introduction," in M. Fransman (ed.) *Global Broadband Battles: why the US and Europe lag while Asia leads*, Stanford, CA: Stanford University Press, 1–57.

Freeman, C. (1996a) "The two-edged nature of technological change: employment and unemployment," in W.H. Dutton (ed.) *Information and Communication Technologies: visions and realities*, Oxford: Oxford University Press, 19–36.

Freeman, C. (1996b) "The factory of the future and the productivity paradox," in W.H. Dutton (ed.) *Information and Communication Technologies: visions and realities*, Oxford: Oxford University Press, 123–41.

Freeman, C. and Louçã, F. (2001) *As Time Goes By: from the industrial revolutions to the information revolution*, Oxford: Oxford University Press.

Freeman, R.B. (2002) "The labour market in the new information economy," *Oxford Review of Economic Policy*, 18: 288–305.

Freeman, R.B. (2005) "What really ails Europe (and America): the doubling of the global workforce," *The Globalist*, June 3, www.theglobalist.com/StoryId.aspx?StoryId=4542 (accessed 19 October 2006)

Frieden, R. (2005) "Lessons from broadband development in Canada, Japan, Korea and the United States," *Telecommunications Policy* 29: 595–613.

Friedmann, J. and Wolff, G. (1982) "World city formation: an agenda for research and action," *International Journal of Urban and Regional Research*, 6: 309–43.

Fromhold-Eisebith, M. (2002) "Regional cycles of learning: foreign multinationals as agents of technological upgrading in less developed countries," *Environment and Planning A*, 34: 2155–73.

Fuentes-Bautista, M. (2001) *Rural Telecommunications Policy Bibliography*, Report for the Rural Policy Research Institute, Austin: University of Texas, www.utexas.edu/research/tipi/reports2/rtbiblio.pdf (accessed 10 May 2007).

Gadrey, J. (2003) *New Economy, New Myth*, London: Routledge.

GAO [General Accounting Office] (2003) *Satellite Communications: strategic approach needed for DOD's procurement of commercial satellite bandwidth* (GAO-04–206), Washington, DC: US General Accounting Office.

GAO [Government Accountability Office] (2006) *Telecommunications broadband deployment is extensive throughout the United States, but it is difficult to assess the extent of deployment gaps in rural areas*, Report to Congressional Committees (GAO-06–426), Washington, DC: Government Accountability Office, www.gao.gov/new.items/d06426.pdf (accessed 10 May 2007).

Gans, J.S., King, S.P. and Wright, J. (2005) "Wireless communications," in S.K. Majumdar, I. Vogelsang and M.E. Cave (eds) *Handbook of Telecommunications Economics*, vol. 2: *Technology evolution and the Internet*, Amsterdam: Elsevier, 241–85.

Garelli, S. (2005) "The world competitiveness landscape in 2005: a higher degree of risk," in International Institute for Management Development (IMD) (2005) *IMD World Competitiveness Yearbook 2005*, Lausanne: IMD, 40–5.

Garrett, R.K. and Danziger, J.N. (2007) "Which telework? Defining and testing a taxonomy of technology-mediated work at a distance," *Social Science Computer Review*, 25: 27–47.

Gaspar, J. and Glaeser, E.L. (1998) "Information technology and the future of cities," *Journal of Urban Economics*, 43: 136–56.

GE (General Electric) (2006) *Driving Efficiency*, www.ge.com/stories/en/20424.html (accessed 5 June 2006).

Ge, L., Konana, P. and Tanriverdi, H. (2004) "Global sourcing and value chain unbundling," working paper, McCombs School of Business, The University of Texas at Austin, www.mccombs.utexas.edu/faculty/prabhudev.konana/globalsourcing.pdf (accessed 21 September 2005).

Gehrig, T. (2000) "Cities and the geography of financial centers," in J. Thisse and J.-M. Huriot (eds) *The Economics of Cities*, Cambridge: Cambridge University Press, 415–45.

Geller, A. (2005) "Homeshoring' means that call center might be in someone's bedroom," *Seattle Post Intelligencer*, 9 May, www.seattlepi.nwsource.com/business/223367_homecall09.html (accessed 19 December 2006).

General Services Administration (2006) "Federal management regulation; Guidelines for alternative workplace arrangements," *Federal Register*, 71 (52) (March): 13845–48.
George, J. (2005) "FTTH design with the future in mind," *Broadband Properties*, September: 30–5.
Gereffi, G. (2006) *The New Offshoring of Jobs and Global Development*, Geneva: International Labour Organization, www.ilo.org/public/english/bureau/inst/download/newoff.pdf (accessed 20 April 2007).
Gershuny, J.I. and Miles, I.D. (1983) *The New Service Economy*, London: Frances Pinter.
Gerst, M. and Bunduchi, R. (2005) "Shaping IT standardization in the automotive industry—the role of power in driving portal standardization," *Electronic Markets*, 15: 335–43.
Gertler, M. (2003) "Tacit knowledge and the economic geography of context, or The undefinable tacitness of being (there)," *Journal of Economic Geography*, 3: 75–99.
Gibson, W. (1984) *Neuromancer*, London: Harper & Collins.
Gillespie, A. and Richardson, R. (1996) "Advanced communications and employment creation in rural and peripheral regions: a case study of the Highlands and Islands of Scotland," *The Annals of Regional Science*, 30: 91–110.
Gillespie, A. and Richardson, R. (2000) "Teleworking and the city," in J. Wheeler, Y. Aoyama and B. Warf (eds) *Cities in the Telecommunications Age: the fracturing of geographies*, London: Routledge, 228–45.
Gillett, S.E., Lehr, W.H., Osorio, C.A. and Sirbu, M.A. (2006) *Measuring Broadband's Economic Impact: final report*, Washington, DC: US Department of Commerce, Economic Development Administration, www.eda.gov/PDF/MITCMUBBImpactReport.pdf (accessed 23 December 2006).
Gilmore, J.H. and Pine, J.B. (eds) (2000) *Markets of One: creating customer-unique value through mass customization*, Boston, MA: Harvard Business School Press.
Gitomer, J. (1998) *Customer Satisfaction is Worthless, Customer Loyalty is Priceless*, Austin, TX: Bard Press.
Glaeser, E.L. (2005) "Reinventing Boston, 1630–2003," *Journal of Economic Geography*, 5: 119–53.
Gleckman, H., Carey, J., Mitchell, R., Smart, T. and Roush, C. (1993) "The technology payoff," *Business Week*, 14 June: 57–79.
Godin, B. (2004) "The New Economy: what the concept owes to the OECD," *Research policy*, 33: 679–90.
Goldman, S.L., Nagel, R.N. and Preiss, K. (1994) *Agile Competitors and Virtual Organizations: strategies for enriching the customer*, New York: John Wiley.
Goldsmith, J. and Wu, T. (2006) *Who Controls the Internet? Illusions of a borderless world*, Oxford: Oxford University Press.
Gordon, G. (2004) "Telecommuting is 'alive and well and growing," *Workforce Management*, 30 September, www.workforce.com/section/00/article/23/84/98.html (accessed 3 April 2006).
Gordon, P., Lee, B. and Richardson, H.W. (2004) "Travel trends in US cities: explaining the 2000 census commuting results," Research paper, Lusk Center for Real Estate, University of Southern California, Los Angeles, www-rcf.usc.edu/~pgordon/pdf/commuting.pdf (accessed 4 October 2006).
Gordon, R.J. (2000) "Does the "new economy" measure up to the great inventions of the past?" *Journal of Economic Perspectives*, 14 (4): 49–74.
Gorman, S.P. (2005) *Networks, Security and Complexity: the role of public policy in critical infrastructure protection*, Cheltenham: Eldward Elgar.
Gorman, S.P. and Malecki, E.J. (2000) "The networks of the Internet: an analysis of provider networks," *Telecommunications Policy*, 24: 113–34.
Gorman, S.P. and Malecki, E.J. (2001) "Fixed and fluid: stability and change in the geography of the Internet," *Telecommunications Policy*, 26: 389–413.
Gottfredson, M. and Aspinall, K. (2005) "Innovation versus complexity: What is too much of a good thing?" *Harvard Business Review*, 83 (11): 62–71.
Gottfredson, M., Puryear, R. and Phillips, S. (2005) "Strategic sourcing: from periphery to the core," *Harvard Business Review*, 83 (2): 132–39.
Gottmann, J. (1961) *Megalopolis: the urbanized northeastern seaboard of the United States*, Cambridge, MA: MIT Press.
Gottmann, J. (1970) "Urban centrality and the interweaving of quaternary functions," *Ekistics*, 29 (174): 322–31.
Grabher, G. (2002) "Fragile sector, robust practice: project ecologies in new media," *Environment and Planning A*, 34: 1911–26.
Graham, S. (1999) "Global grids of glass: on global cities, telecommunications and planetary urban networks," *Urban Studies*, 36: 929–49.

Graham, S. (2004) "Excavating the material geographies of cybercities," in S. Graham (ed.) *The Cybercities Reader*, London: Routledge, 138–42.

Graham, S. and Marvin, S. (2001) *Splintering Urbanism: networked infrastructures, technological mobilities and the urban condition*, London: Routledge.

Graham, S. and Wood, D. (2003) "Digitising surveillance: categorisation, space and inequality," *Critical Social Policy*, 23: 227–48.

Granovetter, M.S. (1985) "Economic action and social structure: the problem of embeddedness," *American Journal of Sociology*, 91: 481–510.

Graves, W. (2004) "The geography of finance and financial services," *The Industrial Geographer*, 2: 1, www.igeographer.lib.indstate.edu/geditorial.pdf (accessed 4 April 2007).

Green, P.E. (2006) *Fiber to the Home: the new empowerment*, New York: John Wiley.

Greenstein, S. (2005) "The economic geography of Internet infrastructure in the United States," in M. Cave, S. Majumdar and I. Vogelsang (eds) *Handbook of Telecommunications Economics*, vol. 2: *Technology evolution and the Internet*, Amsterdam: Elsevier, 289–373.

Grimes, S. (1999) "Rural areas in the information society: diminishing distance or increasing learning capacity?," *Journal of Rural Studies*, 16: 13–21.

Grimes, S. (2003a) "Ireland's emerging information economy: recent trends and future prospects," *Regional Studies*, 37: 3–14.

Grimes, S. (2003b) "The digital economy challenge facing peripheral rural areas," *Progress in Human Geography*, 27: 174–94.

Grimes, S. (2006) "Ireland's emergence as a centre for internationally traded services," *Regional Studies*, 40: 1041–54.

Grossman, G. and Rossi-Hansberg, E. (2006) *Trading Tasks: a simple theory of offshoring*, Princeton: Princeton University, Department of Economics, www.princeton.edu/%7Egrossman/offshoring.pdf (2 May 2007).

Grote, M.H. and Täube, F.A. (2006) "Offshoring the financial services industry: implications for the evolution of Indian IT clusters," *Environment and Planning A*, 38: 1287–1305.

Grover, V. and Ramanlal, P. (2004) "Digital economics and the e-business dilemma," *Business Horizons*, 47 (4): 71–80.

Grubel, H.G. (1987) "All traded services are embodied in materials or people," *The World Economy*, 10: 319–30.

Grubesic, T.H. (2003) "Inequities in the broadband revolution," *Annals of Regional Science*, 37: 263–89.

Grubesic, T.H. (2006) "A spatial taxonomy of broadband regions in the United States," *Information Economics and Policy*, 18: 423–48.

Grubesic, T.H. and Murray, A.T. (2005) "Spatial–historical landscapes of telecommunication network survivability," *Telecommunications Policy*, 29: 801–20.

Guile, B.R. and Quinn, J.B. (eds) (1988) *Technology in Services*, Washington, DC: National Academy Press.

Guislain, P., Ampah, M.A., Besançon, L., Niang, C. and Sérot, A. (2005) *Connecting Sub-Saharan Africa* (World Bank Working Paper No. 51), Washington, DC: World Bank.

Gunasekaran, V. and Harmantzis, F.C. (2007) "Emerging wireless technologies for developing countries," *Technology in Society*, 29: 23–42.

Guru.com (2004) "Emoonlighter announces name change to Guru.com," 8 March, www.guru.com/Press/2004–03–08.cfm (accessed 22 December 2006).

Häcki, R. and Lighton, J. (2001) "The future of the networked company," *McKinsey Quarterly*, 3: 26–39.

Haddad, C. and Ewing, J. (2001) "Ground wars: UPS's rapid ascent leaves FedEx scrambling," *Business Week*, 21 May: 64–8.

Hagel, J. and Singer, M. (1999) "Unbundling the corporation," *Harvard Business Review*, 77 (3): 133–41.

Hall, P. and Preston, P. (1988) *The Carrier Wave: new information technology and the geography of innovation 1846–2003*, London: Unwin Hyman.

Haltiwanger, J. and Jarmin, R. (2003) "A statistical portrait of the new economy," in D.C. Jones (ed.) *New Economy Handbook*, Amsterdam: Elsevier, 3–24.

Hamm, S. (2007) "The trouble with India," *Business Week*, 19 March: 48–58.

Hamrick, K. (ed.) (2003) *Rural America at a Glance*, Rural Development Research Report RDRR 97–1, Washington, DC: US Department of Agriculture, Economic Research Service,www.ers.usda.gov/publications/rdrr97–1/highres_rdrr97–1.pdf (accessed 31 January 2006).

Handy, S.L. and Mokhtarian, P.L. (1996) "The future of telecommuting," *Futures*, 28: 227–40.
Harper, R.A. (1982) "Metropolitan areas as transactional centers," in C. Christian and R.A. Harper (eds) *Modern Metropolitan Systems*, Columbus, OH: Merrill, 87–109.
Harrison, B. (1994) *Lean and Mean: the changing landscape of corporate power in the age of flexibility*, New York: Basic Books.
Hart, J. and Kim, S. (2002) "Explaining the resurgence of US competitiveness: the rise of Wintelism," *The Information Society*, 18 (1): 1–12.
Harvey, D (1989) *The Condition of Postmodernity*, Oxford: Basil Blackwell.
Harvey, D. (2006) *Spaces of Global Capitalism: towards a theory of uneven geographical development*, London: Verso.
Hassan, R. (2003) *The Chronoscopic Society: globalization, time and knowledge in the network economy*, New York: Peter Lang.
Hatfield, D.N., Mitchell, B.M. and Srinagesh, P. (2005) "Emerging network technologies," in S.K. Majumdar, I. Vogelsang and M. Cave (eds), *Handbook of Telecommunications Economics*, Vol. 2: *Technology evolution and the Internet*, Amsterdam: Elsevier, 31–80.
Hayashi, H. and Brown, K. (2005) "Tokyo Electric Power: Japan's FTTH powerhouse," *Broadband Properties*, March: 34–37.
Headrick, D.R. (1991) *The Invisible Weapon: telecommunications and international politics 1851–1945,* New York: Oxford University Press.
Headrick, D.R. and Griset, P. (2001) "Submarine telegraph cables: business and politics, 1838–1939," *Business History Review*, 75: 543–78.
Hecker, D.E. (2005) "Occupational employment projections to 2014," *Monthly Labor Review*, 128 (11): 70–101.
Heinzl, M. and Young, S. (2006) "With rising Internet traffic, spare fiber-optic lines fill up," *Wall Street Journal*, 27 April: B1.
Heitzman, J. (1999) "Corporate strategy and planning in the science city: Bangalore as 'Silicon Valley'?," *Economic and Political Weekly*, 34 (5) 30 January: PE2–PE11.
Helm, B. and Kripalani, M. (2006) "Life on the Web's factory floor. Who do you think turns all those words into an easy click?," *Business Week*, 22 May: 70–1.
Helpman, E. (1998) "Introduction," in E. Helpman (ed.) *General Purpose Technologies and Economic Growth*, Cambridge, MA: MIT Press, 1–14.
Hempel, J. (2005) "The MySpace generation," *Business Week*, 12 December: 86–96.
Hepworth, M. (1990) *Geography of the Information Economy*, New York: Guilford Press.
Hess, M. and Coe, N.M. (2006) "Making connections: global production networks, standards, and embeddedness in the mobile-telecommunications industry," *Environment and Planning A*, 38: 1205–27.
Hindle, T. (2006) "The new organisation: a survey of the company," *The Economist*, 19 January.
Hipp, C. and Grupp, H. (2005) "Innovation in the service sector: the demand for service-specific innovation measurement concepts and typologies," *Research Policy*, 34: 517–35.
Hite, J. (1997) "The Thunen model and the new economic geography as a paradigm for rural development policy," *Review of Agricultural Economics*, 19 (2): 230–40.
Hobday, M., Davies, A. and Prencipe, A. (2005) "Systems integration: a core capability of the modern corporation," *Industrial and Corporate Change*, 14: 1109–43.
Hof, R.D. (1999) "A new era of bright hopes and terrible fears," *Business Week*, 4 October: 84–98.
Hof, R.D. (2005) "Why eBay is buying Skype," *Business Week*, 12 September, www.businessweek.com/the_thread/techbeat/archives/2005/09/why_ebay_is_buy.html (accessed 27 December 2006).
Hof, R.D. (2006) "Jeff Bezos" risky bet," *Business Week*, 13 November: 52–8.
Holmes, E. (2007) "Going online isn't a 'must' move for all entrepreneurs," *The Wall Street Journal*, 24 April: B4.
Holstein, W.J. (2001) "DaimlerChrysler net designs," *Business 2.0*, April 17, 26–8.
Holstein, W.J. (2002) "Middleman becomes master," *Chief Executive*, 182 (October): 53–6.
Hopkinson, P., James, P. and Maruyama T. (2002) *Teleworking at British Telecom: the economic, environmental and social impacts of its Workabout scheme*, Report for the program SUSTEL, University of Bradford, www.btplc.com/Societyandenvironment/PDF/SusTelPilotReport.pdf (accessed 18 October 2006).
Housel, T.J. and Skopec, E.W. (2001) *Global Telecommunications Revolution: the business perspective*, New York: McGraw Hill.

Howells, J. (2002) "Tacit knowledge, innovation and economic geography," *Urban Studies*, 39: 871–84.

Hsu, C. (2004) *2003 Section 43.82 Circuit Status Data, December 2004*, Washington, DC: Federal Communications Commission, www.hraunfoss.fcc.gov/edocs_public/attachmatch/DOC-255737A1.pdf (accessed 9 August 2005).

Hubbard, G. (2006) "The productivity riddle," *Strategy + Business*, 45: 28–33.

Hudson, H.E. (1990) *Communication Satellites: their development and impact*, New York: Free Press.

Hugill, P.J. (1999) *Global Communications since 1844: geopolitics and technology*, Baltimore: Johns Hopkins University Press.

Huggins, R., Izushi, H. and Davies, W. (2005) *World Knowledge Competitiveness Index 2005*, Pontypridd (UK): Robert Huggins Associates.

Hulsink, W. and Davies, A. (2003) "Making markets: telecommunications in Western Europe," in G. Madden (ed.) *World Telecommunications Markets* (*The International Handbook of Telecommunications Economics*, Vol. III), Cheltenham: Edward Elgar: 413–29.

Huseman, R.C. and Goodman, J.P. (1999) *Leading with Knowledge: the nature of competition in the 21st century*, Thousand Oaks, CA: Sage.

Huston, L. and Sakkab, N. (2006) "Connect and develop: inside Procter & Gamble's new model for innovation," *Harvard Business Review*, 84 (3): 58–66.

Huxley, A. (1938) *Brave New World*, New York: Perennial (reprinted 1998).

Iammarino, S. and McCann, P. (2006) "The structure and evolution of industrial clusters: transactions, technology and knowledge spillovers," *Research Policy*, 35: 1018–36.

ICICI OneSource (2006) "ICICI OneSource announces outsourcing centers in Northern Ireland," press release, 13 June, Mumbai: ICICI OneSource.

Ida, T. (2006) "Broadband, information society, and the national system in Japan," in M. Fransman (ed.) *Global Broadband Battles: why the US and Europe lag while Asia leads*, Stanford, CA: Stanford University Press, 65–86.

IDATE (2001) *DigiWorld 2000*, Montpellier: IDATE Foundation.

IDATE (2006) *DigiWorld 2005*, Montpellier: IDATE Foundation.

IEEE-US Committee on Communications and Information Policy (2005) *Providing Ubiquitous Gigabit Networks in the United States*, Washington, DC: IEEE-US, www.ieeeusa.org/volunteers/committees/ccip/docs/Gigabit-WP.pdf (accessed 26 September 2006).

Ihlwan, M. and Hall, K. (2007) "New tech, old habits," *Business Week*, 26 March: 48–9.

Indergaard, M. (2004) *Silicon Alley: the rise and fall of a new media district*, London: Routledge.

India Daily (2005a) "India's infrastructure seen getting major boost from federal budget," *Indiadaily.com*, 1 March, www.indiadaily.com/editorial/1765.asp (accessed 15 September 2006).

India Daily (2005b) "Abbey's call centre recalled," *Indiadaily.com*, 26 October, www.indiadaily.com/breaking_news/49866.asp (accessed 22 September 2006).

Indjikian, R. and Siegel, D.S. (2005) "The impact of investment in IT on economic performance: implications for developing countries," *World Development*, 33: 681–700.

InnoVisions Canada (2007) "CTA members tell their stories: part 12," www.ivc.ca/stories/Stories12.html (accessed 24 April 2007).

Institute of Management and Development (IMD) (2005) *World Competitiveness Index*, Lausanne: IDM.

Intelligent Community Forum (2006) *Intelligent Community Awards*, www.intelligentcommunity.org/displaycommon.cfm?an=1&subarticlenbr=5 (accessed 11 December 2006).

International Customer Management Institute (ICMA) (2006) *Industry Statistics*, www.incoming.com/statistics/demographics.aspx (accessed 16 December 2006).

International Institute for Management Development (IMD) (2005) *IMD World Competitiveness Yearbook 2005*, Lausanne: IMD.

International Telecommunications Union (2003) *ITU Internet Reports 2003: the birth of broadband*, Geneva: International Telecommunications Union.

International Telecommunications Union (2005) *ITU Internet Reports 2005: the Internet of things*, Geneva: International Telecommunications Union.

Internet World Stats (2007) "Usage and population statistics," www.internetworldstats.com/stats.htm (accessed 17 May 2007).

Ishida, T., Aurigi, A. and Yasuoka, M. (2003) "World digital cities: beyond heterogeneity," in P. van den Besselaar and S. Koizumi (eds) *Digital Cities III: information technologies for social capital: cross-cultural perspectives* (Third International Digital Cities Workshop, Amsterdam, The Netherlands, September 18–19, 2003), Berlin: Springer, 188–203.

Isserman, A.M. (2001) "Creating new economic opportunities: the competitive advantages of rural America in the next century," *International Regional Science Review*, 24: 38–58.

ITAC (2003) "Home-based telework by US employees grows nearly 40% since 2001," press release, 4 September, www.workingfromanywhere.org/news/pr090403.htm (accessed 29 September 2006).

ITAC (2005) *Annual survey shows Americans are working from many different locations outside their employer's office*, press release, 4 October, www.workingfromanywhere.org/news/pr100405.htm (accessed 28 September 2006).

IT Maroc.com (2006) "Le Maroc est devenu le havre des centres d'appels francophones," 22 May, www.itmaroc.com/Le-Maroc-est-devenu-le-havre-des.html (accessed 14 December 2006).

ITS America (2006) "Applications overview," www.itsoverview.its.dot.gov (accessed 23 December 2006).

IT Strategic Headquarters (2006) *New IT Reform Strategy : Realizing Ubiquitous and Universal Network Society Where Everyone Can Enjoy the Benefits of IT*, Tokyo: IT Policy Office, Cabinet Secretariat, www.kantei.go.jp/foreign/policy/it/ITstrategy2006.pdf (accessed 23 April 2007).

James, J. and Versteeg, M. (2007) "Mobile phones in Africa: how much do we really know?" *Social Indicators Research*, 84: 117–26.

James, P. (2004) *Is Telework Sustainable? An analysis of its economic, environmental and social impacts*, final report of the SUSTEL project, European Commission's Information Society Technologies (IST) programme 2001–33228, Brussels: European Commision, www.sustel.org/documents/Reports/final%20report%20-%20july%202004%20v2.pdf (accessed 13 August 2007).

Jancovici, J.M. (2001) *Bénéfices environnementaux envisageables liés à l'introduction du télétravail*, Rapport pour l'Observatoire des Stratégies Industrielles, Paris: Ministère de l'Industrie, December, www.manicore.com/documentation/teletravail/OSI_synthese.html (accessed 3 October 2006).

Janelle, D. and Hodge, D. (eds) (2000) *Information, Place and Cyberspace: issues in accessibility*, New York: Springer-Verlag.

Jensen, R. (2007) "The digital provide: information (technology), market performance and welfare in the South Indian fisheries sector," *Quarterly Journal of Economics*, 122 (3): 879–924.

Johnson, B.C. (2002) "Retail: the Wal-Mart effect," *McKinsey Quarterly*, issue 1: 40–3.

Johnson, B.C., Manyika, J.M. and Yee, L.A. (2005) "The next revolution in interactions," *McKinsey Quarterly*, issue 4: 21–33.

Johnson, D.K., Siripong, N. and Brown, A.S. (2006) "The demise of distance? The declining role of physical proximity for knowledge transmission," *Growth and Change*, 37: 19–33.

Johnston, P. and Botterman M. (1998) *Status Report on European Telework—TELEWORK 98*, Report AC981201 for the European Commission, DG Information Society—Unit C1 www.eto.org.uk/twork/tw98/pdf/tw98.pdf (accessed 18 December 2005).

Joint Venture: Silicon Valley Network (2006) *Joint Venture 2006's Index of Silicon Valley*, San Jose, CA: Joint Venture: Silicon Valley Network, www.jointventure.org/PDF/Index%202006.pdf (accessed 20 December 2006).

Josselin, M.-L. (2004) "Dakar se cache au bout du fil du télémarketing," *Libération*, 6 September.

Jupitermedia (2006) "DWDM," in *Webopedia Computer Dictionary*. Darien, CT: Jupitermedia Corporation. www.webopedia.com/TERM/D/DWDM.html (accessed 5 October 2006).

Kaka, N.F. (2006) "Running a customer service center in India: An interview with the head of operations for Dell India," *McKinsey on IT*, Summer, 23–9.

Kanellos, M. (2006) "PC shipments up, but Dell loses ground," *ZDNet News*, 19 April, www.news.zdnet.com/2100–9584_22–6062973.html (accessed 15 June 2006).

Kaplan, S. and Sawhney, M. (1999) "The emerging landscape of business to business E-commerce," *Business 2.0*, September.

Kaplan, S. and Sawhney, M. (2000) "E-Hubs: the new B2B marketplaces," *Harvard Business Review*, 78 (3): 97–106.

Karlsson, M. (2006) "International R&D trends and drivers," in M. Karlsson (ed.) *The Internationalization of Corporate R&D: leveraging the changing geography of innovation*, Stockholm: Swedish Institute for Growth Policy Studies, 55–87.

KarstadtQuelle AG (2006) "Li & Fung assumes worldwide import," press release, 20 May, www.karstadtquelle.com/englisch/ir/1080_8177.asp (accessed 28 May 2006).

Kash, D.E., Auger, R.N. and Li, N. (2004) "An exceptional development pattern," *Technological Forecasting and Social Change*, 71: 777–97.

Katz, R.L. (1988) *The Information Society: an international perspective*, New York: Praeger.

Kearney, A.T. (2000) *Joint Venture's Internet Cluster Analysis 2000*, San Jose, CA: Joint Venture: Silicon Valley Network, www.jointventure.org/PDF/2000internetcluster.pdf (accessed 14 December 2006).

Kearney, A.T. (2004) *AT Kearney's 2004 Offshore Location Attractiveness Index: making offshore decisions*, www.atkearney.com/shared_res/pdf/Making_Offshore_S.pdf (accessed 22 December 2006).

Keivani, R., Parsa, A., Sim, L.L., Ong, S.E. and Younis, B. (2003) *Going Global: a comparison of Singapore and the cities of UAE* (GaWC Research Bulletin 110), Loughborough: University of Loughborough, Geography and World Cities Research Group.

Kellerman, A. (2002) *The Internet on Earth: a geography of information*, Chichester: John Wiley.

Kellerman, A. (2004) "Internet access and penetration: an international urban comparison," *Journal of Urban Technology*, 11 (3): 63–85.

Kellerman, A. (2006) *Personal Mobilities*, London: Routledge.

Kelly, T. (2005) "Redrawing the Jipp curve for Africa," in *The Acacia Atlas 2005: Mapping African ICT Growth*, Ottawa: IDRC, 4.

Kende, M. (2000) *The Digital Handshake: connecting Internet backbones* (OPP Working Paper 32), Washington: Federal Communications Commission, Office of Plans and Policy. www.fcc.gov/Bureaus/OPP/working_papers/oppwp32.pdf (accessed 11 June 2006).

Kenney, M. (1997) "Value creation in the late twentieth century: the rise of the knowledge worker," in J. Davis, T.A. Hirschl, and M. Stack (eds) *Cutting Edge: technology information capitalism and social revolution*, London: Verso, 87–102.

Kenney, M. and Curry, J. (2001) "Beyond transaction costs: E-commerce and the power of the internet database," in T.R. Leinbach and S.D. Brunn (eds) *Worlds of Electronic Commerce: economic, geographical and social dimensions*, New York: John Wiley, 45–65.

King, B. (2001) "Appalachia's blue tech collar," *Wired*, 9 October, www.wired.com/news/culture/0,1284,46562,00.html (accessed 22 December 2006).

King, J.L. and Lyytinen, K. (2005) "Automotive informatics: information technology and enterprise transformation in the automobile industry," in W.H. Dutton, B. Kahin, R. O'Callaghan and A.W. Wyckoff (eds) *Transforming Enterprise: the economic and social implications of information technology*, Cambridge, MA: MIT Press, 283–312.

Kinsman, F. (1987) *The Telecommuters*, New York: John Wiley.

Kirkegaard, J. F. (2004) "Outsourcing—stains on the white collar?" working paper, The Peterson Institute for International Economics, Washington, DC, www.iie.com/publications/papers/kirkegaard0204.pdf (accessed 21 May 2007).

Kistner, T. (2004) "Utah program creates hundreds of rural IT jobs," Networkworld.com, April 19, www.networkworld.com/net.worker/columnists/2004/0419kistner.html (accessed 21 December 2006).

Kitchin, R. (1998) *Cyberspace: the world in the wires*, Chichester: John Wiley.

Kjellerup, N. (2003) "Myth & reality about contact centres in India," www.callcentres.com.au/India_Call_Centres.htm (accessed 20 November 2006).

Klagge, B. and Martin, R. (2005) "Decentralised versus centralised financial systems: is there a case for local capital markets?" *Journal of Economic Geography*, 5: 387–421.

Kluth, A. (2006) "Among the audience: a survey of new media," *The Economist*, 22 April.

Knight, R.V. (1989) "City development and urbanization: building the knowledge-based city," in R.V. Knight and G. Gappert (eds) *Cities in a Global Society*. Thousand Oaks, CA: Sage, 223–42.

Kogut, B. (2003) "The Internet has borders," in B. Kogut (ed.) *The Global Internet Economy*, Cambridge, MA: MIT Press, 1–40.

Komninos, N. (2002) *Intelligent Cities: innovation, knowledge systems and digital spaces*, London: Spon Press.

Komninos, N. (2004) "Regional intelligence: distributed localised information systems for innovation and development," *International Journal of Technology Management*, 28: 483–506.

Kondratieff, N.D. (1935) "The long waves in economic life," *Review of Economic Statistics*, 17 (6): 105–15.

Kotkin, J. (2000) *The New Geography: how the digital revolution is reshaping the American landscape*, New York: Random House.

Kozischek, D.R. (2006a) "Fiber, Coax or DSL? Meeting customer demand for bandwidth," *Broadband Properties*, August: 16–22.

Kozischek, D.R. (2006b) "Fiber, Coax or DSL? Meeting customer upstream bandwidth demand," *Broadband Properties*, October: 60–5.

Kraemer, K.L. and Dedrick, J. (2001) "Dell Computer: using e-commerce to support the virtual company," working paper, CRITO, University of California, Irvine, www.crito.uci.edu/git/publications/pdf/dell_ecom_case_6–13–01.pdf (accessed 6 September 2005).

Kraemer, K.L. and Dedrick, J. (2002) "Dell Computer: organization of a global production network," CRITO, University of California, Irvine, www.crito.uci.edu/GIT/publications/pdf/dell.pdf (accessed 6 September 2005).

Kraemer, K.L. and Dedrick, J. (2004) "Offshoring in Orange County: leader, follower, or mirror of national trends?" Personal Computing Industry Center, Graduate School of Management, University of California, Irvine, http://crito.uci.edu/pubs/2004/OCoffshoring0404.pdf, (accessed 6 September 2005).

Kraemer, K.L., Dedrick, J. and Yamashiro, S. (2000) "Dell Computer: refining and extending the business model with IT," *The Information Society*, 16: 5–21.

Krafft, J. (2006) "Emergence and growth of broadband in the French infocommunications system of innovation," in M. Fransman (ed.) *Global Broadband Battles: why the US and Europe lag while Asia leads*, Stanford: Stanford University Press: 172–94.

Krebs, G. (2006) "A view of the submarine systems supply industry," *Submarine Telecoms Forum*, 26: 18–20.

Kripalani, M. and Burrows, P. (2006) "Why Apple logged out of India," *Business Week*, 19 June: 48.

Kuemmerle, W. (1997) "Building effective R&D capabilities abroad," *Harvard Business Review*, 75 (2): 61–70.

Laguerre, M.S. (2005) *The Digital City: the American metropolis and information technology*, London: Palgrave Macmillan.

Landingin, R. (2005) "Philippines sees call center perks," *Financial Times*, 28 February.

Langdale, J. (1989) "Telecommunications amd international business telecommuications: the role of leased networks," *Annals of the Association of American Geographers*, 79: 501–22.

Langdale, J.V. (2000) "Telecommunications and 24-hour trading in the international securities industry," in M.I. Wilson and K.E. Corey (eds) *Information Tectonics: space, place and technology in an information age*, Chichester: John Wiley, 89–99.

Langdale, J.V. (2001) "Global electronic spaces: Singapore's role in the foreign exchange market in the Asia Pacific Region," in T.R. Leinbach and S.D. Brunn (eds) *Worlds of Electronic Commerce: economic, geographical and social dimensions*, New York: John Wiley, 203–19.

Langley, C.J., van Dort, E., Ang, A. and Sykes, S.R. (2005) *Third-Party Logistics: results and findings of the 10th annual study*, Cambridge, MA: Capgemini US, LLC, www.3plstudy.com (accessed 25 June 2006).

Lau, T.Y., Kim, S.W. and Atkin, D. (2005) "An examination of factors contributing to South Korea's global leadership in broadband adoption," *Telematics and Informatics*, 22: 349–59.

Lauer, S. (2007) "Renault innove en proposant le télétravail à ses salariés," *Le Monde*, 25 January.

Lawton Smith, H. (2003) "Knowledge organizations and local economic development: the cases of Oxford and Grenoble," *Regional Studies*, 37: 899–909.

Lazaric, N. and Thomas, C. (2006) "The coordination and codification of knowledge inside a network, or the building of an epistemic community: the Telecom Valley case study," in W. Dolfsma and L. Soete (eds) *Understanding the Dynamics of a Knowledge Economy*, Cheltenham: Edward Elgar, 129–56.

Leamer, E.E. and Storper, M. (2001) "The economic geography of the Internet age," *Journal of International Business Studies*, 32: 641–65.

Le Blanc, G. (2001) "Les nouveaux districts des technologies de l'information: l'exemple de Denver aux Etats-Unis," in DATAR (Délégation de l'Aménagement du Territoire et de l'Action Régionale), *Réseaux d'entreprises et territoires—regards sur les systèmes productifs locaux*, Paris: La Documentation Française, www.cerna.ensmp.fr/Documents/GLB-Districts.pdf (accessed 24 July 2006).

Lee-Young, J. and Barnett, M. (2001) "Furiously fast fashions," *The Industry Standard*, 11 June, 73–9.

Leinbach, T.R. (2001) "Emergence of the digital economy and E-Commerce," in T.R. Leinbach and S.D. Brunn (eds) *Worlds of Electronic Commerce: economic, geographical and social dimensions*, New York: John Wiley, 3–26.

Leinbach, T.R. and Bowen, J. (2003) "Air cargo services in Asian industrializing economies: electronics manufacturers and the strategic use of advanced producer services," *Papers in Regional Science*, 82: 309–32.

Leinbach, T.R. and Bowen, J. (2004) "Air cargo services and the electronics industry in Southeast Asia," *Journal of Economic Geography*, 4: 299–321.

Leinbach, T.R. and Capineri, C. (2007) *Globalized Freight Transport: intermodality, e-commerce, logistics, and sustainability*, Cheltenham: Edward Elgar.

Le Journal du Net (2004) "Se mettre au vert, un an après, que sont-ils devenus?" 6 April, www.journaldunet.com/management/dossiers/041053vertbis/bernazzani2.shtml (accessed 24 April 2007).

Lemaître, F. and Michel, A. (2006) "Les banques comme HSBC doivent investir dans les pays émergents," *Le Monde*, 8 January.

Levy, F. and Murnane, R.J. (2004) *The New Division of Labor: how computers are creating the next job market*, Princeton, NJ: Princeton University Press.

Lewin, A.Y. and Peeters C. (2006) "The top-line allure of offshoring," *Harvard Business Review*, 84 (3): 22–4.

Lewis, D.E. (2006) "India tech firms seek US talent in offshoring twist," *The Boston Globe*, 30 May, www.boston.com/business/globe/articles/2006/05/30/india_tech_firms_seek_us_talent_in_offshoring_twist/ (accessed 19 October 2006).

Lewis, W.W., Palmade, V., Regout, B. and Webb, A.P. (2002) "What's right with the US economy," *McKinsey Quarterly*, issue 1: 30–40.

Leyshon, A. (2001) "Time–space (and digital) compression: software formats, musical networks, and the reorganisation of the music industry," *Environment and Planning A*, 33: 49–77.

Leyshon, A. and Thrift, N. (1997) *Money/Space: geographies of monetary transformation*, London: Routledge.

Li & Fung Ltd (2006a) *Li & Fung Limited Annual Corporate Report 2005*, http://202.66.146.82/listco/hk/lifung/annual/2005/ar2005.pdf (accessed 9 June 2006).

Li & Fung Ltd (2006b) "Corporate profile," www.lifung.com/business/index.html (accessed 1 December 2006).

Li, F. (1995) "Corporate networks and the spatial and functional reorganization of large firms," *Environment and Planning A*, 27: 1627–45.

Lipsey, R.G., Carlaw, K.I. and Bekar, C.T. (2006) *Economic Transformations: general purpose technologies and long term economic growth*, Oxford: Oxford University Press.

Lo, V. and Grote, M.H. (2003) "Where traders go when stock exchanges go virtual—concentration, dissemination or persistence?" in M. Balling, F. Lierman and A. Mullineux (eds) *Technology and Finance: challenges for financial markets, business strategies and policy makers*, London: Routledge, 190–203.

Long, S. (2006) "Now for the hard part: a survey of India," *The Economist*, 3 June.

Longhi, C. (1999) "Networks, collective learning and technology development in innovative high technology regions: the case of Sophia-Antipolis," *Regional Studies*, 33: 333–42.

Lopez-Claros, A., Porter, M.E. and Schwab, K. (eds) (2005) *The Global Competitiveness Report 2005–2006: policies underpinning rising prosperity*, London: Palgrave Macmillan.

Louçã, F. (2003) "The new economy and economic cycles," in D.C. Jones (ed.) *New Economy Handbook*, Amsterdam: Elsevier, 767–99.

Lüthje, B. (2002) "Electronics contract manufacturing: global production and the international division of labor in the age of the Internet," *Industry and Innovation*, 9: 227–47.

Lüthje, B. (2005) "IT and the changing social division of labor: the case of electronics contract manufacturing," in W.H. Dutton, B. Kahin, R. O'Callaghan and A.W. Wyckoff (eds) *Transforming Enterprise: the economic and social implications of information technology*, Cambridge, MA: MIT Press, 335–57.

Lyne, J. (2001) "Incentives help Nova Scotia land Stream's 900-employee call center," *Site Selection*, November, www.conway.com/ssinsider/incentive/ti0109.htm (accessed 22 December 2006).

Lyons, D. (2000) "Embeddedness, milieu, and innovation among high-technology firms: a Richardson, Texas, case study," *Environment and Planning A*, 32: 891–908.

Lyotard, J.F. (1979) *La condition postmoderne: rapport sur le savoir*, Paris: Minuit.

McAfee, A. (2006) "Mastering the three worlds of information technology," *Harvard Business Review*, 84 (11): 141–9.

Macdonald, S. and Deng, Y. (2004) "Science parks in China: a cautionary exploration," *International Journal of Technology Intelligence and Planning*, 1: 1–14.

McDonough E.F., Kahn K.B. and Griffin A. (1999) "Managing communication in global product development teams," *IEEE Transactions on Engineering Management*, 46: 375–86.

McFarlan, F.W. and Young, F. (2000) "Li & Fung (A): Internet issues," *Harvard Business School Case Studies*, October, 9, 301–9.

Machlup, F. (1962) *The Production and Distribution of Knowledge in the United States*, Princeton, NJ: Princeton University Press.

McGranahan, D. and Wojan, T. (2007) "Recasting the creative class to examine growth processes in rural and urban counties," *Regional Studies*, 41: 197–216.

McKenzie, R.B. (2003) *Digital Economics: how information technology has transformed business thinking*, Westport, CT: Praeger.

McLuhan, M.H. (1962) *The Gutenberg Galaxy: the making of typographic man*, Toronto: University of Toronto Press.

Magretta, J. (1998) "Fast, global and entrepreneurial supply chain management Hong Kong style," *Harvard Business Review*, 76 (5): 102–14.

Maguire, J. (2002) "Walmart.com," *E-commerce guide*, 15 November, www.ecommerce-guide.com/news/trends/article.php/1501651 (accessed 2 December 2006).

Maguire, J. (2005) "The state of E-commerce: online shopping trends," *E-commerce guide*, 2 August, www.ecommerce-guide.com/news/research/article.php/3524581 (accessed 21 May 2007).

Malecki, E.J. (1997) *Technology and Economic Development: the dynamics of local, regional and national competitiveness*, 2nd edn, London: Addison-Wesley Longman.

Malecki, E.J. (2000) "Knowledge and regional competitiveness," *Erdkunde*, 50: 334–51.

Malecki, E.J. (2002) "The economic geography of the Internet's infrastructure," *Economic Geography*, 78: 399–424.

Malecki, E.J. (2003) "Digital development in rural areas: potentials and pitfalls," *Journal of Rural Studies*, 19: 201–14.

Malecki, E.J. (2004a) "Fibre tracks: explaining investment in fibre optic backbones," *Entrepreneurship and Regional Development*, 16: 21–39.

Malecki, E.J. (2004b) "Jockeying for position: what it means and why it matters to regional development policy when places compete," *Regional Studies*, 38: 1101–20.

Malecki, E.J. and Wei, H. (2006) "A wired world: the evolving geography of submarine cables and the shift to Asia," paper presented at the annual meeting of the Association of American Geographers, Chicago, March 2006; currently undergoing revision.

Malhotra, A and Majchrzak, A (2005) "Virtual workspace technologies," *MIT Sloan Management Review*, 46 (2): 11–4.

Marescaux, J., Leroy, J., Gagner, M., Rubino, F., Mutter, D., Vix, M., Butner, S.E., and Smith, M.K. (2001) "Transatlantic robot-assisted telesurgery," *Nature*, 413 (27 September): 379–80.

Mariscal, J. (2005) "Digital divide in a developing country," *Telecommunications Policy*, 29: 409–28.

Markillie, P. (2005) "Crowned at last: a survey of consumer power," *The Economist*, 2 April.

Markillie, P. (2006) "The physical Internet: a survey of logistics," *The Economist*, 17 June.

Markusen, A. (1996) "Sticky places in slippery space: a typology of industrial districts," *Economic Geography*, 72: 293–313.

Marshall, J.N. and Richardson, R. (1996) "The growth of telephone call centres in peripheral areas of Britain: evidence from Tyne and Wear," *Area*, 28: 308–17.

Marshall, J.N. and Richardson, R. (1999) "Teleservices, call centres and urban and regional development," *The Service Industries Journal*, 19: 96–116.

Marshall, S., Taylor, W. and Yu, X. (eds) (2003) *Closing the Digital Divide: transforming regional economies and communities with information technology*, Westport, CT: Greenwood Publishing.

Martin, R. (ed.) (1999) *Money and the space economy*, Chichester: John Wiley.

Martin, R. and Sunley, P. (2003) "Deconstructing clusters: chaotic concept or policy panacea?" *Journal of Economic Geography*, 3: 5–35.

Martin, R., Sunley, P. and Turner, D. (2002) "Taking risks in regions: the geographical anatomy of Europe's emerging venture capital market," *Journal of Economic Geography*, 2: 121–50.

Maskell, P. (1999) "Globalisation and industrial competitiveness: the process and consequences of ubiquitification," in E.J. Malecki and P. Oinas (eds) *Making Connections: technological learning and regional economic change*, Aldershot: Ashgate, 35–59.

Maskell, P., Eskelinen, H., Hannibalsson, I., Malmberg, A. and Vatne, E. (1998) *Competitiveness, Localised Learning and Regional Development: specialisation in small open economies*, London: Routledge.

Maskell, P., Pedersen, T., Petersen, B. and Dick-Nielsen, J. (2005) "Learning paths to offshore outsourcing—from cost reduction to knowledge seeking," DRUID Working Paper 05–17, www.druid.dk/wp/pdf_files/05–17.pdf (accessed 16 June 2006).

Mason, R.O. and Apte, U.M. (2005) "Using knowledge to transform enterprises," in W.H. Dutton, B. Kahin, R. O'Callaghan and A.W. Wyckoff (eds) *Transforming Enterprise: the economic and social implications of information technology*, Cambridge, MA: MIT Press, 131–54.

Matlack, C. (2006) "Airbus: first, blame the software," *Business Week*, 5 October, www.businessweek.com/print/globalbiz/content/oct2006/gb20061005_846432.htm (accessed 26 December 2006).

Matteucci, N., O'Mahony, M., Robinson, C and Zwick, T. (2005) "Productivity, workplace performance and ICT: industry and firm-level evidence for Europe and the US," *Scottish Journal of Political Economy*, 52: 359–86.

Matthiessen, C.W., Schwarz, A.W. and Find, S. (2006) "World cities of knowledge: research strength, networks and nodality," *Journal of Knowledge Management*, 10 (5): 14–25.

Mauldin, A. (2006) "Fueling subsea bandwidth demand: demand drivers and internet traffic growth," *Submarine Telecoms Forum*, 28: 28–30.

Maunula M. (2006) *The Perceived Value-Added of Venture Capital Investors: evidence from Finnish biotechnology industry*, Discussion papers 1030, Helsinki: The Research Institute of the Finnish Economy (ETLA) www.etla.fi/files/1556_Dp1030.pdf (accessed 23 October 2006).

Media Lab Asia (2006) *Research Overview*, www.medialabasia.org/medialabasia/index.php?option=content&task=view&id=28&Itemid=37 (accessed 21 December 2006).

Mehta, S.A. (2006) "Behold the server farm! Glorious temple of the information age!" *Fortune*, 154 (3) 7 August: 69–74.

Melody, W. (1991) "New telecommunications networks and the spatial characteristics of markets," in J. Brotchie, M. Batty, P. Hall and P. Newton (eds) *Cities of the 21st Century: new technologies and spatial systems*, Melbourne: Longman Cheshire, 65–72.

Mensch, G. (1979) *Stalemate in Technology: innovations overcome the depression*, Cambridge, MA: Ballinger.

Mesenbourg, T.L. (2000) "Measuring electronic business: definitions, underlying concepts, and measurement plans," Washington DC: US Census Bureau, www.census.gov/epcd/www/ebusines.htm (accessed 30 March 2007).

Metcalfe, B. (1995) "Metcalfe's Law: a network becomes more valuable as it reaches more users," *Infoworld*, 17 (40) 2 October: 53.

Metcalfe, B. (2006) "Metcalfe's Law recourses down the long tail of social networking," *VCMike's Blog*, 18 August, www.vcmike.wordpress.com/2006/08/18/metcalfe-social-networks/ (28 April 2007).

Meyer-Krahmer, F. and Reger, G. (1999) "New perspectives on the innovation strategies of multinational enterprises: lessons for technology policy in Europe," *Research Policy*, 28: 751–76.

Micklethwait, J. and Wooldridge, A. (2000) *A Future Perfect: the challenge and hidden promise of globalization*, New York: Crown Books.

Miller, R. and Coté, M. (1987) *Growing the Next Silicon Valley*, Lexington, MA: Lexington Books.

Miller, S.K. (2004) "Telecommuting still on the rise 47 million employees work from home at least one day a month," *Processor*, 26 (5) (30 January): 18.

Minges, M. (2006) "Tracking ICTs: World Summit on the Information Society targets," in World Bank, *2006 Information and Communications for Development: global trends and policies*, Washington, DC: World Bank, 125–46.

Mitchell, W.J. (1995) *City of Bits: space, place and the infobahn*, Cambridge, MA: MIT Press.

Mokhtarian, P.L and Bagley, M.N. (2000) "Modeling employees" perceptions and proportional preferences of work locations: the regular workplace and telecommuting alternatives," *Transportation Research Part A—Policy and Practice* 34: 223–42.

Mokhtarian, P.L. and Chen, C. (2004) "TTB or not TTB, that is the question: a review and analysis of the empirical literature on travel time (and money) budgets," *Transportation Research Part A—Policy and Practice*, 38: 643–75.

Mokhtarian, P.L., Collantes, G.O., and Gertz, C. (2004) "Telecommuting, residential location, and commute distance traveled: evidence from State of California employees," *Environment and Planning A*, 36: 1877–97.

Mokhtarian, P.L. and Salomon, I. (1997) "Modeling the desire to telecommute: the importance of attitudinal factors in behavioral models," *Transportation Research Part A—Policy and Practice* 31: 35–50.

Mokhtarian, P., Salomon, I. and Choo, S. (2005) "Measuring the measurable: why can't we agree on the number of telecommuters in the US?" *Quality and Quantity*, 39 (4): 423–52.

Molini, F. (1997) "The migration of teleprofessionals to tourist areas: hypotheses and refutations," *Netcom* 11: 403–16.

Montagne, R. (2006) *FTTH Situation in Europe*, Vienna: IDATE, www.idate.org/pages/download.php?id=307&rub=news_telech&nom=FTTH.pdf (accessed 24 December 2006).

Moriset, B. (2003a) "Rural enterprises in the business intelligence sector: utopia or real development opportunity ?" *Netcom*, 17 (1–2): 3–22.

Moriset, B. (2003b) "Des immeubles intelligents aux hotels de télécommunications: les forteresses de l'économie numérique," *Géocarrefour*, 78: 375–88.

Moriset, B. (2003c) "The new economy in the city: emergence and location factors of Internet-based companies in the metropolitan area of Lyon, France," *Urban Studies*, 40: 2165–86.

Moriset, B. (2004) "Télétravail, travail nomade: le territoire et les territorialités face aux nouvelles flexibilités spatio-temporelles du travail et de la production," *Cybergéo*, 6 February, n° 257, www.193.55.107.45/articles/257res.htm (accessed 6 November 2006).

Moriset, B. and Bonnet, N. (2004) "La géographie des centres d'appel en France," *Annales de Géographie*, 641 (January): 49–72.

Morrill, R. and Sommers, P. (2005) "Seattle as a digital city: unexpected or inevitable?" *Canadian Journal of Regional Science*, 28: 349–68.

Morrissey, W.A. (2005) *Tsunamis: monitoring, detection, and early warning systems*, RL32739, Washington, DC: Congressional Research Service, www.fas.org/sgp/crs/RL32739.pdf (accessed 5 December 2006).

Moulaert, F. and Sekia, F. (2003) "Territorial innovation models: a critical survey," *Regional Studies*, 37: 289–302.

Mowshowitz, A. (1994) "Virtual organization: a vision of management in the information age," *The Information Society*, 10: 267–88.

Mowshowitz, A. (1999) "The switching principle in virtual organization," *Journal of Organizational Virtualness*, 1 (1): 7–18.

Muller, A. (2001) *La net économie* (Que sais-je 3597), Paris: Presses Universitaires de France.

Murphy, A. (2003) "(Re)solving space and time: fulfilment issues in online grocery retailing," *Environment and Planning A*, 35: 1173–1200.

Murphy, A.J. (2007) "Grounding the virtual: the material effects of electronic grocery shopping," *Geoforum*, 38: 941–53.

Nachum, L. and Zaheer, S. (2005) "The persistence of distance? The impact of technology on MNE motivations for foreign investment," *Strategic Management Journal*, 26: 747–67.

Nakamura L. (1999) "Intangibles: what put the new in the new economy?" *Business Review, Federal Reserve Bank of Philadelphia*, July/August: 3–16.

Nana Nzépa, O. and Tankeu Keutchankeu, R. (2005) "Chapter 4: Cameroon," in A. Gillwald (ed.) *Towards an African e-Index: Household and Individual ICT Access and Usage across 10 African Countries*, Johannesburg: Witwatersrand University LINK Center, www.researchictafrica.net/images/upload/Toward2.pdf (accessed 9 May 2007).

NASSCOM (2006) *NASSCOM 2005–2006 Annual Report*, New Delhi: National Association of Software and Service Companies, www.nasscom.in/upload/41527/Annual_report.pdf (accessed 12 September 2006).

National Research Council, Computer Science and Telecommunications Board (2001a) *Embedded, Everywhere: a research agenda for networked systems of embedded computers*, Washington, DC: National Academies Press.

National Research Council, Computer Science and Telecommunications Board (2001b) *The Internet's Coming of Age*, Washington, DC: National Academies Press.

Neef, D. (ed.) (1998) *The Knowledge Economy*, Boston: Butterworth-Heinemann.

Negroponte, N. (1995) *Being Digital*, London: Vintage Books.

Negroponte, N. (1998) "Beyond digital," *Wired*, 6 (12) December, www.wired.com/wired/archive/6.12/negroponte.html (accessed 4 April 2006).

Nelson, R.R. (1998) "The agenda for growth theory: a different point of view," *Cambridge Journal of Economics*, 22: 497–520.

NEPAD e-Africa Commission (2004) *Backbone Telecommunications Infrastructure Development Initiatives in Southern and East Africa*, Washington, DC: World Bank.

Newman, H.B. (2006) "Global networks for high energy physics," presentation at the ICFA Meeting at CERN, Geneva, 17 February 2006, www.monalisa.caltech.edu:8080/Slides/SCIC2006/ICFASCIC Report_hbn020906.ppt (accessed 13 September 2006).

Newman, N.C., Porter, A.L., Roessner, J.D., Kongthon, A. and Jin X.-Y. (2005) "Differences over a decade: high tech capabilities and competitive performance of 28 nations," *Research Evaluation*, 14: 12–18.

Nicolini, D. (2006) "The work to make telemedicine work: a social and articulative view," *Social Science and Medicine*, 62 (11): 2754–67.

Nilles, J. (1976) *The Telecommunications-Transportation Tradeoff: options for tomorrow*, New York: John Wiley.

Nohria, N. and Eccles, R.G. (1992) "Face to face: making network organizations work," in N. Nohria and R.G. Eccles (eds) *Networks and Organizations: structure fom and action*, Boston, MA: Harvard Business School Press, 288–308.

Nordhaus, W.D. (2002) "Productivity growth and the new economy," *Brookings Papers on Economic Activity*, 2: 211–65.

Nunes, M. (1995) "Baudrillard in cyberspace: Internet, virtuality, and postmodernity," *Style*, 29: 314–27.

NYSE Group (2006). *Form 10-K Corporate Report 2005*, Washington, DC: US Securities and Exchange Commission, www.nyse.com/pdfs/2502T05_CNB.PDF (accessed 21 May 2006).

O'Brien, J.M. (2001) "The Making of the Xbox—How did the world's largest software publisher become a hardware manufacturer overnight? One word: Flextronics," *Wired*, 9 (11) November, www.wired.com/wired/archive/9.11/flex.html (accessed 25 June 2006).

O'Brien, R. (1992) *Global Financial Integration: the end of geography*, London: Pinter.

O'Callaghan, R. (2005) "Technological innovation in organizations and their ecosystems," in W.H. Dutton, B. Kahin, R. O'Callaghan and A.W. Wyckoff (eds) *Transforming Enterprise: the economic and social implications of information technology*, Cambridge, MA: MIT Press, 1–12.

Odlyzko, A. (1999) "The current state and likely evolution of the Internet," in *Global Telecommunications Conference, 1999 (GLOBECOM "99)*, vol. 3: 1869–76.

Odlyzko, A. (2000) "The Internet and other networks: utilization rates and their implications," *Information Economics and Policy*, 12: 341–65.

Odlyzko, A. (2003) "Data networks are lightly utilized, and will stay that way," *Review of Network Economics*, 2: 210–37.

OECD (1996) *Employment and Growth in the Knowledge-Based Economy*, Paris: Organisation for Economic Co-operation and Development.

OECD (1999) *OECD Communications Outlook 1999*, Paris: Organisation for Economic Co-operation and Development.

OECD (2001) *Understanding the Digital Divide*, Paris: Organisation for Economic Co-operation and Development.

OECD (2005a) *ICT Use by Businesses: revised OECD model survey*, Paris: Organisation for Economic Co-operation and Development, DSTI/ICCP/IIS(2005)2/FINAL, January 24, www.oecd.org/dataoecd/58/7/35867672.pdf (accessed 9 May 2006).

OECD (2005b) *OECD Communications Outlook 2005*, Paris: Organisation for Economic Co-operation and Development.

OECD (2006) *OECD Broadband Statistics to June 2006*, Paris: Organisation for Economic Co-operation and Development, www.oecd.org/document/9/0,2340,en_2649_34223_37529673_1_1_1_1,00.html (accessed 9 December 2006).

Olavsrud, T. (2002) "Flextronics relocates Xbox manufacturing facility," *Internet News.com*, 15 May, www.internetnews.com/bus-news/article.php/1129171 (accessed 26 June 2006).

Oliner, S.D. and Sichel, D.E. (2002) "Information technology and productivity: where are we now and where are we going?" *Federal Reserve Bank of Atlanta Economic Review*, 87 (3): 15–44.

Olsen, K.B. (2006) *Productivity Impacts of Offshoring and Outsourcing: a review*, Paris: OECD, STI Working Paper 2006/1, www.oecd.org/dataoecd/16/29/36231337.pdf (accessed 20 April 2007).

Orwell, G. (1948; 1998) *1984: a novel*, London: Penguin Books.

Ory, D.T. and Mokhtarian, P.L. (2005) *An Empirical Analysis of Causality in the Relationship Between Telecommuting and Residential and Job Relocation*, research report 05–03, Institute of Transportation Studies, Davis: University of California, February, www.its.ucdavis.edu/publications/2005/UCD-ITS-RR-05–03.pdf (accessed 9 May 2006).

Outsource2india (2006) "Call Centers in India," www.outsource2india.com/why_india/articles/call_centers_india.asp (accessed 14 December 2006).

Oxford Intelligence and IBM Business Consulting Services (2004). *Investment Strategies and Location Benchmarking Study: shared service centres for Europe*, Whadon (UK): Oxford Intelligence.

Oz, E. (2005) "Information technology productivity: in search of a definite observation," *Information and Management*, 42: 789–98.

Pacione, M. (2005) "Dubai," *Cities*, 22: 255–65.

Packet Clearing House (2006) *Internet Exchange Directory*, www.pch.net/ixpdir/Main.pl (accessed 8 December 2006).

Padma, T.V. (2004) "India to build tsunami warning system," *SciDev.Net News*, 31 December, www.scidev.net/News/index.cfm?fuseaction=readNews&itemid=1825&language=1 (accessed 5 December 2006).

Palan, R. (2003) *The Offshore World: sovereign markets, virtual places, and nomad millionaires*, Ithaca, NY: Cornell University Press.

Paltridge, S. (2002) *Internet Traffic Exchange and the Development of End-to-End International Telecommunication Competition* (DSTI/ICCP/TISP(2001)5/FINAL), Paris: Organisation for Economic Co-operation and Development, www.oecd.org/dataoecd/47/20/2074136.pdf (accessed 9 June 2006).

Paltridge, S. (2006) *Internet Traffic Exchange: market developments and measurement of growth* (DSTI/ICCP/TISP(2005)11/FINAL), Paris: Organisation for Economic Co-operation and Development.

Paniccia, I. (2006) "Cutting through the chaos: towards a new typology of industrial districts and clusters," in B. Asheim, P. Cooke and R. Martin (eds) *Clusters and Regional Development: critical reflections and explorations*, London: Routledge, 90–114.

Parayil, G. (2006) *Political Economy & Information Capitalism in India: digital divide, development divide & equity*, Basingtoke: Palgrave Macmillan.

Park, S. and Yoon, S.-H. (2005) "Separating early adopters from the majority: the case of broadband Internet access in Korea," *Technological Forecasting and Social Change*, 72: 301–25.

Parthasarathy, B. (2004) "India's Silicon Valley or Silicon Valley's India? Socially embedding the computer software industry in Bangalore," *International Journal of Urban and Regional Research*, 28: 664–85.

Parthasarathy, B. and Aoyama, Y. (2006) "From software services to R&D services: local entrepreneurship in the software industry in Bangalore, India," *Environment and Planning A*, 38: 1269–85.

Pavitt, K. 2003. "What are advances in knowledge doing to the large industrial firm in the 'new economy'?" in J.F. Christensen and P. Maskell (eds) *The Industrial Dynamics of the New Digital Economy*, Cheltenham: Edward Elgar, 103–20.

Peet, J. (2000) "Shopping around the web: a survey of E-commerce," *The Economist*, 24 February.

Peppard, J. and Rylander, A. (2006) "From value chain to value network: insights for mobile operators," *European Management Journal*, 24: 128–41.

Phillips, F., Ochs, L. and Schrock, M. (1999) "The product is dead—long live the product-service!," *Research-Technology Management*, 42 (4): 51–6.

Phillips, S.-A. M. and Yeung, H. W.-C. (2003) "A place for R&D? The Singapore Science Park," *Urban Studies*, 40: 707–32.

Pinch, S., Henry, N., Jenkins, M. and Tallman, S. (2003) 'From 'industrial districts' to 'knowledge clusters': a model of knowledge dissemination and competitive advantage in industrial agglomerations," *Journal of Economic Geography*, 3: 373–88.

Pine, B.J. and Gilmore, J.H. (1998) "Welcome to the experience economy," *Harvard Business Review*, 76 (4): 97–105.

Pine, B.J. and Gilmore, J.H. (1999) *The Experience Economy: work is theatre and every business a stage*, Boston: Harvard Business School Press.

Pohjola, M. (2002) "The new economy in growth and development," *Oxford Review of Economic Policy*, 18: 380–96.

Pohl, N. (2003) "E-finance: Causing major upheavals in the spatial organization of the financial services sector," in C. Barfield, G. Heiduk and P.J. Welfens, *Internet, Economic Growth and Globalization*, Berlin: Springer, 135–70.

Polèse, M. and Shearmur, R. (2002) *The Periphery in the Knowledge Economy: the spatial dynamics of the Canadian economy and the future of non-metropolitan regions in Quebec and the Atlantic Provinces*, Quebec: Institute national de la recherche scientifique/INRS-Urbanisation, Culture et Société.

Polèse, M. and Shearmur, R. (2006) "Why some regions will decline: A Canadian case study with thoughts on local development strategies," *Papers in Regional Science*, 85: 23–46.

Poon, J. (2003) "Hierarchical tendencies of capital markets among international financial centers," *Growth and Change*, 34: 135–56.

Porat, M.U. (1977) *The Information Economy: definition and measurement*, Washington, DC: US Department of Commerce, Office of Telecommunications.

Port, O., King, R. and Hampton, W.J. (1988) "How the new math of productivity adds up," *Business Week*, 6 June: 103–14.

Porter, M.E. (1985) *Competitive Advantage: creating and sustaining superior performance*, New York: Free Press.

Porter, M.E. (1990) *The Competitive Advantage of Nations*, New York: Free Press.

Porter, M.E. (1998) "Clusters and the new economics of competition," *Harvard Business Review*, 76 (6): 77–90.

Porter, M. (2000) "Location, clusters, and company strategy," in G.L. Clark, M.P. Feldman and M.S. Gertler (eds) *The Oxford Handbook of Economic Geography*, Oxford: Oxford University Press, 253–74.

Porter, M.E. (2001) "Strategy and the Internet," *Harvard Business Review*, 79 (3): 62–78.

Porter, M.E. and Millar, V.E. (1985) "How information gives you competitive advantage," *Harvard Business Review*, 63 (4): 149–60.

Post, J.E., Preston, L.E. and Sachs, S. (2002) "Managing the extended enterprise: the new stakeholder view," *California Management Review*, 45 (1): 6–28.

Powell, W.W. and Snellman, L. (2004) "The knowledge economy," *Annual Review of Sociology*, 30: 199–220.

Power, D. (2001) "Information and communication technologies and the integration of European derivatives markets," in T.R. Leinbach and S.D. Brunn (eds) *Worlds of Electronic Commerce: economic, geographical and social dimensions*, New York: John Wiley, 241–55.

Power, D. and Jansson, J. (2004) "The emergence of a post-industrial music economy? Music and ICT synergies in Stockholm, Sweden," *Geoforum*, 35: 425–39.

Pratt, J. (1999) *Telework America National survey: cost/benefits of teleworking to manage work/life responsibilities*, Washington DC: International Telework Association and Council.

Preparatory Commission for the Comprehensive Nuclear-Test-Ban Treaty Organization (2006) "Comprehensive Nuclear-Test-Ban Treaty Organization: a profile," *Submarine Telecoms Forum*, 29: 11–7.

Press, L., Foster, W., Wolcott, P. and McHenry, W. (2003) "The Internet in India and China," *Information Technologies and International Development*, 1: 41–60.

PriMetrica (2003) *Colocation 2004: colocation centers, carrier hotels and carrier facilities*, Carlsbad, CA: PriMetrica.

Prudential (2007) "Sale of Egg Banking plc," *Group News Releases*, 29 January.

Quah, D. (2003) "Digital goods and the new economy," in D.C. Jones (ed.) *New Economy Handbook*, Amsterdam: Elsevier, 289–321.

Quinn, J.B. Baruch, J.J. and Paquette, P.C. (1987) "Technology in services," *Scientific American*, 257 (6): 50–8.

Qvortrup, L. (1989) "The Nordic telecottages: community teleservice centres for rural regions," *Telecommunications Policy*, 13: 59–68.

Rabinovich, E. and Knemeyer, A.M. (2006) "Logistics service providers in Internet supply chains," *California Management Review*, 48 (4): 84–108.

Rajan, R.G. (2005) "Making India a global hub," *McKinsey Quarterly 2005 special edition: Fulfilling India's promise*: 113–21.

Rallet, A. (2000) "Les deux économies de l'information," *Réseaux*, 100: 299–330.

Ramirez, R. and Richardson, D. (2005) "Measuring the impact of telecommunication services on rural and remote communities," *Telecommunications Policy*, 29: 297–319.

Ramos, J. (2007) "A place in the sun: a special report on offshore finance," *The Economist*, 24 February.

Rantanen, T. (1997) "The globalization of electronic news in the 19th century," *Media, Culture & Society*, 19: 605–20.

Rayburn, J.M. and Conrad, C. (2004) "China's Internet structure: problems and control measures," *International Journal of Management*, 21: 472–80.

Reich, R.B. (1991) *The Work of Nations*, New York: Alfred A. Knopf.

Renault (2006) *Technocentre*, www.renault.com/renault_com/fr/main/Rubrique_Transversale/FAQ/190_Technocentre/index.aspx (accessed 19 December 2006)

Rhoads, C. (2006) "Out of Boston: an entrepreneur has quixotic goal of wiring Rwanda," *The Wall Street Journal*, 17 August: A1.

Richard, P.J. and Devinney, T.M. (2005) "Modular strategies: B2B technology and architectural knowledge," *California Management Review*, 47 (4): 86–113.

Richardson, R. and Gillespie, A. (2000) "The economic development of peripheral rural places in the information age," in M. Wilson and K. Corey (eds) *Information Tectonics: space, place and technology in an electronic age*, Chichester: John Wiley, 199–217.

Richardson, R. and Gillespie, A. (2003) "The call of the wild: call centers and economic development in rural areas," *Growth and Change*, 34: 87–108.

Richardson, R., Belt, V. and Marshall, N. (2000) "Taking calls to Newcastle: the regional implications of the growth in call centres," *Regional Studies*, 34: 357–69.

Richtel, M. (2006) "The long-distance journey of a fast-food order," *The New York Times*, 11 April: A1.

Rifkin, J. (2001) *The Age of Access: the new culture of hypercapitalism, where all of life is a paid-for experience*, New York: Tarcher-Penguin.

Roberts, J.W. and Oueslati-Boulahia, S. (2000) "Quality of service by flow-aware networking," *Philosophical Transactions of the Royal Society of London Series A: Mathematical, Physical and Engineering Sciences*, 358: 2197–2207.

Roberts, S. (2005) (ed.) *Guide to Measuring the Information Society*, Paris: Organisation for Economic Co-operation and Development, www.oecd.org/dataoecd/41/12/36177203.pdf (accessed 9 May 2006).

Rohlfs, J.H. (2005) "Bandwagon effects in telecommunications," in S.K. Majumdar, I. Vogelsang and M.E. Cave (eds) *Handbook of Telecommunications Economics*, volume 2: *Technology evolution and the Internet*, Amsterdam: Elsevier, 79–115.

Roitz, J., Allenby, B., Atkyns, R. and Nanavati, B. (2003) *Organizing Around Networks, Not Buildings: 2002/2003 AT&T employee telework research results*, AT&T Telework white paper, www.telcoa.org/id147.htm (accessed 20 September 2007).

Roitz, J., Nanavati, B. and Levy, G. (2004) *Lessons Learned From The Network-Centric Organization: 2004 AT&T employee telework results*, AT&T Telework Center of Excellence, www.att.com/telework/docs/2004_telework_results.pdf (accessed 18 December 2005).

Romer, P.M. (1990) "Endogenous technological change," *Journal of Political Economy*, 98 (5, part 2): S71–S102.

Rosenthal, B.E. (2004) "Why a Silicon Valley supplier brought its offshore call center home," www.outsourcing-journal.com/may2004-offshore.html (accessed 12 September 2006).

Ross, S.S. (2006) "Wow! 1 million FTTH customers," *Broadband Properties*, October: 19–22.

Rottman, J.W. and Lacity, M.C. (2006) "Proven practices for effectively offshoring IT Work," *MIT Sloan Management Review*, 47 (3): 56–63.

Rutherford, J. (2004) *A Tale of Two Global Cities: comparing the territorialities of telecommunications developments in Paris and London*, Aldershot: Ashgate.

Rutherford, J. (2005) "Networks in cities, cities in networks: territory and globalisation intertwined in telecommunications infrastructure development in Europe," *Urban Studies*, 42: 2389–2406.

Rutherford, J., Gillespie, A. and Richardson, R. (2004) "The territoriality of pan-European telecommunications backbone networks," *Journal of Urban Technology*, 11 (3): 1–34.

Rutherford, J., Gillespie, A. and Richardson, R. (2005) *Technological connectivities of European cities? The potentials and pitfalls of the use of telecommunications data in measurements of world city network formation* (*GaWC Research Bulletin 181*), Loughborough: Globalization and World Cities Research Study Group and Network, www.lboro.ac.uk/gawc/rb/rb181.html (accessed 26 September 2006).

Rycroft, R.W. and Kash, D.E. (1999) *The Complexity Challenge: technological innovation for the 21st century*, London: Pinter.

St George Area Chamber of Commerce (2006) "Population," www.stgeorgechamber.com/EcDev/demographics_population.htm (accessed 22 December 2006).

St John, C.H. and Pouder, R.W. (2006) "Technology clusters versus industry clusters: resources, networks, and regional advantages," *Growth and Change*, 37: 141–71.

Sako, M. (2003) "Modularity and outsourcing: the nature of co-evolution of product architecture and organisation in the global automotive industry," in A. Principe, A. Davies and M. Hobday (eds) *The Business of Systems*, Oxford: Oxford University Press, 229–53.

Salomon, I. (1998) "Technological change and social forecasting: the case of telecommuting as a travel substitute," *Transportation Research Part C—Emerging Technologies*, 6: 17–45.

Salter, C. (2004) "Calling JetBlue," *Fast Company*, 82 (May), www.fastcompany.com/magazine/82/jetblue_agents.html (accessed 13 November 2006).

Sanchez, R. and Mahoney, J. (1996) "Modularity, flexibility and knowledge management in product and organizational design," *Strategic Management Journal*, 17: 63–76.

Sanderson, S. and Uzumeri, M. (1995) "Managing product families—the case of the Sony-Walkman," *Research Policy* 24: 761–82.

Santos, J., Doz, Y. and Williamson, P. (2004) "Is your innovation process global?" *MIT Sloan Management Review*, 45 (4): 31–7.

Sapsted, J., Gann, D., Marshall, N. and Salter, A. (2005) "From here to eternity?: The practice of knowledge transfer in dispersed and co-located project organizations," *European Planning Studies*, 13: 831–51.

Sassen, S. (2001a) "Impacts of information technologies on urban economic and politics," *International Journal of Urban and Regional Research*, 25: 411–8.

Sassen, S. (2001b) *The Global City: New York, London, Tokyo*, 2nd edn, Princeton, NJ: Princeton University Press.

Sassen, S. (2006) *Cities in a World Economy*, 3rd edn, Thousand Oaks, CA: Pine Forge Press.

Saxenian, A. (1994) *Regional Advantage: culture and competition in Silicon Valley and Route 128*, Cambridge, MA: Harvard University Press.

Saxenian, A. (2004) "Taiwan's Hsinchu region: imitator and partner for Silicon Valley," in T. Bresnahan and A. Gambardella (eds) *Building High-Tech Clusters: Silicon Valley and beyond*, Cambridge: Cambridge University Press, 190–228.

Saxenian, A. (2006) *The New Argonauts: regional advantage in a global economy*, Cambridge, MA: Harvard University Press.

Scheleur, S., King, C. and Shimberg, M. (2006) "Quarterly retail E-Commerce sales 1st Quarter 2006," *US Census Bureau News*, CB06–66, 18 May, Washington DC: US Census Bureau, www.census.gov/mrts/www/data/pdf/06Q1.pdf (accessed 21 May 2006).

Schiller, D. (1999) *Digital capitalism: networking the global market system*, Cambridge, MA: MIT Press.

Schilling, M. (2000) "Towards a general modular systems theory and its application to interfirm product modularity," *Academy of Management Review*, 25: 312–34.

Schlender, B. (1999) "The *real* road ahead," *Fortune*, 140 (8) 25 October: 138–52.

Schoales, J. (2006) "Alpha clusters: creative innovation in local economies," *Economic Development Quarterly*, 20: 162–77.

Schneider, F.B. (ed.) (1999) *Trust in Cyberspace*, Washington DC: National Academy Press.

Schwaag Serger, S. (2006) "China: from shop floor to knowledge factory?" in M. Karlsson (ed.) *The Internationalization of Corporate R&D: leveraging the changing geography of innovation*, Stockholm: Swedish Institute for Growth Policy Studies, 227–66.

Schwartz, E. (2004) "Wal-Mart promises RFID will benefit suppliers," *InfoWorld*, 17 June, www.infoworld.com/article/04/06/17/HNwalmart_1.html (accessed 17 November 2006).

Scott, A.J. (1988) *New Industrial Spaces: flexible production, organization and regional development in North America and Western Europe*, London: Pion.

Scott, A.J. (1996) "Economic decline and regeneration in a regional manufacturing complex: Southern California's household furniture industry," *Entrepreneurship and Regional Development*, 8: 75–98.

Scott, A.J. (2000) *The Cultural Economy of Cities: essays on the geography of image-producing industries*, Thousand Oaks, CA: Sage.

Scott, A.J. (2006) *Geography and Economy: three lectures*, Oxford: Oxford University Press.

Scott, A.J., Agnew, J.A., Soja, E.W. and Storper, M. (2001) "Global city-regions," in A.J. Scott (ed.) *Global City-Regions: trends, theory, policy*, Oxford: Oxford University Press, 11–30.

Searle, G. and Pritchard, B. (2005) "Industry clusters and Sydney's ITT sector: northern Sydney as 'Australia's Silicon Valley'," *Australian Geographer*, 36: 145–69.

Sengupta, S. (2006) "Skills gap hurts technology boom in India," *The New York Times*, 17 October, www.nytimes.com/2006/10/17/world/asia/17india.html (accessed 27 December 2006).

Shahin, J. (2006) "A European history of the Internet," *Science and Public Policy*, 33: 681–93.

Shannon, C.E. and Weaver, W. (1949) *The Mathematical Theory of Communication*, Urbana, IL: University of Illinois Press.

Shapero, A. (1985) *Managing Professional People*, New York: Free Press.

Shapiro, C. and Varian, H. (1998) "Versioning: the smart way to sell information," *Harvard Business Review*, 76 (6): 106–14.

Sharma, A. and Searcey, D. (2006) "Cell carriers to web customers: use us, but not too much," *Wall Street Journal*, 11 May: B1.

Shepard, S.B. (1997) "The new economy: what it really means," *Business Week*, 17 November: 38–40.

Sheppard, E. (2002) "The spaces and times of globalization: place, scale, networks, and positionality," *Economic Geography*, 78: 307–30.
Shoup, D. (2005) *The High Cost of Free Parking*, Chicago: American Planning Association.
Sidaway, J.D. and Bryson, J.R. (2001) "Constructing knowledges of "emerging markets": UK-based investment managers and their overseas connections," *Environment and Planning A*, 34: 401–16.
Simmie, J. (2002) Trading places: competitive cities in the global economy," *European Planning Studies*, 10: 201–14.
Sinai, T. and Waldfogel, J. (2005) "Geography and the Internet: is the Internet a substitute or a complement for cities?" *Journal of Urban Economics*, 56: 1–24.
Singhal, A. and Rogers, E.M. (1989) *India's Information Revolution*, New Delhi: Sage.
Smallbone, D. and North, D. (2000) "The innovation and growth of rural SMEs during the 1990s," *Regional Studies*, 34: 145–57.
Smith, M.D., Bailey, J. and Brynjolfsson, E. (2000) "Understanding digital markets: Review and assessment," in E. Brynjolfsson and B. Kahin (eds) *Understanding the Digital Economy*, Cambridge, MA: MIT Press, 99–136.
Soete, L. (2000) "Towards the digital economy: scenarios for business," *Telematics and Informatics*, 17: 199–212.
Solow, R.M. (1987) "We'd better watch out: review of *Manufacturing Matters*," *New York Times Book Review*, 92 (12 July): 36.
Song, M., Berends, H., van der Bij, H. and Weggeman, M. (2007) "The effect of IT and co-location on knowledge dissemination," *Journal of Product Innovation Management*, 24: 52–68.
Spellman, M. (2005) "Broadband, and Google, come to rural Oregon," *Broadband Properties*, December: 28–30.
Standage, T. (2006) "Your television is ringing: a survey of telecoms convergence," *The Economist*, 14 October.
Staple, G.C. (1999) "The soft network: code is a compass," in G.C. Staple (ed.) *TeleGeography 1999*, Washington, DC: TeleGeography, Inc: 3–13.
Steinbock, D. (2003) "Globalization of wireless value system: from geographic to strategic advantages," *Telecommunications Policy*, 27: 207–35.
Steiner, R., Tirivayi, A., Tirivayi, N., Jensen, M., Hamilton, P. and Buechler, J. (2005) *Promoting African Research and Education Networking*, Ottawa: International Development Research Centre, www.idrc.ca/uploads/user-S/11083295861PAREN_Reportv15.doc (accessed 4 September 2006).
Steinfield, C. (2004) "Situated electronic commerce: towards a view as complement rather than substitute for offline commerce," *Urban Geography*, 25: 353–71.
Steinfield, C., Adelaar, T. and Lai, L-j (2002) "Integrating brick and mortar locations with E-commerce: understanding synergy opportunities," *Proceedings of the Hawai'i International Conference on System Sciences*, January 7–10, 2002, Big Island, Hawaii, www.msu.edu/~steinfie/HICSS2002.pdf (accessed 23 November 2006).
Stephan, P.E., Sumell, A.J., Black, G.C. and Adams, J.D. (2004) "Doctoral education and economic development: the flow of new Ph.D.s to industry," *Economic Development Quarterly*, 18: 151–67.
Sternberg, R. (1996) "Reasons for the genesis of high-tech regions—theoretical explanation and empirical evidence," *Geoforum*, 27: 205–23.
Stewart, T.A. (1997) *Intellectual Capital: the new wealth of organizations*, New York: Doubleday Currency.
Stockdale, A. (2006) "Migration: pre-requisite for rural economic regeneration?" *Journal of Rural Studies*, 22: 354–66.
Stone, A. (1998) "Attention netgrocer shoppers! This bargain may be too good to last cheap shipping is burning a hole in its balance sheet," *Business Week*, 13 October, www.businessweek.com/smallbiz/news/coladvice/reallife/rl981013.htm (accessed 3 April 2006).
Storper, M. (1997) *The Regional World: territorial developments in a global economy*, New York: Guilford Press.
Storper, M. and Venables, A.J. (2004) "Buzz: face-to-face contact and the urban economy," *Journal of Economic Geography*, 4: 351–70.
Stratigea, A. and Giaoutzi, M. (2000) "Teleworking and virtual organization in the urban and regional context," *Netcom*, 14: 331–57.
Strom, D. (2000) "What becomes a location most?" *Mappa Mundi Magazine*, 190 (February), www.mappa.mundi.net/inform/inform_0190.html (accessed 24 December 2006).

Stross, R. (2006) "It's not the people you know. It's where you are," *New York Times*, 22 October, www.nytimes.com/2006/10/22/business/yourmoney/22digi.html (accessed 25 October 2006).
Strover, S., Chapman, G. and Waters, J. (2004) "Beyond community networking and CTCs: access, development, and public policy," *Telecommunications Policy*, 28: 465–85.
Stubbings, C. (2004) "Which way round the world?" *Submarine Telecoms Forum*, 17: 29–32.
Sturgeon, T.S. (2002) "Modular production networks; a new model of industrial organization," *Industrial and Corporate Change*, 11: 451–96.
Sturgeon, T.S. (2003) "What really goes on in Silicon Valley? Spatial clustering and dispersal in modular production networks," *Journal of Economic Geography*, 3: 199–225.
Suarez-Villa, L. (2000) *Invention and the Rise of Technocapitalism*, Lanham, MD: Rowman & Littlefield.
Suarez-Villa, L. (2003) "The E-economy and the rise of technocapitalism: networks, firms, and transportation," *Growth and Change*, 34: 390–414.
Sum, N.-L. (2003) "(Re-)Imagining 'Greater China': Silicon Valley and the strategy of siliconization," in C.R. Hughes and G. Wacker (eds), *China and the Internet: Politics of the digital leap forward*, London: RutledgeCurzon, 102–26.
Sun, Y. and Wang, H. (2005) "Does Internet access matter for rural industry? A case study of Jiangsu, China," *Journal of Rural Studies*, 21: 247–58.
Surowiecki, J. (2006) "Philanthropy's new prototype," *Technology Review*, 109 (5): 48–56.
Swyngedouw, E.A. (1997) "Neither global nor local: 'glocalization' and the politics of scale," in K.R. Cox (ed.) *Spaces of Globalization: reasserting the power of the local*, New York: Guilford Press, 137–66.
Symons, F. (1997) "Virtual departments, power and location in different organizational settings," *Economic Geography*, 73: 427–44.
Syrquin, M. (1988) "Patterns of structural change," in H. Chenery and T.N. Srinivasan (eds) *Handbook of Development Economics*, vol. I, Amsterdam: North-Holland, 203–73.
TACTIS and IDATE (2005) "ORTEL—Observatoire Regional des Télécommunications—extraits des résultats 2005," www.ortel.fr (accessed 10 May 2007).
Talbot, J. (2001) "Les déplacements domicile-travail. De plus en plus d'actifs travaillent loin de chez eux," *INSEE Première*, 767, www.insee.fr/fr/ffc/docs_ffc/IP767.pdf (accessed 21 May 2007).
Tamasy, C. (2002) "Are there too many innovation centres in Germany?" in L. Schätzl and J. Revilla Diez (eds), *Technological Change and Regional Development in Europe*, Berlin: Physica-Verlag, 112–31.
Tapscott, D. (1996) *The Digital Economy: promise and peril in the age of networked intelligence*, New York: McGraw-Hill.
Taylor, P.J. (2004) *World City Network: a global urban analysis*, London: Routledge.
Télécoms Sans Frontières (TSF) (2006) "Partnership between Télécoms Sans Frontières and the United Nations: TSF appointed 'First responder' within the Emergency Telecoms Cluster," *News*, 11 October, www.tsfi.org/html_e/index_gb.php (accessed 23 October, 2006).
TeleGeography (2001) *International Bandwidth 2001*, Washington, DC: TeleGeography, Inc.
TeleGeography Research (2004) *Global Internet Geography 2005*, Washington, DC: PriMetrica.
TeleGeography Research (2005a) *International Bandwidth 2005*, Washington, DC: PriMetrica.
TeleGeography Research (2005b) *TeleGeography 2006*, Washington, DC: PriMetrica.
TeleGeography Research (2006a) *Global Internet Geography 2007*, Washington, DC: PriMetrica.
TeleGeography Research (2006b) *World Broadband Yearbook 2006*, Washington, DC: PriMetrica.
TeleGeography Research (2006c) *Global Colocation Database 2006*, Washington, DC: PriMetrica.
The Economist (2003) "Lost in translation," 29 November: 58.
The Economist (2004) "The rise of the green building," 4 December (Technology Quarterly): 17–23.
The Economist (2005a) "Calling across the divide," 12 March: 74.
The Economist (2005b) "The Bangalore paradox," 23 April: 67–9.
The Economist (2005c) "Flying from the computer," 29 September: 65–7.
The Economist (2005d) "Outsourcing in eastern Europe: the rise of nearshoring," 1 December: 65–7.
The Economist (2005e) "India: the next wave," 17 December: 57–8.
The Economist (2005f) "Somalia calling," 24 December: 89.
The Economist (2006a) "Outsourcing to China: Watch out, India," 6 May: 69–70.
The Economist (2006b) "Visions of ecopolis," 23 September (Technology Quarterly): 20–3
The Economist (2006c) "Capital city," 21 October: 83–6.
The Economist (2006d) "The drawbacks of homework," 2 December: 78.
The Economist (2006e) "The phone of the future," 2 December (Technology Quarterly): 18–20.

The Economist (2006f) "Talk is cheap," 9 December: 68.

The Economist (2007a) "A cash call," 17 February, 67–70.

The Economist (2007b) "Hungry tiger, dancing elephant, 7 April: 65–67.

The Telework Coalition (2003) *The Greater Washington DC Telecommuting On-line Survey*, www.telcoa.org/id110.htm (accessed 22 August 2005).

Thomas, D. (2006) "Banks shift more IT offshore," *Computing*, May 11.

Thomas, G. and Wyatt, S. (1999) "Shaping cyberspace—interpreting and transforming the Internet," *Research Policy*, 28: 681–98.

Thorsen, T. (2006) "Newsmaker: Microsoft's entertainment CFO does a 360," *CNET News.com*, February 3.

Thrift, N. (1996) "New urban eras and old technological fears: reconfiguring the goodwill of electronic things," *Urban Studies*, 33: 1463–93.

Thrift, N. (1998) "Virtual capitalism: the globalisation of business knowledge," in J. Carrier and D. Miller (eds) *Virtualism: a new political economy*, Oxford: Berg, 161–86.

Thrift, N. (2000) "Performing cultures in the new economy," *Annals of the Association of American Geographers*, 90: 674–92.

Tickell, A. (2000) "Finance and localities," in G. Clark, M. Feldman and M. Gertler (eds) *The Oxford Handbook of Economic Geography*, Oxford: Oxford University Press, 230–47.

Timmer, M.P. and van Ark, B. (2006) "Does information and communication technology drive EU-US productivity growth differentials?" *Oxford Economic Papers*, 57: 693–716.

Tippins, M.J. and Sohi, R.S. (2003) "IT competency and firm performance: is organizational learning a missing link?" *Strategic Management Journa*l, 24: 745–61.

Tookey, A., Whalley, J. and Howick, S. (2006) "Broadband diffusion in remote and rural Scotland," *Telecommunications Policy*, 30: 481–95.

Törnqvist, G. (1983) "Creativity and the renewal of regional life," in A. Buttimer (ed.) *Creativity and Context* (Lund Studies in Geography, Series B: Human Geography, number 50), Lund: Gleerup, 91–112.

Torre, A. and Rallet, A. (2005) "Proximity and localization," *Regional Studies*, 39: 47–59.

Townsend, A. (2001a) "Network cities and the global structure of the Internet," *American Behavioral Scientist*, 44: 1697–1716.

Townsend, A.M. (2001b) "The Internet and the rise of the new network cities, 1969–1999," *Environment and Planning B: Planning and Design*, 28: 39–58.

Trent, R.J. and Monczka, R.M. (2005) "Achieving excellence in global sourcing," *MIT Sloan Management Review*, 47 (1): 24–32.

Tseng, M.M. and Piller, F.T. (eds) (2003) *The Customer Centric Enterprise: advances in mass customization and personalization*, New York: Springer Verlag.

United Nations Conference on Trade and Development (UNCTAD) (2004) *World Investment Report 2004: the shift towards services*, New York and Geneva: United Nations.

United Parcel Service, Inc. (2006) *Form 10-K Annual Report 2005*, Washington, DC: US Securities and Exchange Commission.

US Census Bureau (2004) "Travel arrangement and reservation services: 2002," September, www.census.gov/prod/ec02/ec0256i05.pdf (accessed 4 April 2006).

US Census Bureau (2006a) "E-commerce 2004" *E-Stats*, 25 May, www.census.gov/eos/www/papers/2004/2004reportfinal.pdf (accessed 24 December 2006).

US Census Bureau (2006b) "Quarterly retail e-commerce sales," www.census.gov/mrts/www/ecomm.html (accessed 22 November 2006).

US Department of Commerce (2002) *Digital economy 2002*, Washington, DC: US Department of Commerce, Economics and Statistics Administration, www.esa.doc.gov/reports/DE2002r1.pdf (accessed 28 September 2005).

US Department of Commerce (2003) *Digital Economy 2003*, Washington, DC: US Department of Commerce, Economics and Statistics Administration, www.esa.doc.gov/2003.cfm (accessed 17 October 2006).

US Department of Transportation, Bureau of Transportation Statistics (2003) *Highlights of the 2001 National Household Travel Survey*, BTS03–05, Washington, DC: US Department of Transportation.

van Ark, B. (2002) Measuring the new economy: an international comparative perspective," *Review of Income and Wealth*, 48: 1–14.

van Ark, B., Inklaar, R. and McGuckin, R.H. (2003) "ICT and productivity in Europe and the United States: where do the differences come from?" *CESifo Economic Studies*, 49: 295–318.

van der Meer, A. and van Winden, W. (2003) "Governance in cities: a comparison of urban information and communication technology policies," *Regional Studies*, 37: 407–19.

Van Marsh, A. (2004) "Outsourcing comes to airlines," CNN.com, December 26, http://edition.cnn.com/2004/TRAVEL/12/23/bt.outsource.airlines/index.html (accessed 31 August 2006).

Van Reenen, J. and Sadun, R. (2006) "Information technology and productivity, or 'It ain't what you do, it's the way that you do I.T.'," in S. Dutta, A. Lopez-Claros and I. Mia (eds) *The Global Information Technology Report 2005–2006: leveraging ICT for development*, London: Palgrave Macmillan, 55–60.

van Welsum, D. and Vickery, G. (2005) *New Perspectives on ICT Skills and Employment*, Paris: Organisation for Economic Co-operation and Development.

van Winden, W. and Woets, P. (2004) "Urban broadband internet policies in Europe: a critical review," *Urban Studies*, 41: 2043–59.

Vara, V. and Mangalindan, M. (2006) "Web pioneers eBay and Amazon face a threat from older retailers," *The Wall Street Journal*, 16 November: A1.

Varian, H. (2000) "Buying, sharing and renting information goods," *Journal of Industrial Economics*, 48: 473–88.

Varian, H. (2003) "Does IT matter?" *Harvard Business Review*, 81 (7): 112.

Vatier, M.S. (2005) "Les garde-fous juridiques du télétravail," *Le Journal du Management*, January, www.management.journaldunet.com/dossiers/050166teletravail/vatier.shtml (accessed 28 September 2006).

Vatne, E. and Taylor, M. (2000) *The Networked Firm in a Global World: the organization of industrial space*, Aldershot: Ashgate.

Veltz, P. (1996) *Mondialisation, villes et territoires: l'économie d'archipel*, Paris: Presses Universitaires de France.

Veltz, P. (2000) *Le nouveau monde industriel*, Paris: Gallimard.

Vernon, R. (1966) "International investment and international trade in the product cycle," *Quarterly Journal of Economics*, 80 (2): 190–207.

Vias, A. C. (1999) "Jobs follow people in the rural Rocky Mountain West," *Rural Development Perspectives*, 14 (2): 14–23.

Vickery, G. and Wusch-Vincent, S. (2006) *The Future Digital Economy: digital content creation, distribution and access. Conference summary*, Paris: Organisation for Economic Co-operation and Development.

Vinodrai, T. (2006) "Reproducing Toronto's design ecology: career paths, intermediaries, and local labor markets," *Economic Geography*, 82: 237–63.

Virtual-Agent Services (2006) "Virtual-Agent Services continues to expand opening 21st North American location," www.vagent.com/061117toronto.aspx (accessed 3 January 2007).

Volvo AB (2006) *The Volvo Group 2006*, annual corporate report, www.volvo.com/NR/rdonlyres/758245BC-1374-4D82–9268-A58E329C4CD2/0/ar2006_volvo_eng.pdf (accessed 30 April 2007).

Volvo IT (2006) "Facts and figures," www.volvo.com/volvoit/global/en-gb/about_us/facts_and_figures (accessed 30 April 2007).

von Hippel, E. (1988) *The Sources of Innovation*, Oxford: Oxford University Press.

von Hippel, E. (2005) *Democratizing Innovation*, Cambridge, MA: MIT Press.

Von Zedtwitz, M. and Gassmann, O. (2002) "Market versus technology drive in R&D internationalization: four different patterns of managing research and development," *Research Policy*, 31: 569–88.

Wachter, R.M. (2006a) "The 'dis-location' of US medicine—the implications of medical outsourcing," *New England Journal of Medicine*, 354: 661–5.

Wachter, R.M. (2006b) "International teleradiology," *New England Journal of Medicine*, 354: 662–3.

Wacker, G. (2003) "The Internet and Censorship in China" in C.R. Hughes and G. Wacker (eds) *China and the Internet: politics of the digital leap forward*, London: Routledge Curzon, 58–82.

Waite, C.A. (1998) "Service sector: its importance and prospects for the future," in W.O. Candilis (ed.) *United States Service Industries Handbook*, New York: Praeger, 1–22.

Walcott, S.M. (2003) *Chinese Science and Technology Industrial Parks*, Aldershot: Ashgate.

Walcott, S.M. and Wheeler, J.O. (2001) "Atlanta in the telecommunications age: the fiber-optic information network," *Urban Geography*, 22: 316–39.

Walker, R. (1985) "Is there a service economy? The changing capitalist division of labour," *Science and Society*, 49: 42–83.

Wallace, B., Duberg, J. and Kirkley, J. (2003) *Dynamics of the Oil and Gas Industry in the Gulf of Mexico: 1980–2000. Final report* (OCS Study MMS 2003–004), New Orleans: US Department of the Interior, Minerals Management Service.

Wallsten, S. (2005a) "Regulation and Internet use in developing countries," *Economic Development and Cultural Change*, 53: 501–23.

Wallsten, S. (2005b) *Broadband Penetration: an empirical analysis of state and federal policies* (Working paper 05–12), Washington, DC: AEI-Brookings Joint Center for Regulatory Studies, www.aei-brookings.org/admin/authorpdfs/page.php?id=1161 (accessed 9 June 2006).

Wal-Mart Stores, Inc. (2006a). *Form 10-K Annual Report for the fiscal year ended January 31, 2006*, Washington, DC: US Securities and Exchange Commission.

Wal-Mart Stores Inc. (2006b) "Continued expansion of radio frequency identification (RFID)," www.walmartstores.com/GlobalWMStoresWeb/navigate.do?catg=339 (accessed 3 January 2007).

Wal-Mart (2006c) *Our Company: The Wal-Mart story: timeline*, www.walmartstores.com/GlobalWMStoresWeb/navigate.do?catg=6&contId=1800 (accessed 5 June 2006).

Wang, W.-C. (2005) "Glob@lizing the network economy: local advantage for high-technology development," in A. Lagendijk and P. Oinas (eds) *Proximity, Distance and Diversity: issues on economic interaction and local development*, Aldershot: Ashgate, 169–89.

Wanyama, W., Ertas, A., Zhang, H-C, and Ekwaro-Osire, S. (2003) "Life-cycle engineering: issues, tools and research," *International Journal of Computer Integrated Manufacturing*, 16: 307–16.

Warf, B. (1999) "The hypermobility of capital and the collapse of the Keynesian State," in R. Martin (ed.) *Money and the Space Economy*, Chichester: John Wiley, 227–40.

Warf, B. (2000) "Telecommunications and economic space," in E. Sheppard and T.J. Barnes (eds), *A Companion to Economic Geography*, Oxford: Blackwell, 484–498.

Warf, B. (2006) "International competition between satellite and fiber optic carriers: a geographic perspective," *Professional Geographer*, 58: 1–11.

Warf, B. and Purcell, D. (2001) "The currency of currency: speed, sovereignty, and electronic finance," in T.R. Leinbach and S.D. Brunn (eds) *Worlds of Electronic Commerce: economic, geographical and social dimensions*, New York: John Wiley, 224–40.

Warnecke, H. (1997) *The Fractal Company: a revolution in corporate culture*, New York: Springer-Verlag.

Warsh, D. (2006) *Knowledge and the Wealth of Nations: a story of economic discovery*, New York: W.W. Norton.

Washington Metropolitan Telework Center (2006) *Telework Center Survey*, March, www.gsa.gov/gsa/cm_attachments/GSA_DOCUMENT/teleworkcenterclientsurveymarch2006_R25-u8R_0Z5RDZ-i34K-pR.doc (accessed 19 December 2006).

Wessel, D. and Davis, B. (2007) "Pain from free trade spurs second thoughts," *The Wall Street Journal*, 28 March: A1.

Whalen, D.J. (undated) "Communications satellites: making the global village possible," www.hq.nasa.gov/office/pao/History/satcomhistory.html (accessed 24 December 2006).

Wheeler, J.O., Aoyama, Y. and Warf, B. (eds) (2000) *Cities in the Telecommunications Age: the fracturing of geographies*, London, Routledge.

Wheelon, A.D. (1988) "The role of satellite communications in the 1990s," *International Journal of Technology Management*, 3: 667–73.

Whinston, A.B., Stahl, D.O. and Choi, S.-Y. (1997) *The Economics of Electronic Commerce*, Indianapolis, IN: Macmillan Technical Publishing.

Whitman, B. (2005) "FTTH expands fast around the world," *Broadband Properties*, September: 42–6.

Williams, C.C. (2004) "A borderless world of hypermobile and homeless money? An evaluation of financial flows in the mutual fund industry," *The Industrial Geographer*, 2 (2): 144–55, www.igeographer.lib.indstate.edu/williams.pdf (accessed 23 October 2005).

Wilson, M. (2000) "The fall of the mall? The impact of online travel sales on travel agencies," *Journal of Urban Technology*, 7 (2): 43–58.

Wójcik, D., Sidaway, J.D. and Beaverstock, J.V. (2007) "European financial geographies," *Growth and Change*, 38: 167–73.

Wolcott, P. and Goodman, S. (2003) "Global diffusion of the Internet I: India: Is the elephant learning to dance?" *Communications of the Association for Information Systems*, 11: 560–646.

Wood, P. (2002a) "Knowledge-intensive services and urban innovativeness," *Urban Studies*, 39: 993–1002.

Wood, P. (2002b) "Services and the 'new economy': an elaboration," *Journal of Economic Geography*, 2: 109–14.

Woodall, P. (2006) "The new titans: a survey of the world economy," *The Economist*, 16 September.

World Trade Organization (WTO) (2006) "Trade by sector: leading exporters and importers of other commercial services, 2005," in *Statistics: International Trade Statistics 2006*, Geneva: World Trade Organization, www.wto.org/english/res_e/statis_e/its2006_e/section4_e/iv93.xls (accessed 13 August 2007).

Wrigley, N. (2000) "The globalization of retail capital: themes for economic geography," in G. Clark, M. Feldman and M. Gertler (eds) *The Oxford Handbook of Economic Geography*, Oxford: Oxford University Press, 292–313.

Wrigley, N. and Currah, A. (2006) "Globalizing retail and the 'new e-conomy': the organizational challenge of e-commerce for the retail TNCs," *Geoforum*, 37: 340–51.

Wusch-Vincent, S. and Vickery, G. (2005) *Digital Broadband Content: music*, Paris: Organisation for Economic Co-operation and Development.

Wyatt, I.D. and Hecker, D.E. (2006) "Occupational changes during the 20th century," *Monthly Labor Review*, 129 (3): 35–57.

Yan, X. and Pitt, D. (2002) *Chinese Telecommunications Policy*, Boston, MA: Artech House.

Yeung, H.W.-C. (2002) "The limits to globalization theory: a geographic perspective on global economic change," *Economic Geography*, 78: 285–305.

Yoffie, D.B. (ed.) (1997) *Competing in the Age of Digital Convergence*, Boston, MA: Harvard Business School Press.

Young, M. and Jude, M. (2004) *The Case for Virtual Business Processes: reduce costs, improve efficiencies, and focus on your core business*, Indianapolis, IN: Cisco Press.

Zaheer, S. (2000) "Time zone economies and managerial work in a global world," in P.C. Early and H. Singh (eds) *Innovations in International Management*, Thousand Oaks, CA: Sage, 339–53.

Zällh, S. (2005) *Significant E-Marketplaces*, Oslo, Norway: eMarket Services, www.emarketservices.com/clubs/ems/artic/SignificanteMarkets.pdf (accessed 30 November 2006).

Zanfei, A. (2000) "Transnational firms and the changing organisation of innovative activities," *Cambridge Journal of Economics*, 24: 515–42.

Zeller, C. (2004) "North Atlantic innovative relations of Swiss pharmaceuticals and the proximities with regional biotech arenas," *Economic Geography*, 80: 83–111.

Zhang, J. (2006a) *Easier Access to Venture Capital in Silicon Valley: some empirical evidence*, San Francisco, CA: Public Policy Institute of California.

Zhang, J. (2006b) "Market transition, state connections and Internet geography in China," *The China Review*, 6: 93–123.

Zhou, Y. (2005) "The making of an innovative region from a centrally planned economy: institutional evolution in Zhongguancun Science Park in Beijing," *Environment and Planning A*, 37: 1113–34.

Zimmermann, H. (1980) "OSI reference model—the ISO model of architecture for open systems interconnection," *IEEE Transactions on Communications*, COM-28: 425–32.

Zittrain, J. and Palfrey, J.G. (2005) *Internet filtering in China 2004–2005: a country study*, www.opennetinitiative.net/studies/china/ (accessed 29 August 2006).

Zook, M.A. (2005) *The Geography of the Internet Industry: venture capital, dot-coms, and local knowledge*, Oxford: Blackwell.

Zook, M.A. (2006) "Geographies of the Internet," in B. Cronin (ed.), *Annual Review of Information Science and Technology*, vol. 40, Medford, NJ: Information Today, 53–78.

Zook, M.A., Dodge, M., Aoyama, Y. and Townsend, A. (2004) "New digital geographies: information, communication, and place," in S.D. Brunn, S.L. Cutter and J.W. Harrington (eds) *Geography and Technology*, Dordrecht: Kluwer, 155–176.

Zucker, L.G. and Darby, M.R. (2005) "Socio-economic impact of nanoscale science: initial results and NanoBank," NBER working paper 11181, Cambridge, MA: National Bureau of Economic Research, http://papers.nber.org/papers/w11181.pdf (accessed 25 October 2006).

Zysman, J. (2005) "Tranforming production in a digital era," in W.H. Dutton, B. Kahin, R. O'Callaghan and A.W. Wyckoff (eds) *Transforming Enterprise: the economic and social implications of information technology*, Cambridge, MA: MIT Press, 257–81.

Zysman, J. and Borrus, M. (1997) "Globalization with borders: the rise of Wintelism as the future of industrial competition," *Industry and Innovation*, 4 (2) 141–66.

INDEX